# COSMIC SECRETS

## Basic Features of Reality

# COSMIC SECRETS

## Basic Features of Reality

WOLFRAM SCHOMMERS

Forschungszentrum Karlsruhe, Germany



World Scientific

NEW JERSEY • LONDON • SINGAPORE • BEIJING • SHANGHAI • HONG KONG • TAIPEI • CHENNAI

*Published by*

World Scientific Publishing Co. Pte. Ltd.
5 Toh Tuck Link, Singapore 596224
*USA office:* 27 Warren Street, Suite 401-402, Hackensack, NJ 07601
*UK office:* 57 Shelton Street, Covent Garden, London WC2H 9HE

**British Library Cataloguing-in-Publication Data**
A catalogue record for this book is available from the British Library.

**COSMIC SECRETS**
**Basic Features of Reality**



ISBN-13 978-981-283-643-4
ISBN-10 981-283-643-8

Typeset by Stallion Press
Email: enquiries@stallionpress.com

Printed in Singapore by Mainland Press Pte Ltd.

# Foreword

Are humans able to formulate the final truth of the physical world? Can we discover, in other words, a complete theory for what is often called basic reality defining the deepest ground of the cosmos? It has been suggested that this should be possible and, furthermore, it is also argued that such an absolute truth will be available soon. However, we have to be careful.

Why do even prominent people believe that we will have the final truth in our hands soon? Here the following aspect seems to be essential. It is often believed that science progresses by approaching the final truth successively. That is, the "final view" of these things are gradually approached by way of an asymptotic approximation, and there is a detailed filling of a given, fundamentally defined theoretical frame of the cosmos. Then, we are approaching the absolute truth by the elimination of secret by secret and, finally, there is no secret left. However, the theory of science demonstrates convincingly that such an asymptotic approximation is not correct. The progress of science does not work in this way. More details are given in Chapter 1.

In this monograph we will demonstrate by reasonable theoretical arguments that the structures of basic (absolute) reality is principally not recognizable for humans. The world that we have in front of us in everyday life cannot be the absolute truth (basic reality). In other words, basic reality hides behind our space–time impressions, and it is principally not possible to say how all these things around us, which are positioned in space and time, are structured in basic reality. Within this theoretical picture the details of basic reality remain hidden for ever and have to be considered as "cosmic secrets". None of these cosmic secrets are eliminatable.

Again, on the basis of the laws of conventional physics it is often believed that the absolute truth (basic reality) will be in our hands soon.

Then, no secret would be left. Is such a view justified? It is not. Let us briefly quote some essential points.

Conventional physics is mainly based on quantum theory and the Theory of Relativity, and these theories lead to sets of laws that work fantastically. Quantum field theory is accurate to about one part in $10^{11}$, General Relativity has been tested to be correct to one part in $10^{14}$. However, if we put them together we inevitably obtain irreconcilable differences; both theories are mutually exclusive. In this connection the cosmological constant is of particular interest because this constant connects quantum phenomena with space–time features of the General Theory of Relativity. However, it turned out that the quantum-theoretical treatment of the cosmological constant obviously leads to a value, which is much too large, and this reflects a big problem, but we have presently no satisfactory solution for it.

Because the cosmological constant is too large and no simple solution for this problem has yet been offered, we may conclude that there is something wrong with the basic concept responsible for the quantum theoretical treatment of the vacuum. The vacuum possibly contains ballast that is not necessary. Concerning this point we find in K. C. Cole's book *The Hole in the Universe* the following interesting comment [29]: "*They have filled the vacuum with so much garbage, there isn't room for the cosmological constant,*" *said Leon Lederman.* "*Einstein freed us from the ether. Now we need to get rid of (today's version of ether) again. We need to sweep the vacuum clean.*"

Here Lederman's statement that "*We need to sweep the vacuum clean*" is of particular relevance because it points to a critical point in connection with space and time. The following question arises: Is the container-principle for space–time really realistic? Here container-principle means that everything (the entire cosmos) is embedded in space–time. Does this view really reflect the basic situation? This question has to be answered, and we come directly to the next question: Is that, which we have in front of us in our everyday life observations, actually the objective, basic reality? It is not. The situation is described by the well-known psychologist C. G. Jung's statement [7]: "*When one thinks about what consciousness really is, one is deeply impressed by the wonderful fact that an event that takes place in the cosmos outside, produces an inner picture, that the event also takes place inside...*"

This statement by C. G. Jung suggests the following: There is a projection of reality onto space–time and the real world outside is also embedded in space–time. After Jung there is a one-to-one correspondence between the inner picture and the structures in the outside world. But why should events in nature occur, so to speak, twice, one outside of us, and again in the form of a picture inside us? This would be against the "principles of evolution" and the "principle of usefulness", respectively. In other words, the principles of evolution dictate, for economic reasons, that the inner picture and the structures in the outside world are different from each other; the inner picture should contain less information than the outside world. It is relatively easy to recognize that the inner picture is just what we have directly in front of us within our observations in everyday life.

Already the philosopher Immanuel Kant (1724–1804) investigated the relation between true and perceived reality. He argued that we cannot make statements about the true reality outside. According to him all things we observe are located within space–time and these elements, space and time, are located inside the observer. According to Kant space and time are not empirical concepts, which are determined by abstraction from experience. Experiences become possible only through the concepts of space and time. According to Kant space and time are not objects, but have to be considered as preconditions for the possibility of all experience.

Then, Lederman's statement "*We need to sweep the vacuum clean*" becomes superfluous because Kant's space–time can hardly be the seat of physically real quantities (quantum fields etc.), and the cosmological constant loses its significance. In connection with Kant's philosophy we have to ask whether the entire material world is actually embedded in space and time. We will recognize that this should not be the case, but it is more realistic to assume that the world is projected onto space and time.

Within conventional physics (classical mechanics, usual quantum theory and the Theory of Relativity) it is assumed, as a matter of course, that the real world is embedded in space and time, and this is possibly the reason for the fact of an unsatisfactory situation. On the one hand, both theories lead to descriptions of observations with almost unimaginable accuracy; on the other hand, both theories are mutually incompatible. Brian Greene's remarked in *The Elegant Universe* [28]: *As they are currently formulated, general relativity and quantum mechanics, cannot both be right.*

In this monograph we will discuss some new basic results that are based on the assumption that the world is not embedded in space and time but, in accordance to Immanuel Kant's ideas, we will treat physics (in particular quantum phenomena) on the basis of the so-called projection principle, which is characterized by the following conception: In basic reality (outside world) there is no space-time, but the processes that take place in the outside world are projected onto the elements space and time and we obtain in this way pictures of reality. We will therefore call this relatively new framework "projection theory". The most remarkable result of projection theory is that time is no longer a classical parameter as in conventional quantum theory, but it is also relevant to mention that we get for each system (object) a certain form (shape), and this effect comes into play by distance-independent interactions that may appear within projection theory in the most natural way; also this effect is not known in conventional quantum theory. Furthermore, it is important to mention that Mach's principle is fulfilled within the framework of projection theory.

Wolfram Schommers
Karlsruhe, Germany

# Contents

# 1

# The Absolute Truth

## 1.1. Final Truth

Can the progress in science, in particular in physics, lead to a final truth of the physical world? In other words, can we discover a complete theory about the cosmos? Are humans able to grasp all the features of the universe? It has often been suggested that this should be possible and, furthermore, it has also often been pointed out that such an absolute truth will be available soon. For example, Stephen Hawking suggested in his famous book *A Brief History of Time* that we should thoroughly be able to discover such a complete theory and also to find out "*why it is why we and the universe exist. If we find the answer to that, it would be the ultimate triumph of human reason — for then we would know the mind of God.*" [1]

The superstring theory is often considered as the final, complete theory of nature. The fundamental point is that here the basic building blocks of nature are not point-like particles, as usually assumed in conventional elementary particle physics, but one-dimensional very short vibrating strings (and other objects: vibrating two-dimensional membranes, three-dimensional blobs and other ingredients). This superstring theory is also known as "Theory of Everything" since the scientific community firmly assumes that this theory is that what we have called above "absolute truth", that is, it is believed that the superstring theory is a complete framework for the explanation of every fundamental feature in the universe.

John Horgan discussed in his book *The End of Science* the situation in superstring theory with Edward Witten, who coined the term superstring theory. In this connection Horgan wrote [2]: "... *acknowledged, even emphasized, that no one had truly fathomed the theory, and that it might be decades before it yielded a precise description of nature. He would not predict, as others had, that superstring theory might bring about the end of physics. Nevertheless, he was serenely confident that it would eventually yield a profound new understanding of reality.*" There is obviously no doubt for Edward Witten that the superstring concept describes the basic and final structure of the cosmos: "Good wrong ideas are extremely scarce ... and good wrong ideas that even remotely rival the majesty of string theory have never been seen." In other words, in the opinion of Witten the superstring concept is too elegant and beautiful to be wrong. But Witten still goes a step further: He is convinced that superstring theory could be discovered by any other advanced civilisations in the universe, that is, independent of any specific culture or biological structure of the observers.

## 1.2. Two Important Questions

In connection with Stephen Hawking's and Edwards Witten's statements and thoughts, respectively, we are concerned with the following important question: How general can a human observer formulate the laws of nature? There are two points of particular relevance:

1. Can a human observer actually formulate the laws of nature independent of his own nature, that is, independent of his biological structure?
2. Can we really find out "why" we and the cosmos exist?

These questions are of central relevance, and they contain secrets of cosmic extent. In this monograph we would like to discuss the essential points in connection with that, what we can know about the universe and what we do not.

## 1.3. Why Does the Cosmos Exist?

In connection with Stephen Hawking's statement "... why we and the universe exists" we have critically to ask the following question: Do the physicists really develop new theories that describe *why* the cosmos exists? Here we have to be careful. No, it is the rule that they do not.

The physical theories describe *how* certain systems work not *why* they work. Modern physical sciences started with Newton's gravitational theory and, as is well known, this theory allows, for example, for us to determine the precise motion of a planet around the sun. Newton's laws tell us *how* the planet moves but gives no explanation of the machinery of gravitation, that is, *why* a planet moves around the sun. "*Newton was originally asked about his theory* — 'But it does not mean anything — it does not tell us anything'. *He said, 'It tells you how it moves. That should be enough. I have told you how it moves, not why'*. [3]

What can we say about basic quantum theory that governs contemporary physics? Do the basic laws of quantum mechanics tell us *why* nature is organized in the way as it is? Here we also have to answer the question of how does it work. What machinery is actually producing these quantum effects? Also here we would like to argue as follows: If we know this machinery, we also know why nature is organized in the way as it is. Can we formulate a basic mechanism (or more than one) that explains quantum phenomena? This question can clearly be answered. Let us give here only some principal remarks:

The understanding in terms of mechanisms or models most often assumes understanding effects on the basis of mechanisms that we observe in everyday life. This is however not possible in the case of quantum mechanics and must be excluded. The modern interpretation of this theory does even not allow us to invent more complex mechanisms that are beyond the usual pictures and mechanisms observed in everyday life. The reason why is that in quantum theory the particle (for example, an electron) motion cannot be analyzed in terms of a continuous motion; we can only say something about probabilities. Here probability is not a mean of approximation, but probability is obviously established at the fundamental scientific level. "*It is not our ignorance of the internal gears, of the internal complications, that makes nature appear to have probability in it. It seems to be somehow intrinsic. Someone has said it this way — Nature herself does not even know which way the electron is going to go.*" [3]

We know the laws of quantum theory but we cannot understand them in the sense that we can analyze them in terms of (intimate) mechanisms and, therefore, if probability is an intrinsic feature of nature we will never know *why* all these things and effects are as they are. (That does not mean that the present quantum laws must represent the final truth in connection with quantum theory.)

In conclusion, as in the case of Newton's gravitational law also quantum mechanics only tells us *how* quantum systems behave, but not *why*. Also within string theory the basic laws of quantum theory are assumed to be valid.

The discussion above has been exclusively based on theoretical physical arguments. Since physics represents that which we call "basic science" we probably never find out at this level *why* we and the universe exist. Clearly, philosophical considerations can help to deepen our knowledge about the universe but philosophical arguments are very often not based on "hard" facts, namely those which are accessible to empirical tests.

## 1.4. Are the Laws of Nature Independent of the Observer's Own Nature?

As we have outlined above, Edward Witten is obviously convinced that the theoretical structures given by superstring theory are quite general and should therefore reflect that which we have called "absolute truth". In particular, Witten is convinced that superstring theory could be discovered by any other advanced civilizations in the cosmos and should therefore be independent of the structure of the individuals forming a certain civilization, that is, the outer shape and the material structure of the brain of the various individuals could in principle be quite different from those of the human observer.

Is such a view realistic? Can the laws of nature be independent of the observers own nature? It is not, as we will recognize below.

From our interaction with the outside world we obtain information about it. We can only say something about a complete theory (about the final truth of the physical world) if this information covers all possible aspects of the universe. Is a human observer really in the position to interact with the outside world in a way that he is able to pick up all aspects of the universe? Is that possible at all? Here we must say that he is definitely not able to. This point will be discussed in more detail in the forthcoming chapters too.

In connection with all these points, the following general question is relevant. How does science proceed? Is there anything like a "final view" of the things? Which tendencies exist historically to categorize a status of knowledge acquired? Let us briefly mention some of the most relevant opinions.

## 1.5. Self-Indulgence was Dominant

Not only Hawking and Witten believe that the present theoretical structures of the physical laws represent the final (absolute) truth, but scientists of all epochs thought that the scientific laws "now" approached their final state of completion, i.e., scientists of all epochs were convinced to have the absolute truth in their hands. However, every time it turned out that this belief was a fallacy. It is a certain kind of self-indulgence.

For instance, Lord Kelvin (1824–1907) thought that the foundations of physics as laid down towards the end of the last century were complete and that only secondary questions were still left to be answered. Berthelot (1827–1907) in 1885 felt that their world no longer concealed any secrets. Haeckel (1834–1919) concluded from his studies (also made towards the end of the 19th century) that all legitimate questions in natural science had essentially been answered. Another example of this tendency is given by Max Planck (1859–1947) [4]:

> *"As I was beginning to study physics (in 1875) and sought advice regarding the conditions an prospects of my studies from my eminent teacher Philipp von Jolly, he depicted physics as a highly developed and virtually full-grown science, which — since the discovery of the principle of the conservation of energy had in a certain sense put the keystone in place — would soon assume its final stable form. Perhaps in this or that corner there would still be some minor detail to check out and coordinate, but the system as a whole stood relatively secure, and theoretical physics was markedly*

> *approaching that the degree of completeness which geometry, for example, had already achieved for hundreds of years. Fifty years ago (as of 1824) this was the view of a physicist who stood at the pinnacle of the times."*

If we jump from the last century to the present, we find a recurrence of the opinions above. For instance, Richard Feynman wrote [4]:

> *"What of the future of this adventure? What will happen ultimately? We are going along guessing the laws; how many laws are we going to have to guess? I do not know. Some of my colleagues say that this fundamental aspect of our science will go on, but I think there will certainly not be perpetual novelty, say for a thousand years. This thing cannot keep on going so that we are always going to discover more and more new laws ... We are very lucky to live in an age in which we are still making discoveries. It is like the discovery of America — you only discover it once. The age in which we live is the age in which we are discovering the fundamental laws of nature, and that day will never come again. It is very exciting, it is marvellous, but this excitement will have to go. Of course in the future there will be other interests ... There will be a degeneration of ideas, just like the degeneration that great explorers feel is occurring when tourist begin moving in on a territory. In this (present) age people are (perhaps for the last time?) experiencing a delight, the tremendous delight that you get when you guess how nature will work in a situation never seen before."*

It is clearly evident from all the examples cited that eminent personalities (and obviously not only these) in different epochs thought that scientific findings "now" approached their final state of completion. Evidently, there have also been epochs in which the knowledge available was judged with less self-indulgence: Nicholas Rescher in his book entitled, "The limits of Science" [4], from which most of the texts of the examples quoted above have been taken, deals with these anti-positions. However, compared with the discussions in present day literature, the opinions of the authors quoted before — especially those expressed by Hawking, Witten and Feynman — seem to dominate for the time being. However, the past showed that this belief was every time a fallacy and only a certain kind of self-indulgence remained.

The facts from the theory of science have taught us that we will never be able to make final and complete statements about the physical universe, i.e., from the point of view of theory of science the "absolute truth" can never be obtained. This is an important statement and will be justified below.

## 1.6. Newton's Mechanics and Its Overestimation

As we have outlined above, scientists of all epochs thought that they have the "absolute truth" in their hands. In other words, they were firmly convinced that the scientific laws "now" approached their final state of completion. However, it was every time a mistake. A typical example for such a delusion is given by the far-reaching conclusion which was drawn from Newton's mechanics. Let us briefly discuss the background of this undesirable development.

### 1.6.1. *Instead of Gods, Capricious Fairies etc. We Have the Equations of Motion*

Before Newton one throughout believed, that the events in field and forest, and of course in heaven, would be influenced decisively by gods, demigods, capricious fairies and other mystical phenomena. This situation changed fundamentally with Newton's new point of view: Newton could state something about the movement of celestial objects without striving gods or capricious fairies. Within the frame of Newton's analysis phenomena in heaven cannot be caused by mystical individuals.

The essential point is that Newton's equations of motion describe the motion of the celestial bodies completely. Besides the gravitation introduced by Newton there are no other influences; in particular, there are no gods, demigods or capricious fairies. The calculations agree with the observations in such a precise way, that it would be absurd to assume that in the motion of the celestial bodies any other elements would be effective — others than those which belong to Newton's equations of motion.

Therefore, the direct experience agrees with the statements of the theory. Capricious fairies do not appear there, and they do not influence the motion of the celestial object arbitrarily, nothing like that, no tremble, no swerve, which would not be under mathematical control. Only the gravitation in the sense of Newton is effective. In other words, Newton's theory eliminates all the gods, capricious fairies, etc.

With Newton's theory people believed to hold the truth in their hands. In its extreme form of application one even believed to hold the truth about man, the truth about life in the hands. In the application of Newton's theory scientists did by no means restrict themselves to appropriate problems, not at all, because also man and life with all its complexity, i.e., with all his bodily, mental and emotional conditions have been considered as a Newtonian system. This appears, however, not appropriate. The reason is easy to recognize and is given by the basic characteristics of Newton's theory.

Within the framework of Newton's world view everything is given by the equations of motion, and these equations have the characteristic that, if one knows the state of a system at a certain time, the states for all future times are exactly determined, that is, man appears here as a certain kind of mechanical machine without spontaneity and creativity. This is certainly an absurd picture of man and is based on the success of Newton's mechanics, and due to this success the laws given by Newtonian theory were seen as the basic laws of nature, and classical mechanics was considered to be the ultimate theory of natural phenomena. In other words, scientists believed to have the "final (absolute) truth" in their hands.

### 1.6.2. *Lamettrie and the Monistic Picture of Man*

It was Lamettrie who required that, in addition, living human bodies (and of course animals and plants) should behave like a machine, i.e., without any spontaneity; in Lamettrie's opinion, the behavior of men was also completely causal and determinate. In other words, it was seen as being possible to predict — in principle — with absolute certainty all the future activities of a man if his present state was known. Within this picture of reality men are merely automata and not free individuals capable of influencing the course of events by their volitions; the world is completely mechanistic in its ultimate nature, and the mind is a production of mechanics. Clearly, all those conclusions about men which were based on religious considerations had no place within such a picture.

### The Cartesian division

The ideas of Newton (1643–1727) and Lamettrie (1709–1751) were influenced by the philosophy of Descartes (1596–1650). Descartes developed the spirit/matter dualism very extensively. He maintained that mind and matter are divided into two separate and independent realms, i.e., the mind should have no contact with matter and vice versa. Mind and matter had entirely different functions to perform: mind to think and matter to occupy space.

Descartes' fundamental division allowed one to treat matter as dead and completely separate from men, and to see the material world as an ensemble of different objects which form a huge machine. Newton constructed his mechanics on the basis of this mechanistic world view. The philosophy of Descartes has had a considerable influence on the general western way of thinking up to the present day.

The Cartesian division guarantees an objective description of the world. On the other hand, the division of mind and matter into two separate and independent realms created a serious problem: if man's will (which can be considered as a result of an operation of the mind) has no contact with the matter of his body, how can it compel this body to turn to the right or to the left as it pleases?

According to Descartes the spirit/matter dualism exists only for men but not for living animals, plants, etc., which was seen as objects without mind, i.e., their behavior was assumed to be completely causal and determinate like to a huge mechanical machine working in accordance with Newton's mechanics. In other words, Descartes assumed for animals, plants, etc. a monistic picture and for men a dualistic one.

However, Descartes world view seems too artificial, and a uniform picture should be the goal. Such a uniform conception of reality was formulated by Lamettrie, who claimed that the monistic picture for animals, etc. is also valid for men. Lamettrie's world view requires that feelings, free will, etc. are effects produced by a complex mechanical system (the human body). Within such a framework, based on Newton's mechanics, all the future feelings, etc., can in principle be predicted if the man's present mechanical state is known. Within Lamettrie's world view all subjective elements are eliminated.

## The influence of Darwin's evolution theory

Lamettrie's radical form of materialistic natural philosophy received little attention in the eighteenth and the first half of the nineteenth century. However, the concept of an evolving world rather than a static one changed the situation. Darwin (1809–1882) proposed in the second half of the nineteenth century that the earth and all living things had evolved through a long history, a history of continual, gradual change. This concept of an evolving world, however, is in contrast to the creation myth of primitive people and of most religions, which have a common, essentially static concept of a world that, once created, has not changed. Darwin's evolution theory was almost universally accepted by serious scientists even before his death in 1882, and the following view became more and more popular: No God had created the earth and everything on it at a certain instant, but all these things had evolved throughout a long history in line with the laws of physics, which were assumed to be Newton's equations of motion.

This view was the reason why Lamettrie's radical form of materialistic natural philosophy became increasingly attractive. For example, within the framework of Haeckel's (1834–1919) monism these tendencies are clearly reflected; there were many editions of his monograph *Die Welträtsel* [5] and it has been translated into more than 25 languages. Haeckel's materialistic world view has had a considerable echo up to the present day.

### 1.6.3. *Conclusion*

Again, with Newton's theory scientists believed to hold the "absolute truth" in their hands. We know that this belief was a fallacy; new theoretical developments (the Theory of Relativity and quantum theory) disproved the assumption made to consider Newton's mechanics as "absolute truth"; on particular, Lamettrie's radical form of materialistic natural philosophy with man as mechanical machine became outdated.

However, the conviction to soon have the "absolute (final) truth" on the table is also believed today in connection with newer theories, which are based on the Theory of Relativity and quantum theory. Many books suggest directly or at least unequivocally, that we will know "God's strategy" soon, that is, the "absolute truth"; with only a few minor steps and everything is known. But here also there are relevant objections against these newer tendencies for the assumption that the final formulation of the physical laws is (almost) to be reached.

There are principal reasons that speak against this assumption. In this connection two points are of particular relevance: (1) The phenomenon of evolution. (2) Basic statements of the theory of science. Before we explain these points in more detail, let us investigate whether Newton's equations of motion describe processes in the absolute reality. We already know that these equations cannot reflect the final, absolute truth, but they could in principle be part of the absolute truth, i.e., an approximate representation of it in absolute (basic) reality.

## 1.7. Scientific Realism

Which kind of truth is ultimately described by theoretical physics? Is physics actually in the position to describe that which we have called "absolute truth"? Let us explain this point by means of Newton's equations of motion.

Let us study the situation by means of an example, namely the earth at its motion around the sun. Earth and sun interact, i.e., they attract each other, and for this attraction Newton could formulate a mathematical relation. Furthermore, if we know the position and the velocity of the earth at a certain but arbitrary time, then the movement of the earth around the sun is determined for all times $\tau$, at least in principle. One knows, in other words, at which position in space the earth will be in the year 2020 or where it was, for instance, in the year 1900. In order to calculate these data, we only have to solve the right equation, which is quite simple in its structure and has the following form: $m_E(d^2\mathbf{r}_\mathrm{E}/d\tau^2) = -Gm_E m_S(\mathbf{r}_E - \mathbf{r}_S)/|\mathbf{r}_E - \mathbf{r}_S|^3$, where $\mathbf{r}_E$ and $\mathbf{r}_S$ are the position vectors of earth and sun, $m_E$ and $m_S$ are their masses, and $G$ is the gravitation constant. More elements are not involved in the equation of motion. All the other details in connection with this equation are not of interest for our discussion. Newton's equation of motion describes the path of a celestial object under the influence of another celestial object, here the path of the earth within the gravitation field of the sun.

Let us assume that the above equation describes the absolute reality. Then, we have to consider the masses $m_E, m_S$ and the gravitational influence between them as really existing in the reality outside. To imagine this, first of all gives us no problem. Also the solutions of Newton's equation of motion (it is a differential equation) have to be considered as really existing, that is, the possible paths of the mass reflect the deepest ground of reality. This assumption too gives us no problems since we actually observe the movements of the celestial objects; we have them directly in front of us. Therefore, we can assume at the outset that these paths also reflect an absolute fact.

Up to this point one can pre-suppose, that the equations of motion with all their elements and solutions reflect the structure of absolute reality and, therefore, actually describe what we have called "absolute truth". In other words, reality at the deepest ground is structured as is formulated by physics.

But we have to be careful, because such a kind of scientific realism would consequently mean that the celestial objects would continuously solve differential equations in their movements through space, since Newton's equations of motion are differential equations, as already mentioned above.

Such an idea must however be considered as absurd. Where is the computer hidden which would do this work? Who does this work with lightning speed and simultaneously everywhere? By the way, the mechanism with which nature solves the differential equations would have to be delivered by the theory itself; in other words, it should be the content of the theory; this is however not the case.

Such a scientific scenario (realism), that the masses solve in their motion through space incessantly differential equations, has to be excluded and must be judged as ridiculous, which is also reflected in the following remark [4]:

> *"As Herschel ruminated long ago, particles moving in mutual gravitational interaction are, as we human investigators see it forever solving differential equations which, if written out in full, might circle the earth."*

This brief analysis makes clear that we affect with Newton's ideas by no means that what we have called "absolute reality" and "absolute truth", respectively, and that is true for all theoretical developments in physics, even for the newer theoretical developments, because everything in natural science is formulated in accordance with Newton's basic way of thinking, for which is characteristic, that the scientific laws are formed on the basis of what is directly in front of the observer. Nicholas Rescher, distinguished epistemologist, expressed this fact by [4]:

> *"Scientific realism is the doctrine that science describes the real world: that the world actually is as science takes it to be, and that its furnishings are as science envisages them to be.... It is quite clear that it is not..."*

The facts (the paths and the physical elements and effects tied with them), which are described by the above introduced equation $m_E(d^2\mathbf{r}_\mathrm{E}/d\tau^2) = -Gm_E m_S(\mathbf{r}_E - \mathbf{r}_S)/|\mathbf{r}_E - \mathbf{r}_S|^3$ only represent a certain kind of truth but cannot be considered as absolute in character. If we accept that (there is obviously no other possibility for that), the consequences are significant for everything which appears directly in front of us: If everything, which is described by the equation of motion $m_E(d^2\mathbf{r}_E/d\tau^2) = -Gm_Em_S(\mathbf{r}_E - \mathbf{r}_S)/|\mathbf{r}_E - \mathbf{r}_S|^3$ (the path of the celestial object, which we have in exactly this form directly in front of us) cannot be the last, absolute truth, then also all the other things, which are in front of us and which we feel by our sense of touch, cannot be the last, absolute truth too. This, of course, is also true for that what we perceive of ourselves and other creatures. This result is of fundamental meaning.

In conclusion, the equations of motion cannot be the last, absolute truth. Since the derivation of these equations have been directly based on that which we have in everyday life in front of us, these optical impressions cannot also be part of an absolute (basic) reality. Thus, a human observer is not able to recognize the absolute truth, that is, the deepest ground of reality. The situation is summarized in Fig. 1.1.

What kind of reality could be involved when we base our descriptions on the equations of motion? Do the impressions that we have directly in front us actually reflect a certain kind of reality at all or is it "only" a symbolic picture of it? When we base our considerations on the usual assumption that everything is embedded in space and time, then we may assume that the impressions in front of us directly reflect a real situation with material objects, i.e., it may not be "basic realty" (see the remarks above in connection with the equations of motion), but it is a certain kind of reality, namely that which is given by material objects, which are embedded in space and time. However, we will recognize below that this should not be the case. It will turn out that it is more realistic to assume that our direct optical impressions are "pictures of reality" but not reality itself, that is, that the objects in space and time are geometrical figures and are not material objects.

## 1.8. An Important Principle: As Little Outside World as Possible

We came to the conclusion that a human observer is not able to grasp the "absolute truth". What we have in front of us in everyday life is obviously not basic reality or absolute truth. Nevertheless, this experienced world could contain the complete information of the

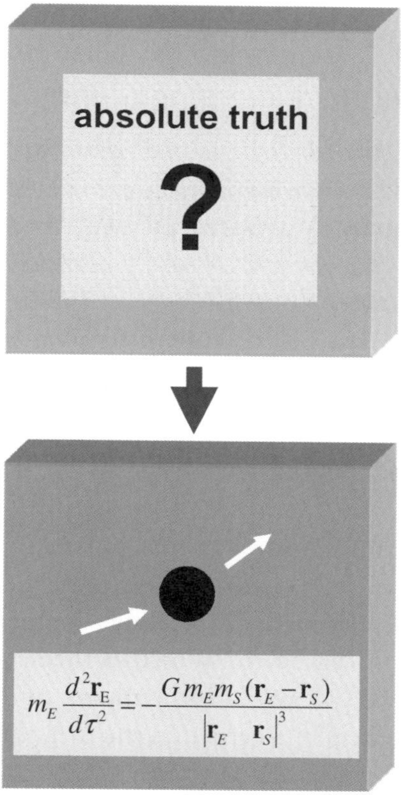


Fig. 1.1. Newton's equations of motion cannot be instruments for the description of the absolute (basic) truth. In particular, we have ascertained that it is hardly realistic to assume that celestial bodies solve differential equations during their motion.

universe, at least in principle. Since the experienced world is essentially relevant for the development of theoretical pictures, these theoretical pictures could also be considered as complete if we were able to observe a complete world. This is however not the case. Why?

The perception of complete reality in the sense of a precise reproduction implies that with growing fine structures, increasing information of the outside world is needed. Then, the evolution would have furnished the sense organs with the property to transmit as much information from the outside world as possible. But the opposite is correct: The strategy of nature is to take up as little information from the reality as possible. Reality outside is not assessed by "complete" and "incomplete" but by "favorable towards life" and "hostile towards life". Concerning this point Hoimar von Ditfurth stated the following [6]:

> *"No doubt, the rule 'As little outside world as possible', only as much as is absolutely necessary is apparent in evolution. It is valid for all descendants of the primeval cell and therefore for ourselves. Without doubt, the horizon of the properties of the tangible environment has been extended more and more in the course of time. But in principle only those qualities of the outside world are accessible to our perception apparatus which, in the meantime, we need as living organisms in our stage of development. Also our brain has evolved not as an organ to understand the world but an organ to survive."*

The principle "as little outside world as possible" can be understood by means of the idea of evolution. The principle of evolution, i.e., the phylogenetic development from simple, primeval forms to highly developed organisms, can be considered as the key for the perception of reality of biological systems. It is the theory of evolution by natural selection which is generally accepted in the meantime. Its foundations have been created by Charles Darwin (1809–1882) more than one hundred years ago. Since then it has been modified and developed further by geneticists. Evolution by natural selection is a two-step process:

*Step 1*: By recombination, mutation, etc., genetic variants are produced at random. Populations with thousands or millions of independent individuals arise.

*Step 2*: Some of these independent individuals will have genes which enable them to manage the predominating situation due to the environment (climate, competition, enemies) better than others. Thus, they have a larger chance for survival than others; they will have, on the statistical average, more descendants than other members of the population. Natural selection takes place in favor of those organisms whose genes have adapted to better cope with the environment.

The number of examples that biological systems have developed in accordance with these criteria is overwhelming. Man and other creatures are characterized by this species-preserving appropriateness. The principle "as little outside world as possible" is compatible with the principles of evolution; it is a succession of evolution. Only those things that are useful for a human and other creatures are relevant. Therefore, the principle "as little outside world as possible" is in a certain kind a "principle of usefulness".

## 1.9. Inside World and Outside World

We experience the world by our sense organs, that is, the observer interacts with realty outside: information about reality outside flow via our senses into the body, and the brain forms a picture of it, and we obtain a "picture of reality". This is definitely a projection. In other words, we have a *reality outside* and an *inner picture* (picture of reality), and we are firmly convinced that the inner picture is identical with reality outside. For example, the well-known psychologist C. G. Jung wrote [7]: "*When one thinks about what consciousness really is, one is deeply impressed of the wonderful fact that an event that takes place in the cosmos outside, produces an inner picture, that the event also takes place inside....*"

This statement by C. G. Jung suggests the following: There is a projection of reality onto space-time and the real world outside is also embedded in space-time. After Jung there is a one-to-one correspondence between the inner picture and the structures in the outside world. Whether or not the outside world is identical with that — which we called above "basic (absolute) reality" — remains an open question here.

### 1.9.1. *One-to-One Correspondence?*

After Jung there is a one-to-one correspondence between the inner picture and the structures in the outside world. However, such a view ignores the principle "as little outside world as possible", that is, the inner picture cannot contain the complete information about the outside world. Again, the statement by C. G. Jung says nothing about whether the outside world is identical with absolute reality and, therefore, it says nothing about the absolute truth with respect to the inner picture. (We know from our analysis in the sections above that it cannot be the absolute truth.)

The common or naive point of view assumes the following: the inside world which we feel to be outside of us actually exists in the outside world in exactly the same form as we perceive. This was also the view of the psychologist C. G. Jung as we have already remarked above. According to this view there is only one difference between the inside world and the outside world: inside there are only geometrical positions, whereas outside there are the real material bodies instead of the geometrical positions. In other words, it is normally assumed that the geometrical positions are merely replaced by material objects.

But why should events in nature occur, so to speak, twice, one outside of us, and again in the form of a picture? This would be against the "principles of evolution" and the "principle of usefulness", respectively.

### 1.9.2. *Cinema and Cinema Ticket*

It would not make much sense if events in a world, which is tailored to fulfil the principle of usefulness, should take place twice. We ascertained in [8, 9] that in nature knowledge for its own sake does not play the major role but rather there is the recognition of the factors "hostile towards survival" and "favorable towards survival". For this purpose a consistent picture of an event occurring in the cosmos must be produced, but not a true reproduction in a one-to-one sense. In particular, such a picture does not need to be complete, because that would unnecessarily burden the biological organism in mastering specific situations in life. The solution of specific problems does not demand a complete knowledge of the world.

From the point of view of evolution the impressions before us are not precise reproductions of reality but merely appropriate pictures of it, formed by the individual from certain pieces of information from the outside world. According to the principles of evolution the central factor is "favorable towards survival" versus "hostile towards survival". The formation of a "true" picture of the world in an absolute sense, which is complete and that represents the absolute truth, is irrelevant. An individual registers situations in the environment in certain patterns which are tailor-made for the particular needs of the species and which are completely free of any compulsion towards precise "objectivity". In particular, we have outlined [8, 9] that we have to assume that an

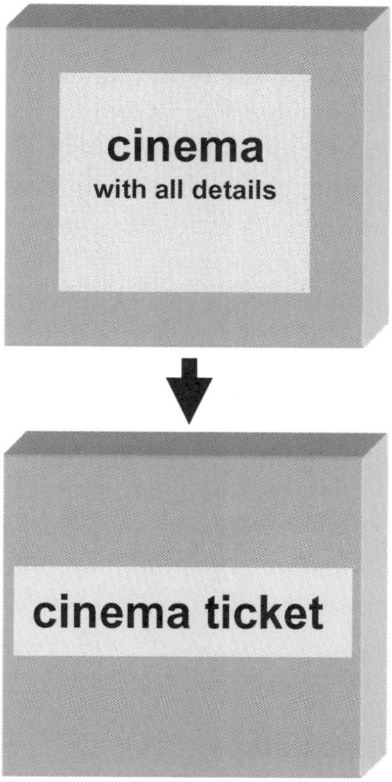


Fig. 1.2. The pictures which are formed by a human observer of the outside world are incomplete and, in particular, they cannot represent elements of the absolute truth. These pictures are primarily useful. So, for example, in order to find a certain place in a cinema, it is not necessary that the visitor gets at the pay desk a small but true model of the cinema, i.e., a precise reproduction of the cinema, which is reduced in size. A simple cinema ticket with the essential information, where one has to go, is more appropriate. In this respect the cinema ticket is the picture of the cinema.

event occurring in the cosmos is portrayed inside a biological system "only" in symbolical form.

Let us quote a simple example (see also [8, 9]). In order to find a certain place in a cinema, it is not necessary that a visitor gets at the pay desk a small but true model of the cinema, i.e., a precise reproduction of the cinema, which is reduced in size; a simple cinema ticket with the essential information is more appropriate. In this respect, the cinema ticket is the picture of the cinema (see also Fig. 1.2).

### 1.9.3. *Summary*

We experience the world by our sense organs and/or its properties are measured with certain measuring instruments. In everyday life the observer interacts with the outside world, that is, information about reality outside flow via our senses into the body and the brain forms a picture of it, and we obtain a picture of the outside world. Due to the principle "as little outside world as possible", the world, which is accessible to a human observer, cannot be complete and does not represent certain elements of the absolute truth.

A lot of scientists believe that the present theoretical structures of the physical laws represent the final and absolute truth. Self-indulgence is dominant! In the light of the biological evolution, the statements by Hawking, and Witten (Section 1.1) appear to be unrealistic; they are in particular misleading and wrong when the principles of evolution are taken into account. In nature cognition does not play the important role but the differentiation between "favorable towards life" and "hostile towards life".

## 1.10. Principal Questions

Now we are able to answer the questions that we have asked in Section 1.2: Can a human observer actually formulate the laws of nature independent of his own nature, that is, independent of his biological structure? Can we really find out "why" we and the cosmos exist? Both questions are of basic relevance because they contain another question: How general can a human observer formulate the laws of nature?

### 1.10.1. *Are Picture-Independent Physical Considerations Possible?*

The observer interacts with the outside world, that is, information about reality outside flow via our senses into the body and the brain forms a picture of it, and we obtain a picture of the outside world. Due to the principle "as little outside world as possible", the world that is reflected by the structures in the pictures must be species-dependent, and this is because the conditions for survival are different for different species.

Only species-dependent information is relevant for the formation of the pictures, and an observer has only access to the outside world over a picture formed by the observers' brain. This in particular means that only picture-dependent statements are possible for man since the species-dependent features exclusively come into existence through the picture. We are imprisoned in this system and cannot escape from it. In other words, a picture-independent point of view must be excluded. Statements about basic reality, which is by definition species-independent, are not possible for man. Thus, a human observer is principally not in the position to make statements about the "absolute truth", which is embedded in basic reality.

Then, we have to answer the question "Can a human observer actually formulate the laws of nature independent of his own nature, that is, independent of his biological structure?" negatively. The laws of nature, formulated by man, are dependent on his biological structure. The laws of basic reality are not accessible to man.

### 1.10.2. *Why Do We and the Cosmos Exist?*

Since man is principally not able to make statement about basic reality, he cannot say "why" we and the cosmos exist. Here the "why" refers to basic reality and, therefore, it is a "why" in the absolute sense. Then, we have also to answer the question "Can we really find out 'why' we and the cosmos exist?" negatively.

When we are able to derive a certain physical law (its mathematical structure) by a mechanism, we may state that this mechanism tells us "why" the effects, which follow from this physical law, happen.

The understanding in terms of mechanisms or models most often assumes an understanding of effects on the basis of mechanisms that we observe in everyday life. However, such a "why", such a mechanism, would be species-dependent and would not be valid for an other species, which have in general other everyday-life experiences than man.

## 1.11. How Does Science Progress?

We can also learn from the theory of science that the "absolute truth" remains principally hidden. To show this we have first to investigate the situation within the so-called "asymptotic convergentism".

### 1.11.1. *Science Progresses by Eliminating the Number of Unanswered Questions*

Let us suppose that there exists a defined set $R_S$ of problems which, in the course of time, are solved, i.e., the questions asked are answered successively. If we denote the number of answered questions at time $\tau$ by $R$ and the number of answered questions at time $\tau'$ ($\tau < \tau'$) by $R'$, we get (see Fig. 1.3)

$$R < R'. \tag{1.1}$$

Accordingly, the number of questions capable of formulation decreases:

$$(R_S - R) > (R_S - R'). \tag{1.2}$$

Within the framework of such a principle all questions will be answered in the course of time, i.e., a maximum of scientific knowledge is achieved.

### 1.11.2. *Principle of Propagation of Questions*

In contrast to that, Immanuel Kant (1724–1804) advocated the "Principle of Propagation of Questions". This principle means that each answer to a scientific question gives rise to new questions. According to Kant this applies because each answer provides new material feeding new questions. But new questions generally change the supposition and this may lead to an extension of the problem

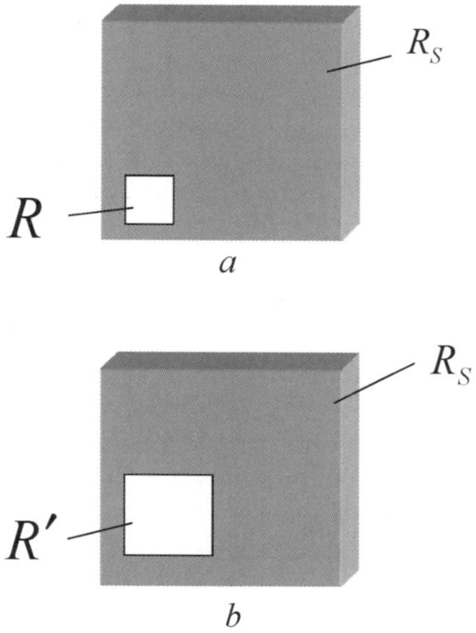


Fig. 1.3. (a) Knowledge at time $\tau$: There is a constant number of $R_S$ problems; the number of answered questions is $R$. (b) Knowledge at time $\tau' > \tau$: There are $R_S$ problems; the number of answered questions is $R' > R$.

horizon: Instead of using Fig. 1.3, we are induced to read Fig. 1.4. Now a variable set of problems is encountered. Then, an increase in the number of unanswered questions can very well accompany an increase in answered questions. Since

$$R_S < R'_S \tag{1.3}$$

(see Fig. 1.4) the following relations

$$R < R' \tag{1.4}$$

$$(R_S - R) < (R'_S - R') \tag{1.5}$$

are possible.

However, the opinions of the scientists quoted above (Lord Kelvin, Feynman, etc., Section 1.5) can well be harmonized with Kant's "Principle of question propagation" provided that not only the number of questions but also their relevance is taken into consideration. As a matter of fact, if the relevance of a problem gradually decreases in the course of time, "later" science must be less important. In that case, it

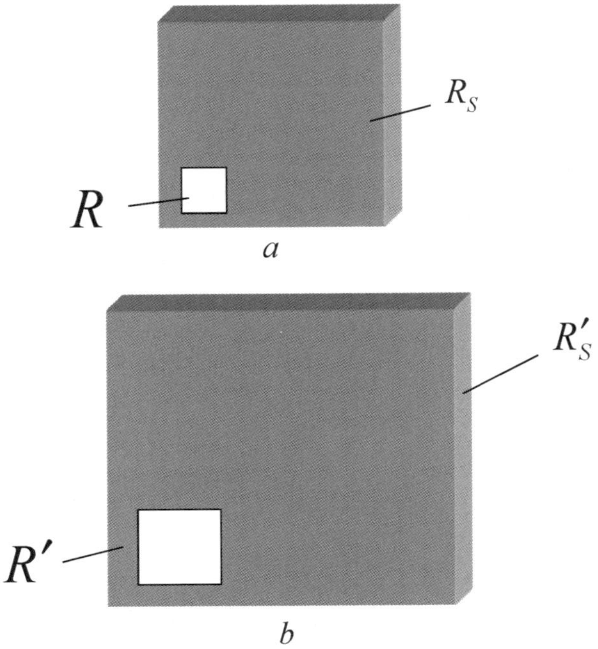


Fig. 1.4. (a) Knowledge at time $\tau$: There are $R_S$ problems; the number of answered questions is $R$. (b) Knowledge at time $\tau' > \tau$: There are $R'_S$ problems; the number of answered questions is $R'$. The number of answered questions increases with time.

can be quite correctly asserted that the basic structure of physics has been worked out although the number of questions to be answered increase with time.

Such a view of the progress of science suggests the analogy with geographic research. In this context, Rescher states among others [4]:

> *"Scientific inquiry would thus be conceived of as analogous to terrestrial exploration, whose product — geography — yields results of continually smaller significance which fill in ever more minute gaps in our information. In such a view, later investigations yield findings of ever smaller importance, with each successive accretion making a relatively smaller contribution to what has already come to hand. The advance of science leads, step by diminished step, toward a fixed and final view of things."*

Accordingly, science progresses by approaching the truth successively: The "final answer" and the "final view" of things, respectively, is gradually approached by the way of an asymptotic approximation. According to Peirce (1839–1914) this truth is obtained in the limit $\tau \to \infty$:

$$R_\infty = \lim_{\tau \to \infty} R_\tau, \tag{1.6}$$

where $R_\tau$ is the time dependent status of knowledge which approaches asymptotically the definite truth $R_\infty$. This detailed filling (accumulation) of given, fundamentally defined patterns resembles greatly the calculation of further decimal points in order to additionally refine a result already roughly estimated, such as in calculating the numerical value of:

$$\begin{aligned} \pi_1 &= 3,1 \\ \pi_2 &= 3,14 \\ \pi_3 &= 3,141 \\ &\vdots \\ \pi &= \lim_{n \to \infty} \pi_n. \end{aligned} \tag{1.7}$$

At least until one generation ago, the opinion was firmly established that science is cumulative and the advocates of the scientific method tried to understand scientific progress to have this cumulative nature (see also [4]). Within this concept the "absolute truth" is set equal to "our ultimate truth". But we will see below that the asymptotic convergentism cannot be upheld in the recognition theory and scientific history.

### 1.11.3. *Substitution Instead of Successive Refinement*

All the statements cited in Sections 1.1–1.6 can be classified on the basis of the asymptotic convergentism. However, essentially two serious objections can be quoted against the statements in Section 1.11.2, i.e., against the asymptotic convergentism:

1) There is no metric to measure the "distance" between bodies of knowledge
The asymptotic convergentism is from the beginning burdened by the great difficulty that it cannot give a metric system allowing us to define the interval concerning

$$R_{\tau} - R_{\tau'}$$

between the status of knowledge $R_{\tau}$ and the status of knowledge $R_{\tau'}$. This means that we are not in the position to decide whether or not we have approached the real truth. How can criteria be formulated which allow such an approximation, that is, to find a metric that is able to measure the "distance" of knowledge? We cannot! There is simply no neutral standpoint in theoretical terms, i.e., no neutral, elevated level (external to science) which could form the basis for a direct comparison between theoretical configurations $R_{\tau}$ and $R_{\tau'}$, etc. Scientific progress can only be measured in connection with the so-called pragmatic level; this point will be outlined in more detail below.

2) There is a fundamental change in perspective
The assumption of a successive approximation cannot be maintained in view of the history of science because the analysis of theories succeeding each other in time shows that the later theory is generally not only supplemented and refined, respectively, but reformulated on the basis of new first principles. A basic change in perspective takes place. Normally the problem does not consist of just adding some further facts but of structuring a new frame of thinking. This situation has been described in a highly instructive manner by Thomas Kuhn who compared Newton's theory to the Theory of Relativity [10] (Einstein' theory), as compared to Newton' theory, provides a basically novel frame of thinking. A change of perspective has taken place, involving the notational decoupling and displacement, respectively, of the network of notions.

## Estimates of the truth

Generally speaking, we have to conclude that a "later" theory became necessary because the "earlier" theory had been limited in its scope. This led to a basically novel concept of the way of seeing nature. Thus, normally, not only improvements and refinements, respectively, are made, but the "earlier" theory is downright replaced by the "later" one. The history of science provides a wealth of examples supporting this statement [4, 10]. Therefore, we have to assume consistently that the "later" theory will also have to be abandoned at some point in time. So, each frame of thinking, independent of the era in which it has been conceived, can never constitute a frame for the "absolute truth". In this context we cannot even provide evidence that we have come closer to the "absolute truth" because no metric system can be defined to measure intervals in recognition. A framework of thinking and a theory, respectively, do not reflect the "absolute truth" but, as formulated by Rescher [4], an "estimate of the truth", which is to be understood as meaning of a tentatively postulated provisional truth.

## Incommensurable structures of thinking

When changing from one to another frame of thinking, generally a conceptual decoupling takes place so that successive structures of thinking may become incommensurable. The advocates of incommensurable theories are basically not in a position to understand each other, because it is not reasonable to make comparisons between incommensurable structures of thinking. Using the following analogy, Rescher reduces the problem to the point [4]: "*One can improve upon one's car by getting a better car, but one cannot improve it by getting a computer or a dishwashing machine.*"

## The pragmatic level

Scientific progress can be defined only if it is possible to project certain tendencies of two incommensurable structures of thinking to one "appropriate" third level (Fig. 1.5). This third level, which is a sort

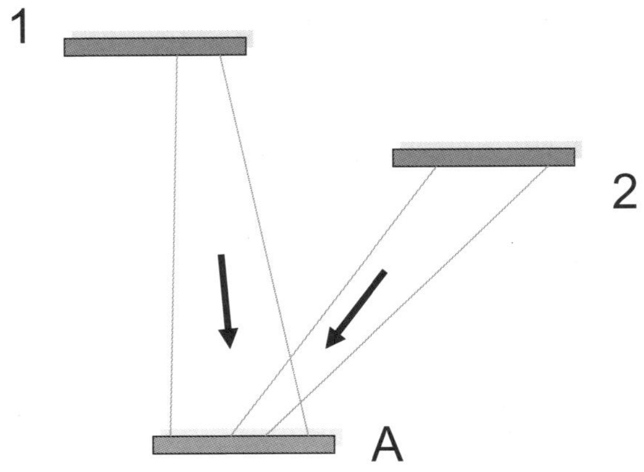


Fig. 1.5.

of reference system (level A in Fig. 1.5) generally will have a "coarser" and "more global" structure than the two incommensurable structures of thinking (levels 1 and 2 in Fig. 1.5). Level A is coarser and more global because, generally, such a projection is not detailed, i.e., a point by point projection and provides only an integral picture (e.g., by averaging). These integral variables of level 1 and level 2 have things in common, if they cover a joint zone on level A. Level A is "appropriate" for comparing two structures of thinking, if it offers to them a finite surface for projection. A structure of thinking is superior to another if it describes in more detail and accuracy the body of facts on level A. In this way it is possible to compare (albeit to a limited extent only) two competing theories. In Fig. 1.5 the structures of thinking underlying level 1 is superior to that of level 2 because the surface (equivalent to the status of recognition which, starting from level 1, explores level A) of level 1 projected onto level A is larger than the surface of level 2 projected onto level A.

Related to the situation of man, level A might constitute the level of everyday life, of technological applications, and of experimental explorations. Thus, technological progress and the elucidation by experiments become the touchstone (albeit to a limited extent) of deviating theoretical positions. Applying this yardstick, a "later" theory must preserve and improve the practical successes of its predecessors;

it is then superior to the "earlier" theories. But it must always be kept in mind that the judgement of a given structure of thinking from the pragmatic level provides only a restricted perspective.

Scientific progress defined in this way will depend in many domains essentially on the technological progress because natural science often needs a more sophisticated technology in order to perform its increasingly complicated interactions with nature. According to what has been said before, from the cognitive point of view, natural science repeatedly starts from its origin; however, given the limits imposed by technology, this process, for merely practical reasons, will proceed at an ever slower rate.

### 1.11.4. *Summary*

Within the asymptotic convergentism it can rather be assumed that the world is actually as science envisages it to be. Here, fundamentally new pictures of the world are not generated again and again, but one frame is filled successively. In that case the concept is justified that theoretical terms like electrons, quarks, etc. are actually existing entities in the world. However, as already outlined in detail above, the asymptotic convergentism can no longer be upheld. The empirical finding (resulting from the analysis of facts provided by the history of science) that science progresses by a sequence of incommensurable schemes of thinking (pictures) must be given a rank equal to that of a relevant experiment in a laboratory. In this connection the following point is relevant: Because a metric for the measurement of the "distance" between bodies of knowledge is not definable, there is no possibility to express certain peculiarities of the absolute truth and basic (absolute) reality, respectively.

## 1.12. Final Remarks

We have asked the following important question: Can the progress in science, in particular in physics, lead to a final truth of the physical world? It cannot! The absolute truth, which is so to speak embedded in basic reality, can never be observed and, therefore, theoretical considerations about it make no sense.

The perception of complete reality in the sense of a precise reproduction implies that with growing fine structures increasing information of the outside world is needed. Then, the evolution would have furnished the sense organs with the property to transmit as much information from the outside world as possible. But the opposite is correct: The strategy of nature is to take up as little information from the reality as possible. Reality outside is not assessed by "complete" and "incomplete" but by "favorable towards life" and "hostile towards life".

The common or naive point of view assumes the following: the inside world which we feel to be outside us, actually exists in the outside world in exactly the same form as we perceive. According to this view there is only one difference between the inside world and the outside world: inside there are only geometrical positions, whereas outside there are the real material bodies instead of the geometrical positions. In other words, it is normally assumed that the geometrical positions are merely replaced by material objects.

But why should events in nature occur, so to speak, twice, one outside of us, and again in the form of a picture? This would be against the "principles of evolution" and the "principle of usefulness", respectively. It would make not much sense if events in a world, which is tailored to fulfil the principle of usefulness, would take place twice. In this connection we stated the following: "*From the point of view of evolution the impressions before us are not precise reproductions of reality but merely appropriate pictures of it, formed by the individual from certain pieces of information from the outside world. According to the principles of evolution the central factor is 'favorable towards survival' versus 'hostile towards survival'. The formation of a 'true' picture of the world in an absolute sense, which is complete and that represents the absolute truth, is irrelevant. An individual registers situations in the environment in certain patterns which are tailor-made for the particular needs of the species and which are completely free of any compulsion towards precise 'objectivity'.*" So, a visitor of a cinema does not get at the pay desk a small but true model of the cinema (a precise reproduction of the cinema,) in order to find a certain place in the cinema, a simple cinema ticket with the essential information is more appropriate. In this respect, the cinema ticket is the picture of the cinema.

The facts from the theory of science have taught us that we will never be able to make final and complete statements about the physical universe, i.e., from the point of view of theory of science the "absolute truth" can never be obtained. Why? Because a metric for the measurement of the "distance" between bodies of knowledge is not definable, there is no possibility to express certain peculiarities of the absolute truth and basic (absolute) reality, respectively. An absolute reality may exist within theory of science but it is not accessible to man.

The results that follow from the theory of science strongly confirm the statements, which we have worked out in connection with the principles of evolution and the principle of usefulness, respectively.

# 2

# The Projection Principle

In Chapter 1, we came to the conclusion that a human observer is not able to grasp the "absolute truth". This deepest ground of reality exists but remains hidden to man and possibly for all biological creatures. The absolute truth can be thought of as embedded in basic reality or can be directly identified with basic reality. In this chapter we would like to analyze space and time, and we will come to features for these elements, which are compatible with what we have extracted in Chapter 1 from the principle of usefulness that is closely linked to the phenomenon of evolution. In this connection, we will introduce the so-called "projection principle" that is obviously of basic relevance for the understanding of physical phenomena.

## 2.1. The Elements of Space and Time

Within the memory of man all essential things are represented within the frame of pictures. This is obviously the most basic statement. Such pictures contain the objects (sun, moon, etc.), the space and the time. We feel the objects, that is, we observe them with our five senses and also with measuring instruments. Thus, we believe in the concrete existence of these things and we call them "material objects". But what can we say about space and time? It is not easy to answer this question but only because these notions, space and time, are burdened with a serious prejudice. We are firmly convinced that all the material objects around us are embedded in space and space–time, respectively. It is normally assumed that the space is a container filled with matter. However, there are some reasons to take another standpoint and to consider this container-conception as unrealistic, as prejudiced, so to speak. Already the philosopher Immanuel Kant pointed out this serious problem [11, 12].

All that we can grasp with our five senses and/or with measuring instruments, can be considered as really existing in the outside reality. However, can we seize space and time with our five senses or with specific measuring instruments? No, we cannot. Therefore, we may conclude that space and time do not exist in the reality outside. Although this conclusion is straightforward, it must be more justified and underpinned with further arguments and more details.

The situation with respect to space and time is very clear: We cannot put a "piece of space" on the table and we can also not put a "piece of time" on the table. Such pieces do not exist. We are not able to observe the elements of space, that is, its coordinates $x, y, z$ and we are also not able to observe the elements of time which we have denoted in Chapter 1 by the letter $\tau$ (normally the letter $t$ is used). We have no senses for that and, furthermore, the development of measuring methods for the detection of $x, y, z$ and $\tau$ are even not imaginable.

Within Newton's mechanics, space and time are independent of each other and also independent of matter. This concept has been developed further by Einstein; within the Theory of Relativity, space and time are no longer independent of each other but form a space–time. Then, the following question arises: Can we put a "piece of space–time" on the table? No, we definitely cannot. As in the cases of space and time, also a piece of space–time does not exist in the sense of an observable quantity.

Nevertheless, we are all familiar with the space–time phenomenon. Thus, we have to conclude that the elements $x, y, z, \tau$ must be inside the observer and, as we will recognize in the course of our debate, it makes no sense to assume that these elements $(x, y, z, \tau)$ are also outside the observer. As we have already mentioned above, already the philosopher Immanuel Kant took this standpoint, but his fundamental ideas have never been seriously used in the formulation of physical theories.

## 2.2. Relationship between Matter and Space–Time

In what form do we experience the space–time phenomenon? The answer is quite clear: We never observe isolated space–time points which we characterize by $x, y, z, \tau$ but without exception only distances of real objects and time intervals in connection with physically real processes. In other words, space and time (or distances of them) never appear without objects and processes. In all cases "matter" and "space and time" are closely linked; neither of them is able to exist without the other. Nobody is able to observe isolated space–time points $x, y, z, \tau$ without objects and processes, respectively.

Is this important experimental finding fulfilled within modern physical theories? No, it is not. Within Newton's mechanics space and time are independent of each other and both elements may exist independent of matter. But also within the Theory of Relativity space–time may appear without material objects (de Sitter already showed in 1917 that an empty space–time is possible within General Theory of Relativity).

## 2.3. Two Relevant Features

In summary, the relationship between "matter" and "space–time" is given by the following facts and features, respectively:

### 2.3.1. *Feature 1*

We definitely cannot see, hear, smell, or taste space and time, that is, space and time (absolute or non-absolute) are not accessible to our senses. Also measuring instruments for the experimental determination of the space–time points $x$, $y$, $z$ and $\tau$ are not known and even not thinkable.

Therefore, space and space–time, respectively, may not be the source of a physically real effect. This requirement is part of Mach's principle which is however not fulfilled within Newton's theory but also not within the Special and General Theory of Relativity. It is obviously not possible to fulfil Mach's principle when we assume that the material bodies are embedded in space (space–time). We will discuss Mach's principle in more detail in Chapter 5.

### 2.3.2. *Feature 2*

We can only say something about *distances in connection with masses*, and time intervals in connection with the physical processes. There is no exception!

Remarks concerning feature 1:

Space and time should never be the source for physically real effects as, for example, inertia. This is not fulfilled within Newton's theory but also not within Einstein's theory. It is still possible within Theory of Relativity to talk about the rotation of the entire mass of the world relative to absolute space; within Gödel's universe (see in particular [11–13]), an absolute space–time is defined. Furthermore, a lone particle in space–time shows the effect of inertia, that is, in this case space–time is the source of inertia.

Remarks concerning feature 2:

a. It follows directly from feature 2 follows that "matter" and "space–time" are closely linked; neither should be able to exist without the other. In other words, an *empty* space–time as physical-theoretical conception should not be existent.
b. When we define a certain distance in space by two material bodies, then there must be a relationship between the two masses and space. The material bodies can have, for example, a *constant* space distance which already expresses a certain kind of relationship. But what kind of relationship? This cannot be due to an interaction between the material bodies and space (space coordinates). The reason is given by feature 1. There can be no interaction between space–time and the sense organs or measuring instruments which are made of material bodies because there can be no interaction between material bodies (masses) and space–time.
c. But what kind of relationship exists between the two material bodies that determine a certain space distance? In the following, we will answer this question.

In conclusion, space and time are not accessible by the human senses, and no measuring instrument is able to make space–time-specific "clicks" but, on the other hand, we experience space–time in connection with material bodies and processes as a concrete phenomenon.

## 2.4. Two Kinds of "Objects"

Due to feature 1 space–time has to be considered as "nothing" in the physical sense but, on the other hand, we experience space and time as a real phenomenon. Clearly, this "nothing" can hardly be occupied by physically real objects. It is problematic to embed the real world into such a metaphysical space–time. Within all modern physical theories (Newton's mechanics, Theory of Relativity, conventional quantum theory, etc.) real material bodies are the contents of an "object" (space–time) that has to be considered as a nothing or as a metaphysical substratum different from matter.

This seems to be a contradiction. Two kinds of objects are mixed which are obviously mutually incompatible. We can overcome this bad situation if we take the position of the philosopher Immanuel Kant: "*Space and time are exclusively features of our brain and the world outside is projected on it*". Then, the material objects that occupy space and time can only be geometrical pictures (Fig. 2.1). In other words, space–time

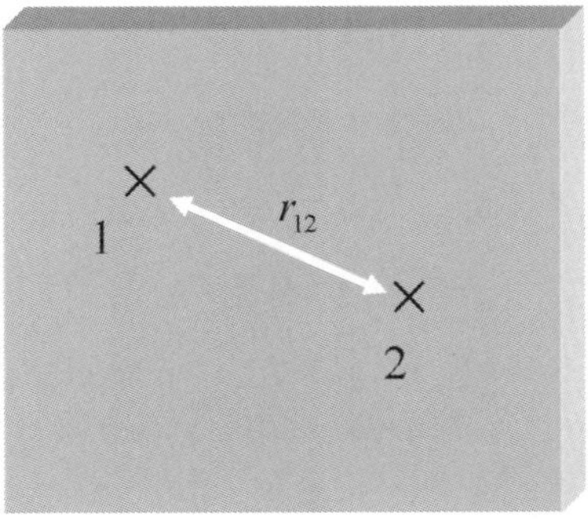


Fig. 2.1. Within the memory of man all essential things are represented within the frame of pictures. This is obviously the most basic statement. What we always have in front of us in our observations in everyday life are geometrical objects which are embedded in space and time. The material objects themselves are not embedded in space and time but only their geometrical positions. In the figure, the geometrical positions $(x_1, y_1, z_1, x_2, y_2, z_2)$ of two objects 1 and 2 are represented by crosses which have the distance $r_{12} = [(x_2 - x_1)^2 + (y_2 - y_1)^2 + (z_2 - z_1)^2]^{1/2}$ at a certain time $\tau$.

has to be considered as an auxiliary element for the representation of physically real processes.

Projection theory (Chapter 3) is based on this conception. If that is correct, the situation would be identical with what we do in connection with our blackboards and notebooks: we draw pictures on them.

## 2.5. Perception Processes

Everything that we see is primarily in our head; it is not outside us. Persons, cars, aeroplanes, the sun, moon and stars are "pictures of reality" in our brain; we have only the impression that all these things are located outside us (see also Fig. 2.2). This conclusion is supported in [14]: "*We have devices in the cerebral cortex which — comparable with a television screen — produce "pictures" in our awareness from the nerve-excitations coming from the retina. It is characteristic for the sight-process that our awareness does not register the picture of a candle on the retina inside the eye, but we have the impression that we are standing opposite the candle-light which is located in the space outside, not standing on its head but upright. We see "real objects" in front of us and around us. Within*

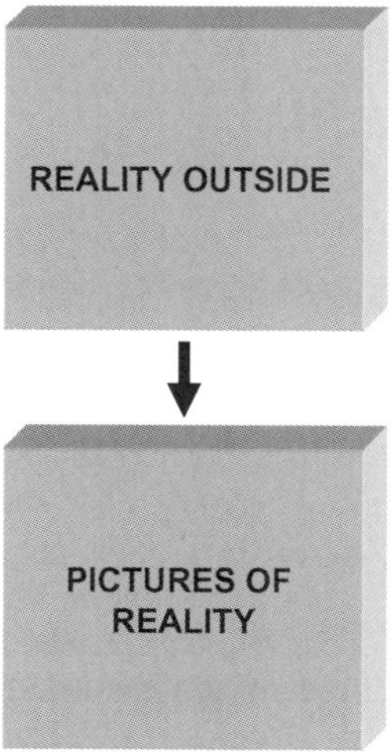


Fig. 2.2. Information of "reality outside" flow via our sense organs into the brain and forms a "picture of reality". We never observe reality outside directly but "only" pictures of it.

*this act of perception, the eye, the optic nerves and the brain work together. To see without the brain is as impossible as to see without eyes.*"

This text means the following: Everything of the observable world is inside the brain, and this situation expresses itself in the fact that the world appears in front of the observer. This situation is particularly shown by an experiment with inverting goggles.

### 2.5.1. *The Experiment with Inverting Goggles*

In Ref. 10, we find the following description of the experiment: "*An experimental subject who puts on goggles fitted with inverting lenses initially sees the entire world upside down. At the start his perceptual apparatus functions as it had been trained to function in the absence of goggles, and the result is extreme disorientation, an acute personal crises. But after the subject has begun to learn to deal with his new world, his entire visual field flips over, usually after an inverting period in which vision is simply confused. Thereafter, objects are again seen as they had been before the goggles were put on. The assimilation of a previously anomalous visual field has reacted upon and changed the field itself. Literally as well as metaphorically, the man accustomed to inverting lenses has undergone a revolutionary transformation of vision.*"

Where is the material world?

All happens in the brain, but it develops in front of us in space, i.e., seemingly outside of us. This is confirmed by the experiment with an observer with inverting goggles. Here we have the following situation: It is the picture inside the brain that is transformed due to the inverting goggles. After the transformation a picture appears which is identical with that before the goggles were put on, although the goggles are still up after the transformation. That is, the observers' brain ignores the inverting goggles. This transformation seemingly happens directly in front of the observer. In other words, that, which is directly in front of the observer cannot be the real, material world because the observers' brain cannot change this material world outside that is apparently embedded in space. In other words, what is directly in front of us, cannot be the material world.

The picture of the world is changed by the brain functions, and this expresses itself by the fact that the world in front of the observer is changed. This is particularly compatible to what we have outlined above about space and time: There can be no space and time in the material world outside, but space and of course time is positioned within the observers' brain.

Most people firmly believe, as a matter of course, that the impressions in front of them directly reflect the real, material world, which is assumed to be embedded in space–time. But this is a fallacy, and this follows directly from the facts outlined above.

Why is there a picture-transformation?

The visual conceptions of the "observer's world" are adapted to what has been developed during the evolution process. Only these visual conceptions are useful for him in the mastering of the species-relevant situations. The brain has obviously stored these conceptions and recognizes that the world with inverting goggles does not belong to these conceptions (developed by the principle of usefulness in the course of evolution) and it simply transforms the picture inside the brain into that before the goggles were put on, where the picture in the brain is nothing more than the impression in front of the observer. Again, the transformation happens although the goggles are still up. In other words, the experiment with inverting goggles confirms the principle of usefulness.

### 2.5.2. *Space and Time Come into Existence by Specific Brain Functions*

We normally assume that these sensations produced by the brain are identical with reality itself (Fig. 2.3), but this should not be the case as we have argued above since space–time cannot be outside the brain because space–time has to be considered as an auxiliary element for the representation of physically real processes. In other words, the outside world, the material bodies, cannot be embedded in space–time. That in particular means that not only the things in front of us (cars, houses, trees, etc.) are in our head but also space–time, where all these

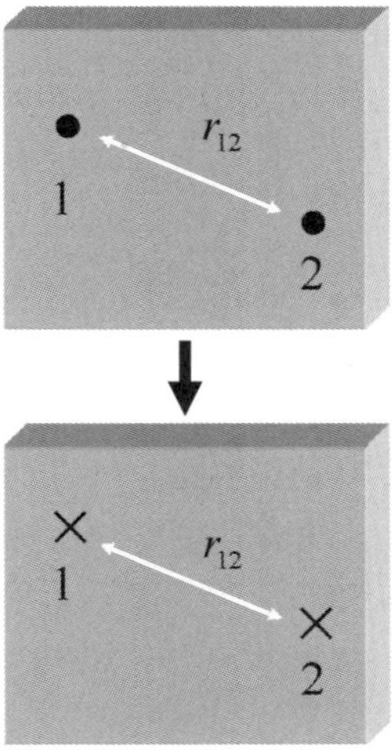


Fig. 2.3. Most people assume *a priori* that the things which they observe and which appear spontaneously in front of them is the material reality itself, and those which are conscious of the fact that it is only a picture normally assume, as a matter of course, that the structures in the picture are identical with the material reality outside, that is, it is assumed that in reality outside the geometrical positions in the picture (crosses in the figure) are replaced by material objects (full points in the figure). This can however not be the case and is obviously a fallacy. This is supported by a large number of indications.

things are positioned. We have only the impression that all these "hard objects", together with space–time, are located outside us. Space and time are obviously elements of the brain; they come into existence due to specific brain functions. This is the reason why we cannot put space and time on the table as we can in the case of matter.

Thus, we have to assume that there is no similarity between the structures and characteristics in the picture (Fig. 2.1) and those in the actual reality outside. How reality outside is constructed can principally not be said because, as is outlined in Chapter 1, a picture-independent point of view does not exist for a human observer. Thus, instead of Fig. 2.3 we obtain Fig. 2.4.

What about feature 2 (Section 2.3)? This feature is expressed by the following fact: *We can only say something about distances in connection with the masses, and time intervals in connection with the physical processes.* The solution for this finding is relatively simple and can be understood within projection theory by the concept that space

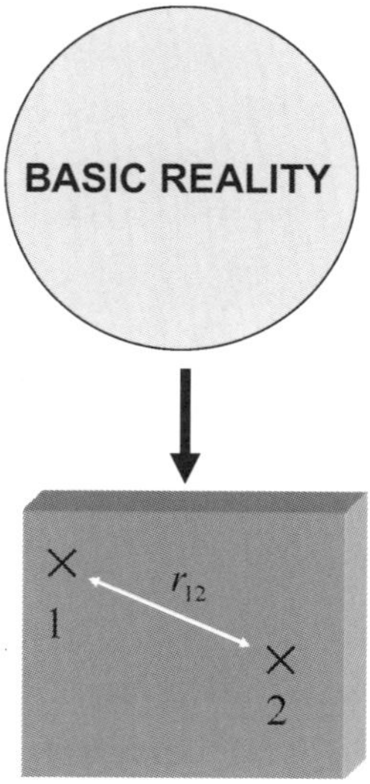


Fig. 2.4. The full circle symbolizes basic (objective) reality. It cannot be observed directly; no statements can be made about basic reality but certain characteristics of it can only appear in the form of pictures in front of us and which are represented in space and time. Space and time do not appear within basic realty.

and time only appear (in the brain) when there is actually something (objects of basic reality) to picture. Then, matter is closely linked to space and time. Why has evolution chosen this way? Here the answer is straightforward: Each species developed in connection with the laws of evolution (Chapter 1, Section 2.7) and these laws are dictated by the "principle of usefulness". An observation of an empty space (space–time) would not be useful because it does not contain any information which the observer needs. The principle of usefulness is obviously only fulfilled if the selected objects are closely linked to space and time.

## 2.6. Inside World and Outside World

We experience the world by our sense organs, that is, the observer interacts with reality outside: information about reality outside flow via our senses into the body, and the brain forms a picture of it. In this way, we obtain a "picture of reality". This is definitely a projection. In other words, we have a *reality outside* and an *inner picture* (picture of reality), and we are firmly convinced that the inner picture is identical with the structures of reality outside. In Chapter 1, we have quoted, as an example, the opinion of the well-known psychologist C. G. Jung who was impressed that an event takes place twice, one in the head of the observer and the other identical event in the world outside.

This view means the following: Although there is a projection of reality onto space–time we may also state that the real world outside is embedded in space–time. But, due to feature 1, this is definitely forbidden, as we have analyzed above. Strictly speaking, there can be no space–time in connection with reality outside because space–time can only be an auxiliary element for the representation of physically real processes that take place outside. It makes no sense to assume that space–time is also an element of reality outside. The *unobserved* reality outside is free of space and time (Fig. 2.4 is correct and not Fig. 2.3).

Thus, the structure of reality outside must be quite different from the structure of the inner picture that is always based on space and time. However, the inner picture is the most direct impression we have of reality outside. Arthur Eddington wrote (Swarthmore Lecture (1929), *Science and the Unseen World*): "*In comparing the certainty of things, spiritual and things temporal, let us not forget this — Mind is the first and most direct thing in our experience; all else is remote inference ...*".

Again, the inner picture must be quite different from the structures in the outside world and this is because there can be no space–time in the outside world.

## 2.7. The Influence of Evolution

All the statements we have made in connection with space–time and its relationship to reality are completely in accord to that what we have outlined in connection with "basic reality" and the "picture of reality" (see Chapter 1). Let us briefly discuss this point.

In Chapter 1, we came to the conclusion that the "absolute truth" (basic reality) must be quite different from the "picture of reality", which the observer directly experiences (which is directly in front of us). In particular, we pointed out that there should be no point-to-point correspondence between the structures in basic reality and those in the picture of reality. We come exactly to the same conclusion when we consider the features of space–time, which we have outlined so far in this chapter.

The fact that space and time are not elements of the world outside means the following: The structures in the world outside (basic reality) without space–time must be different from those in the picture of reality, which is formed on the basis of space and time. In other worlds, there can be no point-to-point correspondence between the structures in basic reality and those in the picture of reality. As said, this fact is in accordance with that, which we have worked out in Chapter 1 on the basis of the principles of evolution. Therefore, the reason for this specific space–time behavior must be given by the principles of evolution. Thus, we may conclude that the space–time elements have been developed in accordance with the principles of evolution. Exactly the same arguments (used in Chapter 1) are valid for space and time; we only have to repeat what we have outlined in Chapter 1 about the relationship between basic reality and the picture of reality.

The perception of true reality in the sense of a precise reproduction (one-to-one correspondence) implies that we need as much information from the actual reality outside as possible; in the case of a precise reproduction we need the complete information about the outside world. Then, evolution would have developed sense organs with the property to transmit as much information from reality as possible. However, the opposite view is correct: The strategy of nature is to take up as little information from the outside world as possible. The reason is relatively simple.

The solution of specific problems does not demand a complete knowledge of the world. From the point of view of evolution, the impressions before us are not precise reproductions of reality but merely appropriate pictures of it, formed by the individual from certain pieces of information from the outside world. According to the principles of evolution, the central factor is "favorable towards survival" versus "hostile towards survival". The formation of a "true" picture of the world in an absolute sense is irrelevant. An individual registers situations in the environment in certain patterns which are tailor-made for the particular needs of the species (principle of usefulness) and which are completely free of any compulsion towards precise "objectivity". In particular, we have outlined in Chapter 1 that we have to assume that an event occurring in the cosmos is portrayed inside a biological system "only" in symbolical form. This is consistent with the fact that there is no space–time in basic reality but only in the picture leading to completely different configurations.

In this connection, we discussed in Chapter 1 a simple example. In order to find a certain place in a cinema, it is not necessary that a visitor gets at the pay desk a small but true model of the cinema, i.e., a precise reproduction of the cinema, which is reduced in size; a simple cinema ticket with the essential information is more appropriate. In this respect, the cinema ticket is the picture of the cinema.

We ascertained in Chapter 1 that in nature knowledge for its own sake does not play the major role but rather the recognition of the factors "hostile towards survival" and "favorable towards survival". For this purpose a consistent picture of an event occurring in the cosmos must be produced, but not a true reproduction in a one-to-one sense. In particular, such a picture does not need to be complete, because that would unnecessarily burden the biological organism in mastering specific situations in life.

## 2.8. Information in the Picture Versus Information in Basic Reality (Outside Reality)

From the principles of evolution we have to conclude that basic reality contains much more information than an observer can depict in space and time (Fig. 2.5). This space–time information is mainly that part of the outside world which a human observer needs for survival.

The contents of the space–time information, selected from basic reality, are dictated by the principles of evolution. The unselected information, on which the selected picture of reality is based, remains principally unknown. For example, we know the observers shape in space and time but not his structure in basic reality (reality outside). However, the projection principle also says that the observer must be more than what is pictured in space and time which exclusively

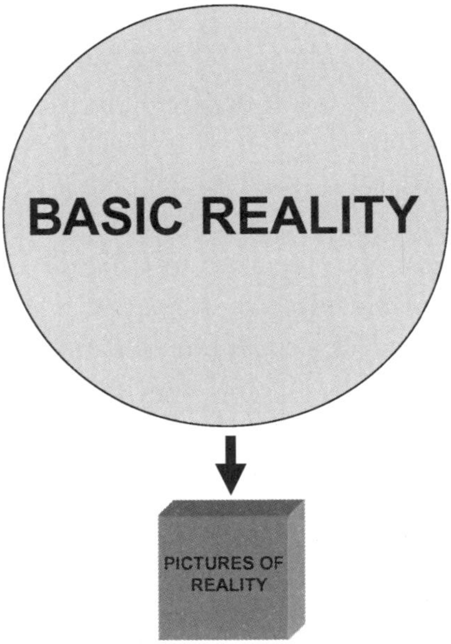


Fig. 2.5. Basic reality (symbolized by the full circle) in relation to the part of information (box with "picture of reality") an observer can experience of basic reality. The circle is much larger than that in Fig. 2.4 indicating that the basic reality contains more information (possibly much more information) than an observer can depict in space and time. It is mainly the information of the outside world which a human observer needs for survival.

shows the material part of the observer. But this material part is only the selected information about the observer, and the unselected information must necessarily be "richer" than the material part which is pictured in space and time (see also Figs. 2.6 and 2.7), that is, we may not conclude that the space–time pictures contain the complete information about the reality outside. In fact, we know that a human observer has mind and feelings; his intellectual abilities are accompanied by logical laws that have to be considered as real extensions of the material world. However, these logical laws etc. are not depictable in space and time.

Why can certain things be represented in space and time and others not? The answer is simple and is also given here by the principles of evolution. Certain facts are relevant and other facts are less relevant. Those things and processes, respectively, which are depictable in space

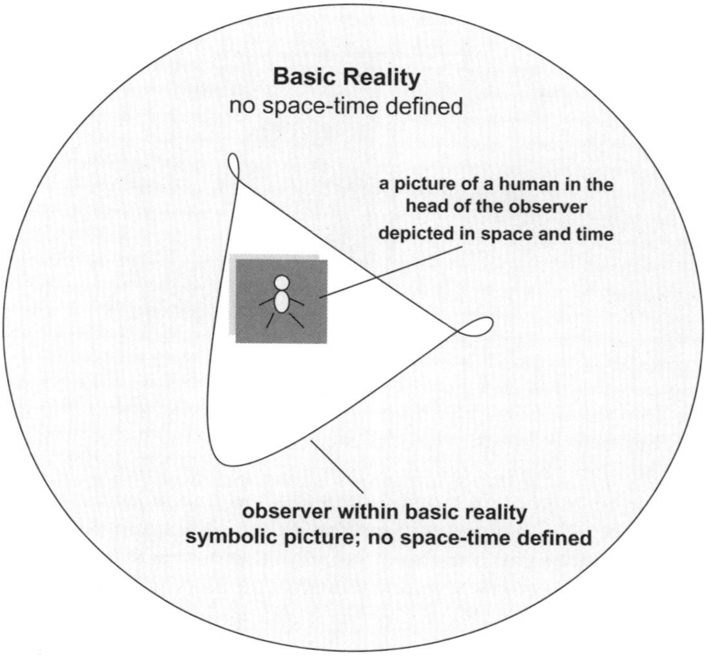


Fig. 2.6. A human observer is able to make statements about himself only in connection with space and time, but he has not the ability to conclude from the space–time picture what exists in basic reality. Therefore, the form of the observer in basic reality has been chosen arbitrarily. This arbitrarily chosen shape expresses that the space–time picture of the observer in his head must be different from what exists within basic reality.

and time, are of particular importance for the survival of a human observer. Intellectual statements, that is, the products of mind, are also relevant but not in connection with processes which touch specific questions of survival. The reason for that is given in detail in [8, 9]. As said, here the principles of evolution are relevant, and in Chapter 4 we will introduce "levels of reality" for the unique description of all these phenomena.

## 2.9. Other Biological Systems

The picture of reality must be species-dependent. In other words, we have to conclude that the actions of other biological systems are in

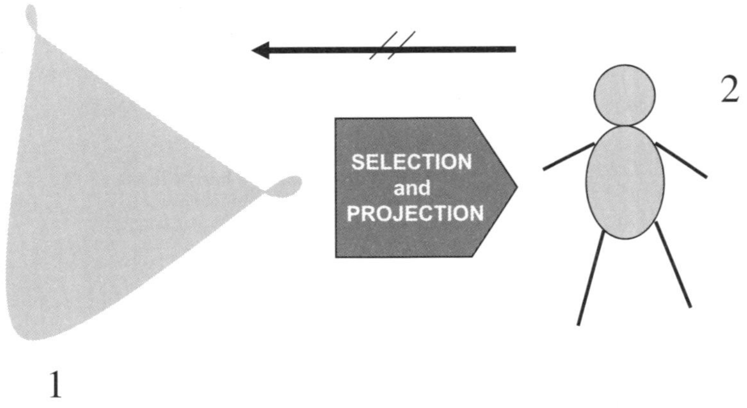


Fig. 2.7. The observer in basic reality is characterized by structure 1. His shape has been chosen arbitrarily because we can say nothing about the observer in basic reality; a picture-independent point of view is not possible within projection theory (see in particular Chapter 1). The observer in space and time is characterized by the familiar structure 2 as he appears after a selection and projection process. The observer in basic reality (structure 1) contains more information than the part of him which is depictable in space and time (structure 2). The observer is principally not able to conclude from structure 2, familiar to all observers, to structure 1 (observer in basic reality).

general based on a picture of reality that is different from that of the human observer. How can we verify that? Wolfgang Schleidt performed some interesting experiments using a turkey, its chick and a weasel which is the turkeys' deadly enemy, and he studied the behavior of the turkey in order to learn something about the perception apparatus of the turkey.

Schleidt worked with more or less everyday methods. However, his experiments are of such importance as certain key experiments in physics which have fundamentally changed the scientific world view. Schleidt demonstrated convincingly that the perception apparatus of a turkey must be quite different from that of humans. These experiments led to dramatic and unexpected results and demonstrate that the turkey must experience the world optically quite differently from the way we do, even though the eyes of the turkey are similar to ours. There is obviously no similarity between what the turkey experiences and what a human being sees in the same situation.

Both systems, man and turkey, react correctly in the normal case because both species are able to exist in the world, which can only be possible from the point of view of modern principles of evolution if their particular views of the world are correct. Therefore, although the conceptions of the world of man and turkey are on the one hand different from each other, they are on the other hand correct in each case. This means that neither of these two conceptions of the world can be true in the sense that they are a faithful reproduction of nature: objective realty (basic reality) must be different from the images which biological systems construct from it. We already came to a similar conclusion on the basis of general space–time arguments (Sections 2.1–2.5). A detailed description of these experiments together with conclusions and interpretations is given in Refs. 8 and 9.

In summary, these experiments by Schleidt deliver essential contributions about our understanding of that what we call reality and they can help to learn something about the relationship between reality and any kind of observer (man, turkey, etc.). However, we can only say that the perception apparatus of the turkey is different from that of humans; the details are not accessible in this way but we do need more information for the answering of such principal questions. We do not know even if such biological systems (turkeys) experience "their world" within the framework of space and time.

These results can be generalized because there is no reason to believe that turkeys have to be considered as an exception. Without doubt Schleidt's results support strongly our view concerning space and time developed above.

## 2.10. How Many (Geometrical) Objects can be in Space–Time?

The *pictures of reality* are species-dependent. On the other hand, the ability of an observer to interact usefully (species-preserving) with his environment is essentially based on the *pictures of reality*. Both, the abilities of the observer for the useful dealing with his environment as well as the observers' pictures of reality have been developed in the course of evolution with a definite interdependence. In this way the observers' form (shape) with arms, legs, eyes, etc. has been developed, but also his brain with its specific functions. So, the selected structures in the picture of reality are strongly correlated with the observers' abilities in dealing with them.

Observation of the world means that a certain kind of information has been selected by the brain from basic reality and this information appears in front of us. It is actually in the brain and we have only the impression that the world is in front of us (Section 2.5). We know that the infinite, in whatever form it appears, is never a result of an observation; only quantities with a finite set of values can be processed with established measuring methods and the five senses. The infinite is not only quantitatively inaccessible, but it can also not be experienced qualitatively by the imagination.

The selected part of information is by definition identical with the observed part. Otherwise the principle of selection would make no sense. We have outlined above that the observer is not able to process an infinite number of objects. Then, also the selection of an infinite number of objects should be impossible, because it would be against the principle of usefulness (principles of evolution).

Why should nature do that when only a finite number of elements can be processed by the human observer? This would make no sense. As we have outlined above, the abilities of the observer for the useful dealing with his environment as well as the observer's pictures of reality have been developed in the course of evolution with a definite interdependence. Therefore, the selection of an infinite number of objects from basic reality would make no sense at all. It would be superfluous because an observer could not process an infinite number of objects, and evolution does not allow us to develop superfluous things.

### Situation in conventional physics

Within conventional physics it is assumed that an outside world (basic reality) is recognizable for human observers, that is, the impression in front of the observer is assumed to be identical with that, which we have called basic reality. The material objects of basic reality are embedded in space and time, just as the impression in front of the observer suggests.

Moreover, the space–time structure of the material objects within this universe is in a good approximation independent of an observer (within the Copenhagen interpretation of quantum the observer has a certain option by the act of observation, but the observer does not influence its possible structures), and there is no connection between these space–time structures of the outside world and the observers biological structure. Within conventional physics no evolutionary effects have been taken into account in the basic formulation of the laws. Thus, within conventional physics it is not forbidden that the universe contains an infinite number of objects, although a human observer is principally not able to process such a scenario.

## 2.11. Two Types of Space–Time?

Space and time are not accessible by the human senses, and no measuring instrument is able to make space–time-specific "clicks". This in particular means that space and time can never be transmitted from the world outside to the brain of the human observer. Thus, space and time are independent elements of the brain, and this statement is close to the ideas of the philosopher Immanuel Kant, which we will discuss in more detail in Chapter 5.

Whether or not space and time are also elements of the world outside (basic reality) becomes then a question without sense because we can never prove this point. Furthermore, our theoretical system does not need a space–time that is also an element of basic reality, because basic reality is in any case not accessible to a human observer. Besides, the situation would require two types of space–time. (i) The space–time of basic reality would be a container for matter. (ii) The space–time of brain would be a screen for the representation of geometrical structures. Both types of space–time would require properties quite different from each other; the one space–time would have nothing to do with the other type.

Such a specific construction can be excluded. We have one type of space–time for one task, namely for the representation of geometrical structures in the observer's brain. We can principally not prove whether or not the other space–time type exists. In particular, we do not need it.

In conclusion, to assume that there is also a space–time in the outside world, so to say, would not be scientific.

## 2.12. Summary

The material objects which we observe in everyday life are not embedded in space (space–time). Space and time have to be considered as auxiliary elements for the representation of the selected information about the outside world (basic reality). The chick experiment demonstrated that these pictures must be species-dependent, and this is confirmed by the principles of evolution. In other words, the physical reality is projected onto space and time, and the space–time elements $x, y, z, \tau$ cannot be seized with our five senses or with specific measuring instruments. These features of space and time must be considered in the formulation of physical theories. In the next chapter, we will treat quantum phenomena on the basis of the projection principle that we have introduced in this chapter.

3

# Fictitious Realities

In Chapters 1 and 2 we came to the following conclusion. (1) A human observer is principally not able to recognize the absolute truth (basic reality). (2) Due to the principles of evolution, a human observer never grasps the complete information of the world outside. (3) Space and time are not elements of basic reality; space and time are exclusively auxiliary elements in the brain of the observer for the geometrical representation of species-relevant parts of the world outside.

From these three statements, it follows directly that the pictures in front of us are the most direct impression of the world, and we may state the following. Within the memory of man all essential things are represented within the frame of pictures that appear spontaneously in front of us. It is the challenge of physics to explain these pictures (which has to be considered as the basic information) theoretically, whereby the conditions in connection with absolute truth, completeness, role of space and time have to be fulfilled in detail.

The cause for the appearance of the pictures are processes in basic reality that are projected onto space and time (see Chapter 2). In other words, in order to be able to explain and to find a picture, we have to describe these processes, which take place in basic reality, and the result must be projected onto space–time.

However, we have also outlined in Chapters 1 and 2 that basic reality is not accessible to man. Therefore, we have to construct alternative realities which have called in [12] "fictitious realities". In this chapter we will summarize and supplement the results given in [12]. Before we start with our considerations, let us briefly discuss the situation in conventional quantum theory with respect to the conditions quoted above: absolute truth, completeness, role of space and time.

## 3.1. Conventional Quantum Theory: Critical Remarks

Conventional quantum theory massively violates these conditions. As in Newton's mechanics, within this theory, matter is still embedded in space, and the phenomenon of evolution (with the principle of usefulness) has not been considered in the basic formulation of conventional quantum theory. It is therefore not surprising, from the point of view of projection theory [12], that there are some critical points in connection with conventional quantum theory. Let us quote a few of them here.

### 3.1.1. *A Diversity of Opinion*

The essential features of the mathematical formalism of (non-relativistic) quantum theory were constructed by Heisenberg and Schrödinger in 1925/1926. On the basis of this formalism, an enormous number of effects in atomic physics, chemistry, solid state physics, etc. could be predicted and explained. However, more than 80 years after its formulation the interpretation of conventional quantum theory is "*by far the most controversial problem of current research in the foundations of physics and divides the community of physicists and philosophers of science into numerous opposing schools of thought*". [15] There is an immense diversity of opinion and a huge variety of interpretations. This is not convincing at all. Most of these interpretations lead to completely different "pictures of reality".

For example, the Copenhagen interpretation of conventional quantum theory (proposed by Bohr in 1927) is quite different from the many-worlds theory (proposed by Everett III in 1957). Both, the Copenhagen interpretation and the many-worlds theory, give predictions in accord with experience and, therefore, they are equivalent from the experimental point of view.

The question is then whether a formulation (the form of conventional quantum theory) with a diversity of completely different interpretations can be accepted as the final truth. We can certainly not.

### 3.1.2. *Some Specific Problems within Conventional Quantum Theory*

In fact, there are obviously some specific serious problems in connection with conventional quantum theory. These problems are often ignored, and this is mainly due to the fact that the empirical success of usual quantum theory is impressive. In [12], we already have quoted some critical points which appear in connection with conventional quantum theory. The main facts can be summarized as follows:

1. Schrödinger's equation could not be deduced in conventional quantum theory but was assumed on the basis of "reasonable arguments". In other words, we do not really know the physical reality on which Schrödinger's equation is based.
2. The *collapse of the wavefunction* cannot be explained in usual quantum theory; one would expect that Schrödinger's equation (the fundamental tool within non-relativistic quantum theory) were able to describe such a process but this is unfortunately not the case. The collapse of the wave function and the assignment of statistical weights are not explained. Within usual quantum theory "*they are consequences of an external a priori metaphysics, which is allowed to intervene at this point and suspend the Schrödinger equation.*" [16] In other words, the situation in connection with the collapse of the wavefunction is unsatisfactory and has to be solved.
3. Time $\tau$ is still a classical parameter in conventional quantum theory. While the coordinates are *statistical* quantities, time $\tau$ does not behave statistically. Time $\tau$ remains unchanged when we go from classical mechanics to quantum theory; this is clearly reflected in the fact that the coordinates can be *operators*; time is always a simple *parameter*. This is the reason why there is no uncertainty relation for the time $\tau$ and the energy $E$ which would agree in its physical content with the position–momentum uncertainty relation. (More details concerning this point will be discussed in Chapter 5).
4. The use of a particle defined as a local existent (which is assumed within the frame of usual quantum theory) seem also to be

> problematic, and this is probably due to the fact that we transfer a picture used in everyday life to the microscopic realm. Can a particle (point-like individual) exist without trajectory (usual quantum theory)? We argued that such a concept is problematic because an individual demands a minimum description since this notion has been chosen with respect to our intuitive demands for *visualizability* (due to our experiences in everyday life). The concept of a point-like individual without trajectory obviously leads to a paradoxical situation. [12]

In the next sections, we will discuss quantum phenomena in connection with the so-called "projection theory". It will turn out that the point of view of conventional quantum theory cannot simply be extended, but we have to include further fundamental facts, so far not directly considered in the basic description of physical systems. We will in particular see that we can learn something new about the relationship between the "observer" and that what we call "reality" when we consider certain facts from biological evolution and behavior research.

First, we apply the so-called projection principle. The projection principle is based on the insight that the material objects (sun, moon, etc.) are not embedded in space (space–time), but the physical reality is projected onto space (space–time) within the frame of the projection principle. After that, we use this principle for the mathematical formulation of the relevant quantum laws. In this way, new features come into play. In particular, we get a completely new formulation for the notion "time" within projection theory. Time becomes a real quantum variable, and it is no longer an external parameter as in classical physics and conventional quantum theory.

All these shortcomings that appear in conventional quantum theory are probably due to the fact that the conditions in connection with absolute truth, completeness as well as the specific features of space and time are not fulfilled within conventional quantum theory. In the following, we will introduce a physical framework that fulfils all these conditions and requirements, respectively.

## 3.2. The Projection Principle in Connection with Fictitious Realities

In Chapter 2, we have discussed the so-called projection principle. We already remarked above that the projection principle is based on the insight that the material objects (sun, moon, etc.) are not embedded in space (space–time), but the physical reality is projected onto space (space–time) within the frame of the projection principle. Space and time have to be considered as auxiliary elements for the representation of selected information that is part of the outside world (basic reality). In other words, the selected part of physical reality is projected onto space and time. Selection processes come into play through evolution. Within all forms of conventional physics, the real world is embedded in space and time.

### 3.2.1. *Alternative Realities*

Basic reality is not accessible to human observers and we have to introduce alternative realities for the description of projection effects. This point has been discussed extensively in [12]. Alternative realities play the role of fictitious realities.

It turned out in [12] that the variables of fictitious reality must be given by the momentum $\mathbf{p}$ and the energy $E$, that is, fictitious reality is embedded in the so-called $(\mathbf{p}, E)$-space. The information that is embedded in $(\mathbf{p}, E)$-space is completely projected onto space and time. Since the variables $\mathbf{p}$ and $E$ reflect reality, where reality has to be identified with a certain system, the variables $\mathbf{p}$ and $E$ have to be considered as system-specific quantities. Thus, the projected information onto space and time must also be system-specific. This however inevitably means that the space–time frame, on which the information of $(\mathbf{p}, E)$-space is projected, may not contain the time $\tau$. $\tau$ is an external parameter and is in all cases measured by our clocks and, therefore, it cannot reflect system-specific properties. Thus, we have to introduce a system-specific time, and we would like to use the letter $t$ for it. In other words, the space–time frame on which the (fictitious) realty is projected is given by the variables $\mathbf{r}$ and $t$. In the following, we want to use for the space–time frames, which are involved in the projections, the marking $(\mathbf{r}, t)$-space.

In summary, we have the outside world (basic reality) which is however not accessible to the human observer. Therefore, the pictures in front of us are the most direct impression of the outside world; these impressions (pictures) have to be considered as the most essential information for a human observer. Thus, we may state the following. Within the memory of man all essential things are represented within the frame of pictures that appear spontaneously in front of us. This is the most basic statement. These pictures of reality are represented in $(\mathbf{r}, t)$-space and are described on the basis of processes which take place in fictitious reality. Each process is characterized by $(\mathbf{p}, E)$-fluctuations in $(\mathbf{p}, E)$-space (fictitious reality). The processes in $(\mathbf{p}, E)$-space are projected onto $(\mathbf{r}, t)$-space (see also Fig. 3.1). As we have outlined in Chapter 2, also the information transfer from basic reality to the picture of reality has to be considered as a projection. Since basic reality (outside world) is not accessible to man, we are principally not able to describe the information transfer from basic reality to the picture

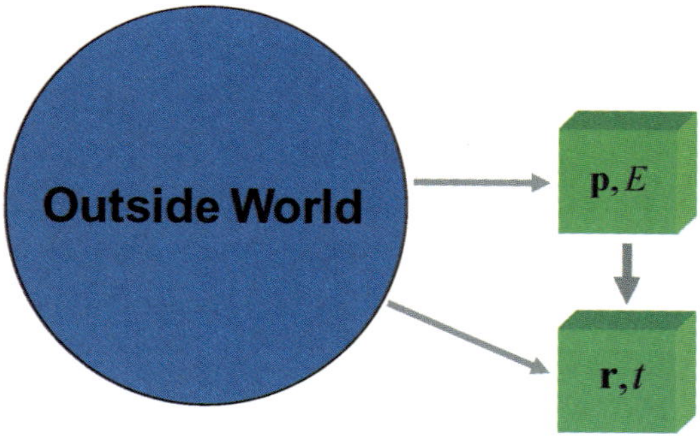


Fig. 3.1. Basic reality (outside world), fictitious reality and their connection to the pictures of reality. Since basic reality (outside world) is not accessible to man, we are principally not able to describe the information transfer from basic reality to the picture of reality. Therefore, we have to construct fictitious realities. More details are given in the text.

of reality, and this is the reason why we have to construct fictitious realities, that is, alternative realities.

### 3.2.2. *Relationships*

In classical mechanics (Newton's mechanics, Theory of Relativity) we have deterministic laws. We observe the motion of objects in space and time (for example, celestial bodies), that is, the position vector $\mathbf{r} = (x, y, z)$ of an object is observed as a function of time $\tau$. Here $\tau$ is the time, which we measure with our clocks and is therefore an external parameter. In this way we obtain a trajectory of an object, and the solution of the equations of motion leads to expressions for $x(\tau), y(\tau), z(\tau)$.

The transition from Newton's mechanics to usual quantum theory is essentially influenced by Planck's constant $\hbar$. We have no longer deterministic laws but the objects behave statistically. The situation is described by the scheme represented in Fig. 3.2.

In conventional quantum theory, trajectories there are not defined, but we observe here a statistical behavior of the objects. However, as in classical mechanics, the geometry of the objects has to be postulated. For example, in the case of an electron it is assumed that it forms a point-like unit. Furthermore, the material objects are assumed to be embedded in space, that is, we have exactly the same situation as in classical mechanics. Within conventional quantum theory, the

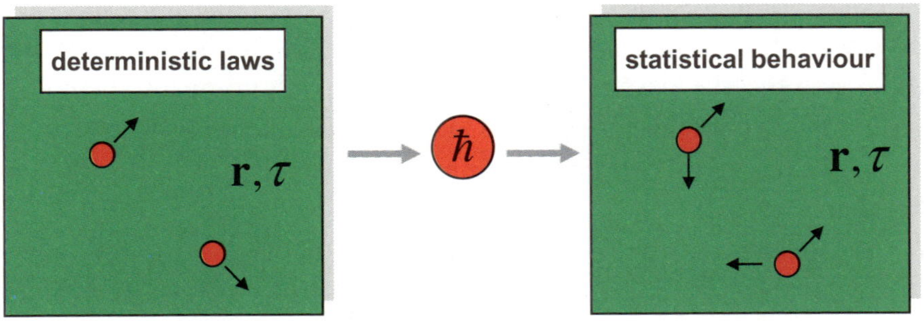


Fig. 3.2. The transition from Newton's mechanics to usual quantum theory is essentially influenced by Planck's constant $\hbar$. Instead of trajectories, the material particles behave statistically but are embedded in space as in the classical case. The time $\tau$ remains an external, classical parameter, i.e., $\tau$ is not influenced by $\hbar$.

statistical behavior of the system is restricted to the coordinates $x, y, z$. The time $\tau$ is not changed by the transition from classical mechanics to conventional quantum theory, i.e., $\tau$ remains an external parameter and has nothing to do with the process under investigation.

The statistical behavior of the particle (embedded in space) is described by the wave function $\psi(\mathbf{r}, \tau)$. According to Born, the probability of finding a real quantum object (assumed to be a point-like particle) in the volume element $dV = dxdydz$ at a certain position $\mathbf{r} = (x, y, z)$ and at a certain time $\tau$ is given by $\psi^*(\mathbf{r}, \tau)\psi(\mathbf{r}, \tau)dxdydz$, where the quantity $\psi^*\psi$ plays the role of a probability density. Again, the particle behaves statistically with respect to the variable $\mathbf{r}$ but not in connection with the time $\tau$.

The transition from classical mechanics to projection theory is described by the scheme represented in Fig. 3.3. Here the influence of $\hbar$ leads to two spaces: $(\mathbf{p}, E)$-space and $(\mathbf{r}, t)$-space. Fictitious reality is embedded in $(\mathbf{p}, E)$-space, and the $(\mathbf{p}, E)$-fluctuations in $(\mathbf{p}, E)$-space are projected onto $(\mathbf{r}, t)$-space. The $(\mathbf{r}, t)$-space is not a container for material objects; in connection with $(\mathbf{r}, t)$-space only geometrical structures are defined. We experience basic reality as a geometrical $(\mathbf{r}, t)$-structure and we describe this $(\mathbf{r}, t)$-structure by projections of $(\mathbf{p}, E)$-fluctuations (interactions in $(\mathbf{p}, E)$-space); basic reality is principally not accessible to a human observer.

If a certain $(\mathbf{r}, t)$-structure is identified as a particle, then the particle can only be the result of certain $(\mathbf{p}, E)$-fluctuations, i.e., such a particle can only come into being by interaction processes that are expressed by the variables $\mathbf{r}$ and $t$. Independent of the particles interaction with other systems, the existence of the particle can only come into being by distance-independent interactions. Distance-independent interactions are not known in conventional physics. More details are given in Section 3.10.

As already mentioned above, projection theory has the following further peculiarity: Each process and system, respectively, defines its own time-structure where the letter $t$ is the variable for describing these system-specific time-structures. The quantity $t$ does not replace the time $\tau$, but the external parameter $\tau$ also exists in projection theory as reference time. In other words, projection theory extends

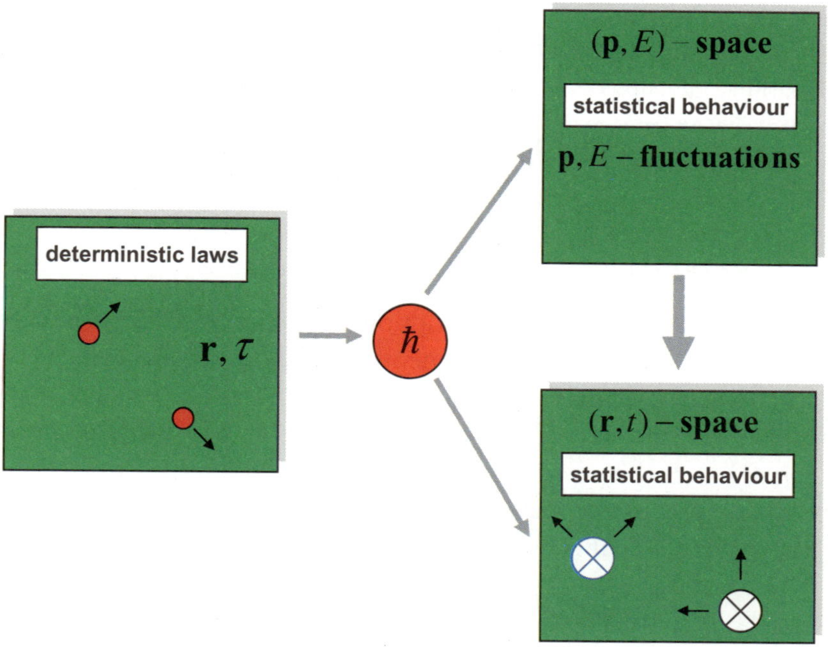


Fig. 3.3. Also the transition from Newton's mechanics to projection theory is essentially influenced by Planck's constant $\hbar$. Here we have two spaces, $(\mathbf{p}, E)$-space and $(\mathbf{r}, t)$-space. The variables $\mathbf{p}, E, \mathbf{r}, t$ of both spaces behave statistically. In contrast to conventional physics, the $(\mathbf{r}, t)$-space is not a container for material objects. In connection with $(\mathbf{r}, t)$-space only geometrical structures are defined. The $(\mathbf{p}, E)$-fluctuations in $(\mathbf{p}, E)$-space are projected onto $(\mathbf{r}, t)$-space. Particles [more or less localized geometrical structures in $(\mathbf{r}, t)$-space] cannot be static units but must come into play through distance-independent interactions. Within projection theory each system defines its own time-structure and $t$ is the variable for describing these system-specific time-structures. The quantity $t$ does not replace the time $\tau$, but the external parameter $\tau$ also exists in projection theory as reference time.

the frame of description by $t$; instead of $\mathbf{r}, \tau$ (conventional physics), within projection theory we have $\mathbf{r}, t, \tau$.

Not only $\mathbf{r}$ acts as a statistical variable but also $t$. Thus, the wavefunction of a system is defined in terms of the variables $\mathbf{r}$, $t$, i.e., we have $\Psi = \Psi(\mathbf{r}, t)$, and $\Psi(\mathbf{r}, t)$ can be dependent on $\tau$ or not. The probability distributions for $\mathbf{r}$ and $t$ can only be determined in connection with $\mathbf{p}$ and $E$ [the variables of $(\mathbf{p}, E)$-space], and we come to the following statement (see in particular [12]): "The measurement of one of the possible values for **p** and for $E$ at time $\tau$ is done in the space–time intervals $\mathbf{r},\ \mathbf{r}+d\mathbf{r}$ and $t,\ t+dt$ with the probability density of $\Psi^*(\mathbf{r},t)\Psi(\mathbf{r},t)$." Here measurement means to use a detector (measuring instrument) that is adjustable to the specific **r**, **t**-configurations. Clearly, this instruction is more general than Born's probability description.

### 3.2.3. *Sequences*

Within Born's probability description only the variable **r** behaves statistically at time $\tau$. Projection theory extends usual quantum theory by the system-specific time $t$ and therefore, here the variables **r** and $t$ behave statistically at time $\tau$ and not only **r** as in usual quantum theory. One of the possible values for **r** and $t$ are given with a certain probability, and this probability is determined by the probability density $\Psi^*(\mathbf{r},t)\Psi(\mathbf{r},t)$. That is, we have the following scheme:

$$\tau : \Psi^*(\mathbf{r},t)\Psi(\mathbf{r},t), \ -\infty \leq \mathbf{r}, t \leq \infty. \tag{3.1}$$

Here $\Psi^*(\mathbf{r},t)\Psi(\mathbf{r},t)$ is assumed to be independent of time $\tau$.

While the reference time $\tau$ goes strictly from the past to the future, the system-specific time $t$ jumps in general arbitrarily from one $t$-position to another, and we cannot know when we observe — with respect to time $\tau$ — an event in connection with the system under investigation in the past, present or in the future. The following example illustrates this.

When we measure at time $\tau_i$ one space–time point of the configuration $\Psi^*(\mathbf{r},t_i)\Psi(\mathbf{r},t_i)$, then the result of such measurement could be given for example by

$$\begin{aligned} \tau_1 &\rightarrow t_2 \\ \tau_2 &\rightarrow t_4 \\ \tau_3 &\rightarrow t_1 \\ \tau_4 &\rightarrow t_3 \end{aligned} \tag{3.2}$$

in the case of $i = 1,\ldots,4$, that is, the sequence with respect to the reference time is given by

$$t_2, t_4, t_1, t_3 \tag{3.3}$$

for

$$\tau_1 < \tau_2 < \tau_3 < \tau_4 \tag{3.4}$$

and

$$t_1 < t_2 < t_3 < t_4. \tag{3.5}$$

Because of the statistical behavior of the system-specific time $t$ the same measurement could have led also to the following result:

$$\begin{aligned} \tau_1 &\rightarrow t_3, \\ \tau_2 &\rightarrow t_1, \\ \tau_3 &\rightarrow t_2, \\ \tau_1 &\rightarrow t_4. \end{aligned} \tag{3.6}$$

In this case, we have the sequence

$$t_3, t_1, t_2, t_4 \tag{3.7}$$

which is different from (3.3). Clearly, here (3.4) and (3.5) are also valid.

In conclusion, the system-specific time $t$ jumps in general arbitrarily from one $t$-position to another, and we cannot know when we observe an event in the past, present or in the future. The reference time $\tau$, measured with our clocks, goes by definition always strictly from the past to the future.

This feature of time is completely new. Even the notion "system-specific time" $t$ is not known in conventional quantum theory. Here only the reference time $\tau$ appears which plays the role of an external parameter, as we have already mentioned above. More details in connection with $t$ will be discussed in the forthcoming sections.

## 3.3. Distribution of Information

There are no selections from $(\mathbf{p}, E)$-space to $(\mathbf{r}, t)$-space. Only a transformation takes place. This is exactly what we need because we do not know the contents of basic reality. The only thing we can say is that the observed data in $(\mathbf{r}, t)$-space can only be a fraction of that what is actually in basic reality, and this is due to the observed evolution in nature. However, we have not the slightest knowledge about the contents of basic reality and, therefore, the selection process cannot be described by a human observer, but only the selected fraction. Since the projection from $(\mathbf{p}, E)$-space to $(\mathbf{r}, t)$-space takes place without selection, we have just found the solution that we need. Only a transformation takes place. Just this transformation justifies using the projection principle.

$(\mathbf{r}, t)$-space and $(\mathbf{p}, E)$-space are equivalent as far as their information is concerned. Both spaces are connected by a Fourier transform. [12] In the case of a Fourier transform, the information of $(\mathbf{p}, E)$-space is transformed to $(\mathbf{r}, t)$-space without any loss of information.

In connection with the Fourier transform, the following feature is relevant. In order to get the information of the system at a certain space–time point, say $\mathbf{r}_1, t_1$, we need all the information of $(\mathbf{p}, E)$-space. This can symbolically expressed by

$$\mathbf{r}_1, t_1 \leftarrow (\mathbf{p}, E). \tag{3.8}$$

Thus, each point of $(\mathbf{r}, t)$-space contains in principle the complete information about the system. (In this case, it is the $(\mathbf{p}, E)$-information).

On the other hand, in order to get the information of the system at a certain point in $(\mathbf{p}, E)$-space, say $\mathbf{p}_1, E_1$, we need all the information of $(\mathbf{r}, t)$-space and, instead of (3.8), we get

$$\mathbf{p}_1, E_1 \leftarrow (\mathbf{r}, t). \tag{3.9}$$

Thus, also each point of $(\mathbf{p}, E)$-space contains in principle the complete information about the system. (It is the $(\mathbf{r}, t)$-information in this case.)

In connection with the corresponding wavefunction, we in particularly get for the Fourier transform [12]

$$\Psi(\mathbf{r}_1, t_1) = \frac{1}{(2\pi\hbar)^2} \int_{-\infty}^{\infty} \Psi(\mathbf{p}, E) \exp\left\{ i \left[ \frac{\mathbf{p}}{\hbar} \cdot \mathbf{r}_1 - \frac{E}{\hbar} t_1 \right] \right\} dp_x dp_y dp_z dE \tag{3.10}$$

with $\mathbf{p} = (p_x, p_y, p_z)$. The inverse transformation is given by

$$\Psi(\mathbf{p}_1, E_1) = \frac{1}{(2\pi\hbar)^2} \int\limits_{-\infty}^{\infty} \Psi(\mathbf{r}, t) \exp\left\{-i\left[\frac{\mathbf{p}_1}{\hbar} \cdot \mathbf{r} - \frac{E_1}{\hbar} t\right]\right\} dtdxdydz \tag{3.11}$$

with $\mathbf{r} = (x, y, z)$.

If the cosmos could be treated as a unified whole, the complete $(\mathbf{r}, t)$-structure of its picture could be described by one wavefunction $\Psi_{\cos}(\mathbf{r}, t)$ and by the probability density $\Psi^*_{\cos}(\mathbf{r}, t)\ \Psi_{\cos}(\mathbf{r}, t)$, respectively. That is, all details of the cosmos (humans, animals, son, moon, galaxies, etc.) would be expressed by $\Psi^*_{\cos}(\mathbf{r}, t)\ \Psi_{\cos}(\mathbf{r}, t)$. Then, in analogy to (3.8) we have the following situation:

$$[\mathbf{r}_1, t_1]_{\cos} \leftarrow [(\mathbf{p}, E)]_{\cos}. \tag{3.12}$$

In other words, each point of $(\mathbf{r}, t)$-space contains in principle the complete information about the cosmos. Each $(\mathbf{r}, t)$-point with $\Psi^*_{\cos}(\mathbf{r}, t)\Psi_{\cos}(\mathbf{r}, t) \neq 0$ contains the information of all the other $(\mathbf{r}, t)$-points. More descriptively: The complete information about the cosmos is already condensed in a grain of sand. We will come back to this point below.

Remark

From Eq. (3.10), it follows that we need all the information of $(\mathbf{p}, E)$-space in order to be able to create the wavefunction $\Psi(\mathbf{r}_1, t_1)$ at the space–time position $\mathbf{r}_1, t_1$, that is, we have

$$\Psi(\mathbf{r}_1, t_1) \leftarrow \Psi(\mathbf{p}, E), -\infty \leq \mathbf{p}, E \leq \infty. \tag{3.13}$$

But it is not possible to determine from $\Psi(\mathbf{r}_1, t_1)$ the function $\Psi(\mathbf{p}, E)$, i.e., the inverse transition to (3.13)

$$\Psi(\mathbf{r}_1, t_1) \rightarrow \Psi(\mathbf{p}, E), -\infty \leq \mathbf{p}, E \leq \infty \tag{3.14}$$

is not possible. The definition of the Fourier transform does not allow the transition (3.14). The point $\mathbf{r}_1, t_1$ contains the information of the entire system but it cannot be retrieved.

Clearly, exactly the same is the case in connection with point $\mathbf{p}_1, E_1$ of $(\mathbf{p}, E)$-space. From Eq. (3.11) it follows that we need all the information of $(\mathbf{r}, t)$-space for the creation of $\Psi(\mathbf{p}, E)$ at point $\mathbf{p}_1, E_1$, i.e., we have

$$\Psi(\mathbf{p}_1, E_1) \leftarrow \Psi(\mathbf{r}, t), -\infty \leq \mathbf{r}, t \leq \infty. \tag{3.15}$$

But here it is not possible to determine from $\Psi(\mathbf{p}_1, E_1)$ the function $\Psi(\mathbf{r}, t)$, i.e., the inverse transition to (3.15)

$$\Psi(\mathbf{p}_1, E_1) \rightarrow \Psi(\mathbf{r}, t), -\infty \leq \mathbf{r}, t \leq \infty \tag{3.16}$$

is not possible. The point $\mathbf{p}_1, E_1$ contains the information of the entire system but it cannot be retrieved.

## 3.4. Basic Transformation Effects

Due to the coexistence of spaces and the connection of their variables $\mathbf{r}$, $t$ and $\mathbf{p}$, $E$ by a Fourier transform, remarkable properties come into play, and this is not a matter of interpretation. In other words, these specific properties can principally not be eliminated by the use of another interpretation but have to be considered as basic facts. In this section, let us briefly discuss non-local effects, the role of time, and we also want to investigate whether point-like particles are definable within the frame of projection theory.

### 3.4.1. *Particles*

Within usual quantum theory, a material particle having the energy $E_0$ and the momentum $\mathbf{p}_0$, is *embedded* in $(\mathbf{r}, t)$-space. As we have outlined above, the situation is different in projection theory. The material particle (if it exists) is embedded in $(\mathbf{p}, E)$-space and not in $(\mathbf{r}, t)$-space. Only its picture is represented in $(\mathbf{r}, t)$-space. The picture contains the maximum information about the particle because both spaces are equivalent concerning their information content.

Let us consider a simple system (particle) in $(\mathbf{p}, E)$-space and let the $(\mathbf{p}, E)$-distribution, say $A(\mathbf{p}, E)_\tau$, be characterized at time $\tau$ by delta-functions:

$$A(\mathbf{p}, E)_\tau = \delta(E - E_0)\delta(p - p_0), \tag{3.17}$$

where $E_0$ and $p_0$ are constant parameters for the energy and momentum at time $\tau$ ($\tau$ is again the time of our clocks which we use in everyday life). In other words, the system under investigation is defined in reality by the distribution $A(\mathbf{p}, E)_\tau$. The picture of reality [represented in $(\mathbf{r}, t)$-space] contains, at time $\tau$, exactly the same information, say $A(\mathbf{r}, t)_\tau$, and $A(\mathbf{r}, t)_\tau$ is given by the projection of $A(\mathbf{p}, E)_\tau$ onto $(\mathbf{r}, t)$-space. This projection can be performed by using the usual Fourier transform and we obtain immediately [12]

$$A(\mathbf{r}, t)_\tau = B \exp\left\{\frac{i}{\hbar}(\mathbf{p}_0 \cdot \mathbf{r} - E_0\, t\right\}, \quad B = \frac{1}{(2\pi\hbar)^2}. \tag{3.18}$$

From this equation, it directly follows that the system, having at time $\tau$ the energy $E_0$ and the momentum $\mathbf{p}_0$, is completely *delocalized* in $(\mathbf{r},\ t)$-space, that is, it takes at time $\tau$ all possible values for $\mathbf{r} = (x, y, z)$ and $t$:

$$-\infty \leq \mathbf{r} \leq \infty,$$

$$-\infty \leq t \leq \infty.$$

In other words, within the theoretical picture given here a system with a definite energy $E_0$ and a definite momentum $\mathbf{p}_0$ can never be *localized* in space, that is, its picture can never represent a point-like particle in $(\mathbf{r}, t)$-space, but just the opposite is correct, i.e., it is completely delocalized, and this is due to the mathematical fact that the Fourier transform of a delta-function does not lead to a delta-function.

As already pointed out several times, within usual quantum theory material particles are embedded in $(\mathbf{r}, t)$-space and, furthermore, they are assumed to be strict localized, that is, they are point-like in character.

In conclusion, the system (particle) is localized in $(\mathbf{p}, E)$-space [see (3.17)], but this space has nothing to do with the space of our experience and this is exclusively given by $(\mathbf{r}, t)$-space. A *localized* object with definite energy $E_0$ and definite momentum $\mathbf{p}_0$ is a requisite of classical mechanics and this is based on our experiences in everyday life. Cars, trees etc. are localized in space in a good approximation.

### 3.4.2. *Role of Time t*

The time $t$ is defined by the system under investigation. In our example the system is characterized by Eq. (3.17). In other words, time $t$ is a *system-specific time*; its characteristics are different for different systems. This is a new feature. But how are these characteristics fixed? What is the relation between the various time scales and the $\tau$-time scale? Our simple example cannot give the answer, but we will answer these questions below with a more detailed analysis using equations that are similar to Schrödinger's equation, which are however generalized with respect to the system-specific time $t$.

It should be emphasized once more (see [12]) that in usual quantum theory there is only one time scale and this is exclusively given by the time $\tau$ of our clocks used in everyday life. However, $\tau$ merely plays the role of an external parameter and has nothing to do with the system under investigation. The situation within projection theory is entirely new, and we will see below that the existence of a system-specific time has serious consequences.

### 3.4.3. *Non-Local Effects*

Let us consider the same system but now at time $\tau + \Delta\tau$, where $\Delta\tau$ is an arbitrarily chosen interval. In principle, $\Delta\tau$ can be infinitesimal but should be different from zero. We would like to assume that, due to an interaction process, the system has no longer the energy $E_0$ and the momentum $\mathbf{p}_0$ but $E'_0$ and $\mathbf{p}'_0$ at $\tau + \Delta\tau$. Then, we get instead of $A(\mathbf{p}, E)_\tau$ [see (3.17)] the following expression for $A(\mathbf{p}, E)_{\tau+\Delta\tau}$:

$$A(\mathbf{p}, E)_{\tau+\Delta\tau} = \delta(E - E'_0)\delta(\mathbf{p} - \mathbf{p}'_0). \tag{3.19}$$

Using this distribution and the usual Fourier transform, we obtain

$$A(\mathbf{r}, t)_{\tau+\Delta\tau} = B \exp\left\{\frac{i}{\hbar}(\mathbf{p}'_0 \cdot \mathbf{r} - E'_0\, t\right\}. \tag{3.20}$$

By comparison of (3.18) with (3.20), we immediately recognize that the properties at all points $\mathbf{r}$, $t$ $(-\infty \leq \mathbf{r} \leq \infty, -\infty \leq t \leq \infty)$ at $\tau + \Delta\tau$ are different from those at time $\tau$. These global changes take place instantaneously since $\Delta\tau$ can in principle be infinitesimal. In other words, we are confronted with non-local effects.

These *non-local effects* are entirely due to the projection process from reality [$(\mathbf{p}, E)$-space] onto $(\mathbf{r}, t)$-space, like a flashlight on a screen. There is no preferred space–time point in $(\mathbf{r}, t)$-space, say $\mathbf{r}_1$ and $t_1$, which would influence instantaneously all the other points $\mathbf{r}, t$ $(-\infty \leq \mathbf{r} \leq \infty, -\infty \leq t \leq \infty)$ by actions through space–time. Since there are only projections, no information or material objects move through space–time. Therefore, the instantaneous transition from one global structure [described by (3.18)] to the other [described by (3.20)] is not in contradiction to the Theory of Relativity.

### 3.4.4. *Conclusion*

In this section, we have pointed out that point-like particles with definite energy and definite momentum are not definable within the frame of the investigation given here. This "point-model", used in usual quantum theory, is obviously a holdover from classical mechanics and is mainly based on our observations in everyday life.

Furthermore, there are non-local effects, that is, we observe in $(\mathbf{r}, t)$-space global instantaneous changes in the structure, and it is a feature of the projection that all space–time points $\mathbf{r}, t$ $(-\infty \leq \mathbf{r} \leq \infty, -\infty \leq t \leq \infty)$ are instantaneously involved. This effect is comparable with a flashlight on a screen.

Note, that this non-locality is not restricted to space, but time $t$ is involved too. We have no longer one time $\tau$ measured with our clocks, but also a system-specific time $t$. Its characteristics are different for different systems. More details are given below.

There still remains of course a lot of questions. For example, we do our measurements in $(\mathbf{r}, t)$-space but how is the contact to reality [$(\mathbf{p}, E)$-space] organized? There is no particle or any other material object embedded in $(\mathbf{r}, t)$-space but everything real is located in $(\mathbf{p}, E)$-space and we have to clarify how we can fix the properties of a system in $(\mathbf{r}, t)$-space, which are expressed within projection theory as $(\mathbf{p}, E)$-distribution. These points will also be discussed below.

Remark

We know that the variables $\mathbf{r}$ and $t$ ($\mathbf{p}$ and $E$) are not simultaneously existent, but play the role of *random variables*, that is, at time $\tau$ there is a certain probability to observe one of all possible $(\mathbf{r}, t)$-points [$(\mathbf{p}, E)$-points]. Within the frame of the example discussed in this subsection, these points are defined at time $\tau$ by the range of $A(\mathbf{r}, t)_\tau$ [$A(\mathbf{p}, E)_\tau$].

## 3.5. Pictures within Projection Theory

Again, within the memory of man all essential things are represented within the frame of pictures. This is the most basic statement from the point of view of projection theory. An example is given in Fig. 3.4(a). In everyday life, we have such or similar pictures spontaneously in front of us and, due to this impression, it is normally assumed in conventional physics that real matter is embedded in space and time. Already Immanuel Kant recognized that this should be a fallacy. Within projection theory is what we have in everyday life spontaneously in front of us, given by geometrical structures [as, for example, the two systems represented in Fig. 3.4(a)]. We have outlined in Chapter 2 that these geometrical structures are exclusively elements (states) of the brain. This was exactly the standpoint of the philosopher Immanuel Kant.

The real world (real matter) remains principally hidden to the human observer. He feels it but he cannot recognize it directly. All the real actions take place in basic reality and the entire complex, which is relevant in connection with these real actions, is projected onto $(\mathbf{r}, t)$-space [see Fig. 3.4(b)]. We describe the geometrical structures in Fig. 3.4(a) by means of fictitious realties, that is, by $(\mathbf{p}, E)$-fluctuations (interactions) in $(\mathbf{p}, E)$-space [Fig. 3.4(c)].

The probability density $\Psi^*(\mathbf{r}, t)\Psi(\mathbf{r}, t)$ defines the original $(\mathbf{r}, t)$-size of the system under investigation. In Fig. 3.4(a), two systems are

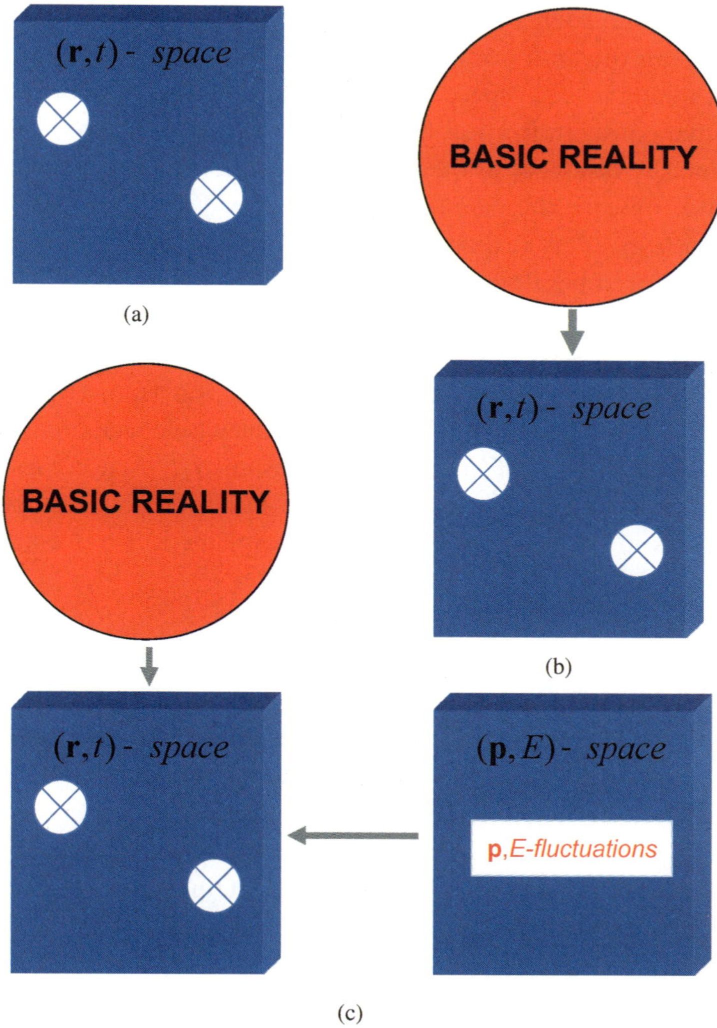


(a) (b) (c)

Fig. 3.4. (a) In everyday life, we have certain pictures spontaneously in front of us. (Here the picture consists of two systems.) In conventional physics, it is normally assumed that real matter is embedded in space and time. Within projection theory there can only be geometrical structures in the picture and are exclusively elements (states) of the brain. (b) All the real actions and processes, respectively, take place in basic reality and the entire complex, which is involved, is projected onto $(\mathbf{r}, t)$-space. (c) We describe the geometrical structures by means of fictitious realities, that is, by $(\mathbf{p}, E)$-fluctuations (interactions) in $(\mathbf{p}, E)$-space. The $(\mathbf{p}, E)$-structures produced by these fluctuations are projected onto $(\mathbf{r}, t)$-space.

represented and each of them is shown with the original $(\mathbf{r}, t)$-size. That is, from the basics of projection theory follows directly the true size of a system in $(\mathbf{r}, t)$-space. The pictures, which we have in everyday life spontaneously in front of us, are made of such systems. However, in general the distances of the $N$ systems from the observer vary and we do not observe the true $(\mathbf{r}, t)$-sizes of the $N$ systems that fill the picture in front of us at time $\tau$. Therefore, for the description of such pictures we need beside the functions $\Psi^*(\mathbf{r}, t)_i \Psi(\mathbf{r}, t)_i, i = 1, \ldots, N$, a further tool which considers the distances of the $N$ systems from the observer. These corrections could be done by geometrical optics. The whole scenario in the picture is a projection from $(\mathbf{p}, E)$-space onto $(\mathbf{r}, t)$-space. Therefore, the source for the whole scenario in $(\mathbf{r}, t)$-space is given in terms of $(\mathbf{p}, E)$-fluctuations in $(\mathbf{p}, E)$-space. However, it is probably easier to use geometrical optics, that is, to correct the pictures in $(\mathbf{r}, t)$-space. This point is explained in more detail in Section 5.7.4.2.

## 3.6. Auxiliary Constructions

As we have outlined in [12], the variable $\mathbf{p}$ and $E$ are constructions on the basis of the space–time variables $\mathbf{r}$ and $t$. It follows immediately from this, that all the physical formulas for the description of the structures in $(\mathbf{r}, t)$-space as well as in $(\mathbf{p}, E)$-space are auxiliary constructions and their elements cannot have real existence in this form. Let us discuss these points in a little more detail.

### 3.6.1. *Energy within Conventional Physics*

We have stated above that we cannot make any statement about basic (absolute) reality. This is true even in connection with its variables. On the other hand, within projection theory the variables $\mathbf{p}$ and $E$ represent reality but this reality is a fictitious reality and has nothing to do with basic (absolute) reality.

The variables $\mathbf{p}$ and $E$ have to be considered as abstract ideas, as products of the human mind so to speak. Thus, reality that is embedded in $(\mathbf{p}, E)$-space also has to be considered as an abstract idea, i.e., it cannot be identified with basic reality. Therefore, (**p**, $E$)-space represents a "fictitious reality". The energy is introduced on the basis of (**r**, $t$)-space to which our intuition is adapted. In fact in the unit of energy ($g \times \text{length}^2/\text{time}^2$) the units of space and time explicitly occur. In other words, **p** and $E$ do not occur in nature. This point has already been discussed in literature with respect to conventional physics (see, for example, [9]). A good statement in connection with energy $E$ is given by Christian von Baeyer in his interesting book *Information*: "*The gradual crystallization of the concept of information during the last hundred years contrasts sharply with the birth of the equally abstract quantity called energy in the middle of the nineteenth century. Then, in the brief span of twenty years, energy was invented, defined and established as a cornerstone, first in physics, then of all science. We don't know what energy is, any more than we know what information is, but as a now robust scientific concept we can describe it in precise mathematical terms, and as a commodity we can measure, market, regulate and tax it.*"

Concerning energy, Edgar Lüscher remarked the following in his book *Experimentalphysik*: "*Energy is not a quantity that actually occurs in nature, but it is an abstract idea, a product of the human mind who tries to understand nature within his capabilities.*"

In conclusion, the statements by Christian von Bayer and Edgar Lüscher about the nature of energy have been formulated on the basis of conventional physics. Projection theory confirms that impressively. Here the energy $E$ (and also the momentum **p**) plays the role of auxiliary constructions, that is, within projection theory the variables **p** and $E$ have to be considered as abstract ideas, as products of the human mind, which do not appear in nature.

### 3.6.2. *The Physical Laws of Basic Reality*

What about the physical laws? How a physical law, deduced by a human observer, is actually formed within basic reality remains principally hidden. In other words, we come exactly to the same conclusion, which we have already drawn in Chapter 1 from the structure of Newton's equations of motion. As an example, we discussed in Chapter 1 the equation of motion for the earth around the sun. We will come back to this point in the next two sections when we discuss the physical laws within the framework of projection theory.

### 3.6.3. *Remark*

Since the $(\mathbf{p}, E)$-values fluctuate for each system there must be an interaction [$(\mathbf{p}, E)$-fluctuations] between the system under investigation and the environment. In principle, the fluctuations may cover the range $-\infty \leq \mathbf{p}, E \leq \infty$. The $(\mathbf{p}, E)$-values of the interacting systems are strongly correlated at each time $\tau$. Only in this way the conservation laws for momentum and energy can be fulfilled at each time $\tau$. If the system under investigation would perform $(\mathbf{p}, E)$-fluctuations independently from other systems of the environment, these conservation laws would be violated.

Therefore, the systematic $(\mathbf{p}, E)$-fluctuations in $(\mathbf{p}, E)$-space reflect processes and we have to consider $(\mathbf{p}, E)$-space as reality. However, because the variables $\mathbf{p}$ and $E$ are auxiliary elements this reality is a "fictitious reality". Therefore, also the $(\mathbf{p}, E)$-fluctuations (the interactions of the systems with other systems) are fictitious in character and merely simulate the real, but unknown processes in basic reality.

Due to the conservation laws for the momentum $\mathbf{p}$ and the energy $E$ we must have systematic interaction processes [$(\mathbf{p}, E)$-fluctuations] of the system under investigation with other systems of the environment and, therefore, we have to consider $(\mathbf{p}, E)$-space as (fictitious) reality. However, there do not exist such conservation laws for the variables $\mathbf{r}$ and $t$, i.e., the realized $(\mathbf{r}, t)$-point at time $\tau$ of the system under investigation is independent of the realized $(\mathbf{r}, t)$-points at the same time $\tau$ of the other systems of the environment. There is no exchange of certain "space–time pieces" between the pictures. In other words, there are no correlations between the $(\mathbf{r}, t)$-points at time $\tau$. This indicates that $\mathbf{r}$ and $t$ are in fact elements of a picture. In other words, the information of fictitious reality [$(\mathbf{p}, E)$-space] is projected onto $(\mathbf{r}, t)$-space by a Fourier transform and we get the "picture of reality".

The variables $\mathbf{p}$ and $E$ cannot exist in basic realty because the variables $\mathbf{r}$ and $t$ do not exist there. (We have outlined above that $\mathbf{p}$ and

$E$ are introduced on the basis of the variables $\mathbf{r}$ and $t$ of $(\mathbf{r}, t)$-space to which our intuition is adapted.) In particular, we have discussed in Chapter 2 that the variables $\mathbf{r}$ and $t$ are observer-specific and direct statements about basic reality are by definition not observer-specific and picture-specific, respectively, i.e., an observer-independent (picture-independent) point of view is not defined in projection theory.

## 3.7. Basic Laws

How can we determine the wavefunctions $\Psi(\mathbf{r}, t)$ and $\Psi(\mathbf{p}, E)$? Here we need certain basic equations. In [12], we have deduced such equations for the $(\mathbf{r}, t)$-space as well as for the $(\mathbf{p}, E)$-space. Here we do not want to repeat the procedure for the deduction of these equations. In this section let us discuss only the main results and the specific structure of these formulas. For comparison, the results of conventional quantum theory (Schrödinger's equations) will be quoted too.

### 3.7.1. *Stationary Case*

Within usual (conventional) quantum theory the equations for the determination of the wave function (Schrödinger's equation) cannot be deduced [12] but had to be assumed. For the stationary case, we have (Schrödinger's equation for the stationary case)

$$i\hbar\frac{\partial}{\partial\tau}\psi(\mathbf{r}, \tau) = -\frac{\hbar^2}{2m_0}\Delta\psi(\mathbf{r}, \tau) + U(x, y, z)\psi(\mathbf{r}, \tau). \qquad (3.21)$$

The corresponding result of projection theory is given by [using the variables of $(\mathbf{r}, t)$-space with $\mathbf{r} = (x, y, z)$],

$$i\hbar\frac{\partial}{\partial\tau}\psi(\mathbf{r}, t) = -\frac{\hbar^2}{2m_0}\Delta\psi(\mathbf{r}, t) + V(x, y, z, t)\psi(\mathbf{r}, t). \qquad (3.22)$$

In contrast to usual quantum theory, where Schrödinger's equation (3.21) could not be deduced, the basic equation of projection theory [Eq. (3.22)] could be derived on the basis of operator rules. (More details are given [12] and also in Chapter 5.)

The equivalent law to (3.22), expressed by the variables of $(\mathbf{p}, E)$-space, has the form (see in particular [12])

$$E\Psi(\mathbf{p}, E) = \frac{\mathbf{p}^2}{2m_0}\Psi(\mathbf{p}, E) + V\left(i\hbar\frac{\partial}{\partial p_x}, i\hbar\frac{\partial}{\partial p_y}, i\hbar\frac{\partial}{\partial p_z} - i\hbar\frac{\partial}{\partial E}\right)\Psi(\mathbf{p}, E). \quad (3.23)$$

As in the case of Schrödinger's equation of usual quantum theory, Eq. (3.22) is the stationary case with respect to time $\tau$. A stationary solution with respect to the system-specific time $t$ is not defined here [12], i.e., within projection theory there do not exist solutions for the case

$$V(x, y, z, t) = V(x, y, z). \quad (3.24)$$

This feature is entirely due to the structure of Eqs. (3.22) and (3.23), respectively, and the fact that $(\mathbf{r}, t)$-space and $(\mathbf{p}, E)$-space are connected by a Fourier transform.

### 3.7.2. *Non-Stationary Case*

The non-stationary cases within conventional quantum theory and projection theory both refer to the reference time $\tau$. As is well known, the non-stationary Schrödinger equation of conventional quantum theory is given by

$$i\hbar\frac{\partial}{\partial\tau}\psi(\mathbf{r}, \tau) = -\frac{\hbar^2}{2m_0}\Delta\psi(\mathbf{r}, \tau) + U(x, y, z, \tau)\psi(\mathbf{r}, \tau). \quad (3.25)$$

The corresponding result of projection theory is expressed by [using the variables of $(\mathbf{r}, t)$-space]

$$i\hbar\frac{\partial}{\partial t}\Psi(\mathbf{r}, t)_\tau = -\frac{\hbar^2}{2m_0}\Delta\Psi(\mathbf{r}, t)_\tau + V(x, y, z, t)_\tau\Psi(\mathbf{r}, t)_\tau. \quad (3.26)$$

The basic, non-stationary law, expressed by the variables of $(\mathbf{p}, E)$-space, has the form [12]

$$E\,\Psi(\mathbf{p}, E)_\tau = \frac{\mathbf{p}^2}{2m_0}\Psi(\mathbf{p}, E)_\tau + V\left(i\hbar\frac{\partial}{\partial p_x}, i\hbar\frac{\partial}{\partial p_y}, i\hbar\frac{\partial}{\partial p_z}, -i\hbar\frac{\partial}{\partial E}\right)_\tau \Psi(\mathbf{p}, E)_\tau. \quad (3.27)$$

The stationary behavior with respect to the wavefunctions $\Psi(\mathbf{r}, t)_\tau$ and $\Psi(\mathbf{p}, E)_\tau$ comes into play through the potential, that is, we obtain the stationary wavefunctions $\Psi(\mathbf{r}, t)$ and $\Psi(\mathbf{p}, E)$ for the replacements

$$V(x, y, z, t)_\tau \rightarrow V(x, y, z, t), \quad (3.28)$$

$$V\left(i\hbar\frac{\partial}{\partial p_x}, i\hbar\frac{\partial}{\partial p_y}, i\hbar\frac{\partial}{\partial p_z}, -i\hbar\frac{\partial}{\partial E}\right)_\tau \rightarrow V\left(i\hbar\frac{\partial}{\partial p_x}, i\hbar\frac{\partial}{\partial p_y}, i\hbar\frac{\partial}{\partial p_z}, -i\hbar\frac{\partial}{\partial E}\right) \quad (3.29)$$

which leads to the transitions

$$\Psi(\mathbf{r}, t)_\tau \rightarrow \Psi(\mathbf{r}, t), \quad (3.30)$$

$$\Psi(\mathbf{p}, E)_\tau \rightarrow \Psi(\mathbf{p}, E). \quad (3.31)$$

This of course also holds for conventional quantum theory, i.e., here the stationary case is defined by the replacement

$$U(x, y, z, \tau) \rightarrow U(x, y, z). \quad (3.32)$$

In the following section, we will briefly discuss the physics behind the functions $V(x, y, z, t)$ [$V(x, y, z, t)_\tau$] and $U(x, y, z)$ [$U(x, y, z, \tau)$].

### 3.7.3. *Discussion*

Schrödinger's equations (3.21) and (3.25) as well as the basic equations of projection theory [(3.22) and (3.23), (3.26) and (3.27)] are equally based on the well-known classical equations $E = \mathbf{p}^2/(2m_0) + U(x, y, z)$ and $E = \mathbf{p}^2/(2m_0) + U(x, y, z, \tau)$. Within conventional quantum theory the potential $U(x, y, z)$ [$U(x, y, z, \tau)$] remains classical when we go from classical mechanics to the quantum-theoretical description. The situation is quite different in projection theory. Here the function $V(x, y, z, t)$ $[V(x, y, z, t)_\tau]$ is a quantum-theoretical quantity, that is, in the transition from classical physics to quantum physics the classical function $U(x, y, z)$ $[U(x, y, z, \tau)]$ is automatically replaced within projection theory by the quantum-theoretical expression $V(x, y, z, t)$ $[V(x, y, z, t)_\tau]$. In other words, the transitions

$$U(x, y, z) \rightarrow V(x, y, z, t) \tag{3.33}$$

and

$$U(x, y, z, \tau) \rightarrow V(x, y, z, t)_\tau \tag{3.34}$$

really reflect a transition from the classical to the quantum-theoretical aspect.

## 3.8. Extension of Conventional Quantum Theory

By comparison of the Eqs. (3.21)–(3.27), we have to conclude that there is a real extension of quantum theory when we go from conventional quantum theory to projection theory. While in conventional quantum theory we have only the variable $\mathbf{r} = (x, y, z)$ (and of course the external parameter $\tau$), within projection theory we have not only the variable $\mathbf{r} = (x, y, z)$ but in addition we have here the system-specific time $t$. (Clearly, we also have the external parameter $\tau$.) In other words, when we go from usual quantum theory to projection theory we have the transition

$$\mathbf{r} \rightarrow \mathbf{r}, t. \tag{3.35}$$

The additional variable $t$ is important for the basic understanding of that which we call time. In Section 3.16, we will discuss this point in more detail. The reference time $\tau$ directly reflects our time-feeling which we have in everyday life, and $\tau$ has been chosen by man to meet this time-feeling, but $\tau$ appears merely as an external parameter and is measured by clocks. But this ad hoc-construction cannot explain the source of the notion "time". Clearly, within theory of relativity $\tau$ becomes flexible but the principal problem remains here.

Equation (3.22) describes the possible geometrical structures in $(\mathbf{r}, t)$-space, i.e., (3.22) allows the determination of the wavefunction $\Psi(\mathbf{r}, t)$ if the function $V(x, y, z, t)$ is known. The probability density $\Psi^*(\mathbf{r}, t)\Psi(\mathbf{r}, t)$ directly defines the observed structure in space and time [Fig. 3.4(a)]. The picture of reality does not contain any material object but exclusively geometrical positions. These geometrical structures have their seat in the human brain.

This in particular means that the potential $V(x, y, z, t)$ cannot describe the potential energy between certain structures in $(\mathbf{r}, t)$-space as, for example, between the two systems represented in Fig. 3.4(a). It makes no sense to define a potential energy, say $V(x, y, z, t)$, between geometrical positions. In conclusion, Eq. (3.22) can only be an auxiliary construction for the determination of $\Psi(\mathbf{r}, t)$ and $\Psi^*(\mathbf{r}, t)\Psi(\mathbf{r}, t)$, respectively. We do not know at all how the auxiliary process described by (3.22) can be expressed in terms of the features of basic reality where the real world is actually embedded. Figure 3.5 reflects this situation. We come exactly to the same conclusion, which we have drawn from the structure of Newton's equations of motion where we discussed, for an example, the equation for the motion for the earth around the sun: $m_E(d^2\mathbf{r}_E/d\tau^2) = -Gm_E m_S(\mathbf{r}_E - \mathbf{r}_S/|\mathbf{r}_E - \mathbf{r}_S|^3$ (Chapter 1, Fig. 1.1).

Within conventional quantum theory (physics), the situation is different. Here we have no projections but the real systems (masses) are embedded in space and the function $U(x, y, z)$ can therefore be considered as potential energy between certain systems which are separated by the distance between them. However, as we will still discuss below, the function $U(x, y, z)$ is in a certain sense an occult quantity and its introduction without the possibility to model it seems to be artificial. Then, the following question arises: Is the classical conception "the world (the real masses) is embedded in space" a realistic conception? As we have stated in Chapter 2, this conception should be in fact questionable. Thus, also the classical equation $E = \mathbf{p}^2/(2m_0) + U(x, y, z)$ which has been used for the formulation of Schrödinger's equation should not be interpreted in terms of a reality that is embedded in space (space–time). Note, that Schrödinger's equation of conventional quantum theory had to be assumed, that is, it

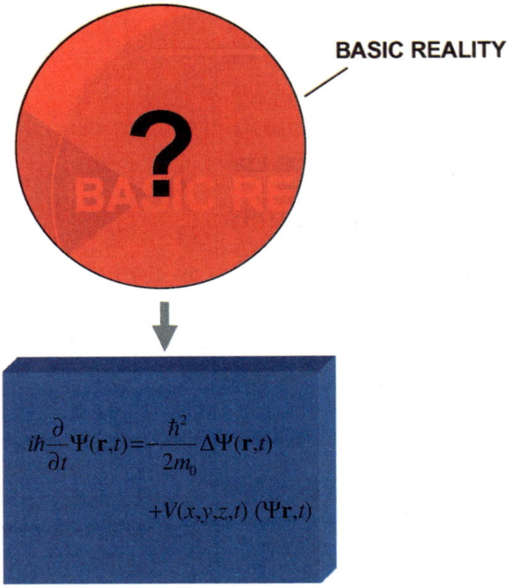


Fig. 3.5. The equation for the determination of the wavefunction $\Psi(\mathbf{r}, t)$ in $(\mathbf{r}, t)$-space [Eq. (3.22)] describes the structural states in the picture but not the events in basic reality. A human observer is principally not able to make statements about basic reality. How basic reality is structured remains hidden.

could not be derived. The reason for this shortcoming is probably the questionable basic concept of conventional physics. The world (the real masses) is embedded in space. In Chapter 5, we will discuss another interpretation, which is compatible with the projection principle.

## 3.9. Only Processes are Relevant!

### 3.9.1. *Free Systems*

It can be shown within projection theory that free (non-interacting) systems cannot exist if the system is elementary in character [12]. Elementary in character means that the system cannot be divided into subsystems. A system can be considered as free if the potential function becomes zero: $V(x, y, z, t) = 0$; then, also the operator $V(i\hbar\partial/\partial p_x, \ldots, -i\hbar\partial/\partial E)$ (Section 3.7) is not definable.

In the case of free systems, the momentum $\mathbf{p} = \mathbf{p}_0$ and the energy $E = E_0$ remain constant in the course of time $\tau$. There is no interaction between the system under investigation and other systems (environment), and there are no $(\mathbf{p}, E)$-fluctuations in the case of free systems: $\Delta\mathbf{p} = 0, \Delta E = 0$. What do we get within projection theory for the wavefunctions $\Psi(\mathbf{r}, t)$ and $\Psi(\mathbf{p}_0, E_0)$ in this specific case?

The outcome can be summarized as follows [12]:

1. The wavefunctions $\Psi(\mathbf{r}, t)$ and $\Psi(\mathbf{p}_0, E_0)$ become zero:

$$\Psi(\mathbf{r}, t) = 0, \tag{3.36}$$

$$\Psi(\mathbf{p}_0, E_0) = 0. \tag{3.37}$$

   Thus, also the probability densities are zero:

$$\Psi^*(\mathbf{r}, t)\Psi(\mathbf{r}, t) = 0, \tag{3.38}$$

$$\Psi^*(\mathbf{p}_0, E_0)\Psi(\mathbf{p}_0, E_0) = 0. \tag{3.39}$$

2. The analysis showed [12] that the systems momentum $\mathbf{p}_0$ and its energy $E_0$ become zero too: $\mathbf{p}_0 = 0,\ E_0 = 0$. That is, nothing has been taken away from the cosmic $(\mathbf{p}, E)$-pool. Therefore, from the point of view of projection theory free elementary systems cannot exist in nature (see also Fig. 3.6).

### 3.9.2. *Principle of Usefulness*

In conclusion, projection theory requires that non-interacting systems cannot exist in nature. Why does nature not admit such systems, i.e., free elementary systems without internal structure? Such free systems with constant momentum $\mathbf{p}_0$ and constant energy $E_0$ are in a certain sense "dead systems" because they are not involved in the processes in nature. There are no $(\mathbf{p}, E)$-fluctuations between a free system and its surroundings, and such systems would be useless. In other words, such systems with

$$\mathbf{p}_0 = \text{const.} \neq 0, \tag{3.40}$$

$$E_0 = \text{const.} \neq 0 \tag{3.41}$$

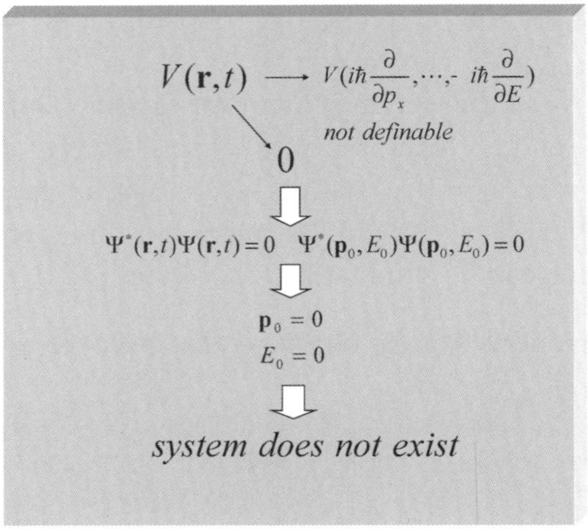


Fig. 3.6. In the case of non-interacting elementary systems we have $V(x, y, z, t) = 0$ and, therefore the operator $V(i\hbar\partial/\partial p_x, \ldots, -i\hbar\partial/\partial E)$ is not definable. Then, the wavefunctions $\Psi(\mathbf{r}, t)$ and $\Psi(\mathbf{p}_0, E_0)$ and the corresponding probability distributions become zero and, furthermore, the systems momentum $\mathbf{p}_0$ and its energy $E_0$ become zero too: $\mathbf{p}_0 = 0$, $E_0 = 0$. That is, nothing has been taken away from the cosmic $(\mathbf{p}, E)$-pool. Therefore, from the point of view of projection theory free elementary systems cannot be realized in nature.

have no place in nature because they would be completely detached from the scenario and this is against the principle of usefulness. Thus, from this point of view free systems should not exist in nature. In fact, projection theory excludes such states explicitly because we have $\mathbf{p}_0 = 0$ and $E_0 = 0$. In particular, we have $\Psi(\mathbf{r}, t) = 0$ and $\Psi(\mathbf{p}, E) = \Psi(\mathbf{p}_0, E_0) = 0$), i.e., the properties given by the Eqs. (3.40) and (3.41) do not exist within projection theory. In other words, projection theory supports the existence of the principle of usefulness.

In fact, the world of man is formed in accordance with the principle of usefulness, and a lot of examples, in particular in connection with evolution, support this view (see also Chapter 1). It is astonishing that this principle is obviously even reflected at the basic level of theoretical description (projection theory). However, this is not a description of the *objective reality* but it is "merely" man's view. Man is principally not able to recognize the properties of that which we have called "objective (basic) reality".

### 3.9.3. *Real Situation*

Nevertheless, the world seems to be built up of subsystems which are normally assumed to be independent from the interaction between them. In other words, we suppose anyhow that there are really certain subsystems possible without mutual interactions. It is assumed that each of the subsystems can exist in an isolated state and that its existence is not dependent on certain external units and parameter, respectively. Our observations in everyday life suggest such a model. However, we have to be careful.

Such free (non-interacting) systems could be considered as useful (in the sense of evolution) if, at a certain time $\tau_{\text{begin}}$, an interaction could be switched on and, in principle, this interaction could be switched off again at time $\tau_{\text{end}}$. However, we do not observe such a switching-on/switching-off scenario in nature. In particular, it assumes a mechanism that is hostile against projection theory because it can obviously not be based on $(\mathbf{p}, E)$-fluctuations. In conclusion, free (non-interacting) systems should not exist in nature, and projection theory should be able to proof that if the projection principle contains the "principle of usefulness" (see Chapter 2).

The results of this section actually show that this is the case. In [12] we showed mathematically that free (non-interacting) systems without further substructure cannot exist within projection theory. It turned out that the relevant quantities, the probability distributions $\Psi^*(\mathbf{r}, t)\Psi(\mathbf{r}, t)$ and $\Psi^*(\mathbf{p}_0, E_0)\Psi(\mathbf{p}_0, E_0)$, are exactly zero.

### 3.9.4. *Summary*

Only interacting systems with non-vanishing $(\mathbf{p}, E)$-fluctuations $(\Delta\mathbf{p} \neq 0, \Delta E \neq 0)$ are relevant when we base our considerations on the "principle of usefulness". That is also a requirement of projection theory because within this theory free (non-interacting) systems without substructure cannot exist. Thus, projection theory obviously contains the "principle of usefulness". Only processes are relevant because static systems are dead systems and cannot be involved in the scenario in nature. Like Whitehead and Bergson, we argue for the primacy of process.

## 3.10. Interactions

In this section, we would like to give some specific statements about that what we call "interaction" and, in particular, its peculiarities in connection with projection theory. In contrast to conventional quantum theory, interactions within projection theory are not restricted to the space variable **r** but it is always an interaction with respect to space–time, i.e., the interaction is described in projection theory by the variables **r** and $t$ and is not restricted to the space variable **r**. Let us mention once more that the system-specific time $t$ is not defined in conventional physics but only a certain kind of external time $\tau$ which we have called above "reference time". Again, $\tau$ is the time that we measure with our clocks and reflects the observers' time feeling. In projection theory beside the variables **r** and $t$ also the reference time $\tau$ plays a particular role. (More details are given in [12] and in Section 3.16.2.)

As we have already pointed out above, the interaction takes place in $(\mathbf{p}, E)$-space and not in $(\mathbf{r}, t)$-space. (In usual quantum theory the real interaction processes take place in space because here it is always assumed that the bodies are really embedded in space–time, and the interaction strength is dependent on the distance between the bodies.) As we have also outlined several times, this picture of usual quantum theory cannot be used in projection theory. Within the framework of projection theory, the interaction does not take place in space $[(\mathbf{r}, t)$-space$]$, and there are no material systems (objects) embedded in $(\mathbf{r}, t)$-space. In projection theory reality has to be identified with states in $(\mathbf{p}, E)$-space. The interactions between individual systems exclusively take place in $(\mathbf{p}, E)$-space and are exclusively expressed by $(\mathbf{p}, E)$-fluctuations (Section 3.2.2). Therefore, the interaction mechanisms which are used in conventional physics cannot be applied here. All these peculiarities change the situation fundamentally and have to be analyzed further.

### 3.10.1. *What Does Interaction Mean within Projection Theory?*

In contrast to the interaction-concept in projection theory, interactions in usual quantum theory are treated on the basis of the classical conception of physics: Everything is embedded in space and the processes, which we observe, are considered to be real material processes taking place in space. In fact, Born's interpretation of the wavefunction is based on the assumption that there are material objects embedded in space. According to Born, the probability to find at time $\tau$ a real quantum object (assumed to be a point-like particle) in the volume element $dV = dxdydz$ at a certain position $\mathbf{r} = (x, y, z)$ is given by $\psi^*(\mathbf{r})\psi(\mathbf{r})dV$ (Section 3.2.2). Within projection theory the situation is quite different: Here measurement means that the properties of $(\mathbf{r}, t)$-space and $(\mathbf{p}, E)$-space are connected and, as we have already outlined in Section 3.2.2, this leads to the following picture. The measurement of one of the possible values for $\mathbf{p}$ and for $E$ is done at time $\tau$ in the space–time intervals $\mathbf{r}, \mathbf{r} + d\mathbf{r}$ and $t,\ t + dt$ with the probability density of $\Psi^*(\mathbf{r}, t)\Psi(\mathbf{r}, t)$ (Section 3.2.2). Again, this statement means that we are working simultaneously on the basis of two spaces [$(\mathbf{r}, t)$-space and $(\mathbf{p}, E)$-space].

Within projection theory we have to distinguish between basic reality, fictitious realty, and the picture of reality. In this connection, it is important to mention that also the measuring instruments (detectors) are positioned in basic reality, and the registration processes take place in basic reality too. However, this situation is exclusively observed by the observer as a picture in $(\mathbf{r}, t)$-space: A measuring instrument and the movements of its pointer appear as a "picture of reality" and has nothing to do with material processes in $(\mathbf{r}, t)$-space but actually take place in basic reality.

Let us underline once more that basic (absolute) reality cannot be experienced by humans. The starting-point of projection theory are the pictures that we have spontaneously in front of us. These pictures have to be considered as "transformed geometrical structures" and they reflect directly certain situations of basic (absolute) reality. In this connection, the following point is important: *Within the memory of man, all essential things are represented within the frame of pictures.* Therefore, we have chosen in our theoretical conception the pictures, which we have spontaneously in front of us, as the most basic facts, and we have used them as starting-point for the development of a theoretical framework (projection theory).

An essential point within projection theory is the feature that (fictitious) reality and the pictures are connected by a Fourier transform. Just this peculiarity makes it necessary to discuss the notion "interaction" in a new light because all interactions exclusively take place in $(\mathbf{p}, E)$-space and not in $(\mathbf{r}, t)$-space (in conventional physics interactions exclusively take place in space). In other words, due to the Fourier transform we are confronted with a completely new situation when we try to analyze interaction processes in projection theory.

In connection with the projection principle, the following feature is important. A system that is localized in $(\mathbf{r}, t)$-space is delocalized in $(\mathbf{p}, E)$-space. This property follows directly from the Fourier transform. For example, let us consider two systems, say $i$ and $j$, which have sharp space–time positions $\mathbf{r}_i, \mathbf{r}_j$ and $t_i, t_j$ in $(\mathbf{r}, t)$-space. However, these two systems are uniformly distributed in $(\mathbf{p}, E)$-space, that is, we have $-\infty \leqslant \mathbf{p}, E \leqslant \infty$ in this specific case (in analogy to that which we have outlined in Section 3.4.1).

The interaction scenario in projection theory is characterized by the following feature: The relative changes in the space–time positions of system $i$ and system $j$ are not due to interaction processes through space–time [$(\mathbf{r}, t)$-space] but come into play by $(\mathbf{p}, E)$-fluctuations which take place in $(\mathbf{p}, E)$-space, and the projection of these fluctuations leads to specific structures in $(\mathbf{r}, t)$-space. If the law for the $(\mathbf{p}, E)$-fluctuations is changed, the Fourier transformation leads automatically to the effect that the relative space–time distances between system $i$ and system $j$ and also the shapes (forms) are changed [12].

The space–time structure, which appears in $(\mathbf{r}, t)$-space, is a projected structure and does not come into existence by an exchange of a certain kind of information through $(\mathbf{r}, t)$-space. Therefore, any specific law $\Psi(\mathbf{p}, E)$ [$\Psi^*(\mathbf{p}, E)\Psi(\mathbf{p}, E)$], which reflects specific $(\mathbf{p}, E)$-fluctuations (interactions) in $(\mathbf{p}, E)$-space, leads to correlated space–time configurations ($(\mathbf{r}, t)$-correlations); there is nothing else.

When we talk in the following about interactions with respect to space–time positions in $(\mathbf{r}, t)$-space, we always mean that there are correlations between the various $(\mathbf{r}, t)$-positions. In other words, "interaction" in $(\mathbf{r}, t)$-space always means in projection theory that there are correlated $(\mathbf{r}, t)$-positions, and there is definitely no exchange of a certain kind of information through $(\mathbf{r}, t)$-space between the various $(\mathbf{r}, t)$-positions.

### 3.10.2. *Delocalized Systems in* $(\mathbf{p}, E)$*-Space*

The whole interaction scenario in physics (classical physics, usual quantum theory, quantum field theory, string theory) is based on processes between localized systems in space (space–time). Since the interactions in projection theory take place in $(\mathbf{p}, E)$-space, none of these specific processes can be applied here. Furthermore, in contrast to the conventional conceptions (here localized systems are used), within projection theory we are confronted with delocalized systems in $(\mathbf{p}, E)$-space, if the same systems (their geometrical structure) are localized in $(\mathbf{r}, t)$-space (see also [12]).

### 3.10.3. *Abstract Interaction Laws*

The description of the interaction between delocalized systems in $(\mathbf{p}, E)$-space is not possible on the basis of interpretations and mechanisms, respectively, which we are using in classical physics or usual quantum theory. A mechanism is normally based on our intuition, and man's intuition is tailor-made to localized bodies that are embedded in space [$(\mathbf{r}, t)$-space] and not to delocalized systems that are positioned in $(\mathbf{p}, E)$-space. Our intuition is adjusted to the structures in space which we experience in everyday life, and these space structures are composed of localized bodies that are embedded in space [$(\mathbf{r}, t)$-space]. Our complete intuitive abilities have been developed in connection with situations which we experience in everyday life, and we call a certain process "imaginable" when it can be explained by the means of everyday life. A "mechanism" is normally defined on the basis of an imaginable process. However, interacting delocalized systems which are positioned in $(\mathbf{p}, E)$-space do not belong to the category of imaginable processes. Thus, it does not make much sense to explain these $(\mathbf{p}, E)$-fluctuations by means of such "mechanisms" but we have to accept that they simply "emerge." In conclusion, the interaction processes (the $(\mathbf{p}, E)$-fluctuations in $(\mathbf{p}, E)$-space) have to be considered as abstract laws.

## 3.11. Distance-Independent Interactions

In [12], we have treated a system consisting of $N$ interacting subsystems, and we assumed that a distance-dependent interaction between the subsystems is effective, that is, the mutual interactions between the subsystems are dependent on the space–time distances $\mathbf{r}_{ki} - \mathbf{r}_l, t_k - t_l$ with $k, l = 1, \ldots, N, k \neq l$ where $\mathbf{r}_1, \mathbf{r}_2, \ldots, \mathbf{r}_N, t_1, t_2, \ldots, t_N$ are the positions and system-specific times in $(\mathbf{r}, t)$-space at time $\tau$. However, projection theory opens the possibility for another kind of interaction that is not dependent on the space–time distances and, therefore, we would like to call it "distance-independent interactions". In other words, there can be interactions between two systems, say $i$ and $j$, whose strength is not dependent on the space–time distances $\mathbf{r}_i - \mathbf{r}_j, t_i - t_j$.

This type of interaction reflects an interesting physical situation, not merely as a formal aspect but in particular with respect to many scientific real effects as, for example, non-local events with respect to the various $(\mathbf{r}, t)$-states. However, the projection principle not only emerges new aspects in connection with the notion "interaction" but it influences the entire range of physical reality as, for example, Mach's principle. First, we will discuss some principal features in connection with "distance-independent interactions". After that we will briefly analyze the relative motion of certain systems that interact by this type of interaction, and Mach's principle comes inevitably into play.

### 3.11.1. *General Remarks*

Does such kind of interaction make sense? Yes, it definitely does. These distance-independent interactions furnish a system with a specific form (shape). Therefore, in the following we will call this kind of interaction also "form interaction". If the systems (particles) $i$ and $j$ are elementary in character this form interaction creates their shape. In conventional physics, we also work with certain forms of elementary systems: We have point-like particles, strings and branes. However, these specific forms had to be assumed in conventional physics and could not be derived. In contrast to these developments, projection theory opens the possibility to explain (derive) certain elementary forms in nature by means of this new kind of interaction, i.e., by distance-independent (form) interactions. In the next sections, we will discuss some basic features of distance-independent interactions. But let us first show some further arguments for the introduction of distance-independent interactions in projection theory.

### 3.11.2. *Principal Analysis*

In Section 3.4, we have discussed Fourier-transform effects. In particular, we have studied simple space–time configurations at time $\tau$ and $\tau + \Delta\tau$. Equation (3.18) is the space–time configuration at time $\tau$, and Eq. (3.20) is the corresponding space–time configuration at time $\tau + \Delta\tau$ where $\Delta\tau$ is an arbitrarily chosen interval. In principle, $\Delta\tau$ can be infinitesimal but should be different from zero.

The following has been stated in Section 3.4: "*By comparison of* (3.18) *with* (3.20), *we immediately recognize that the properties at all points* $\mathbf{r}, t(-\infty \leq \mathbf{r} \leq \infty, -\infty \leq t \leq \infty)$ *at* $\tau + \Delta\tau$ *are different from those at time* $\tau$*; these global changes take place instantaneously since* $\Delta\tau$ *can in principle be infinitesimal. In other words, we are confronted with non-local effects. These non-local effects are entirely due to the projection process from reality* [$(\mathbf{p}, E)$*-space*] *onto* $(\mathbf{r}, t)$*-space, like a flashlight on a screen.*"

These global changes in the $(\mathbf{r}, t)$-structure come into being through the $(\mathbf{p}, E)$-fluctuations

$$\Delta E = E_0 - E_0', \Delta \mathbf{p} = \mathbf{p}_0 - \mathbf{p}_0' \tag{3.42}$$

in $(\mathbf{p}, E)$-space [see (3.18) and (3.20)]. However, due to the projection effect from $(\mathbf{p}, E)$-space onto $(\mathbf{r}, t)$-space there is no privileged space–time point, say $\mathbf{r}_P, t_P$, from which the information in connection with the $(\mathbf{p}, E)$-fluctuations $\Delta E = E_0 - E_0', \Delta\mathbf{p} = \mathbf{p}_0 - \mathbf{p}_0'$ would be distributed to all the other space–time positions (within a certain time interval). The information transfer to a certain space–time point $\mathbf{r}, t$, which causes the transition from (3.18) to (3.20) just at this $(\mathbf{r}, t)$-point, is independent of all the other space–time points $\mathbf{r}', t'$ with $\mathbf{r}' \neq \mathbf{r}$, $t' \neq t$. In other words, due the non-local effects (projections), distance-dependent properties are not defined in such cases. This can easily be verified when we use the expressions (3.17) and (3.18) and their connection with the Fourier transform. In the case of point $\mathbf{r}, t$ we have

$$B \exp\left\{\frac{i}{\hbar}(\mathbf{p}_0 \cdot \mathbf{r} - E_0\, t\right\}$$
$$= \frac{1}{(2\pi\hbar)^2} \int_{-\infty}^{\infty} \delta(\mathbf{p} - \mathbf{p}_0)\delta(E - E_0)$$
$$\times \exp\left\{i\left[\frac{\mathbf{p}}{\hbar} \cdot \mathbf{r} - \frac{E}{\hbar} t\right]\right\} dp_x dp_y dp_z dE \qquad (3.43)$$

and in the case of the points $\mathbf{r}', t'$, we obtain exactly the same expression:

$$B \exp\left\{\frac{i}{\hbar}(\mathbf{p}_0 \cdot \mathbf{r}' - E_0\, t'\right\}$$
$$= \frac{1}{(2\pi\hbar)^2} \int_{-\infty}^{\infty} \delta(\mathbf{p} - \mathbf{p}_0)\delta(E - E_0)$$
$$\times \exp\left\{i\left[\frac{\mathbf{p}}{\hbar} \cdot \mathbf{r}' - \frac{E}{\hbar} t'\right]\right\} dp_x dp_y dp_z dE. \qquad (3.44)$$

That is, the properties of space–time point $\mathbf{r}, t$ is determined by the $(\mathbf{p}, E)$-distribution alone, and this is valid for all the other points $\mathbf{r}', t'$. In other words, no space–time point influences the others, and there is no exchange of information between the space–time points. This in particular means that the space–time structure exclusively comes into play by projection effects and distance-dependent properties are not defined in such cases.

This is of course also true for the wavefunctions $\Psi(\mathbf{r}, t)$, $\Psi(\mathbf{p}, E)$ and, instead of (3.43) and (3.44), we get the original formulations [12]. For the space–time point $\mathbf{r}, t$ we have

$$\Psi(\mathbf{r}, t) = \frac{1}{(2\pi\hbar)^2} \int_{-\infty}^{\infty} \Psi(\mathbf{p}, E) \exp\left\{ i \left[ \frac{\mathbf{p}}{\hbar} \cdot \mathbf{r} - \frac{E}{\hbar} t \right] \right\} dp_x dp_y dp_z dE \tag{3.45}$$

and in the case of the points $\mathbf{r}', t'$, we obtain in analogy to (3.44):

$$\Psi(\mathbf{r}', t') = \frac{1}{(2\pi\hbar)^2} \int_{-\infty}^{\infty} \Psi(\mathbf{p}, E) \times \exp\left\{ i \left[ \frac{\mathbf{p}}{\hbar} \cdot \mathbf{r}' - \frac{E}{\hbar} t' \right] \right\} dp_x dp_y dp_z dE. \tag{3.46}$$

That is, also here the properties of space–time point $\mathbf{r}, t$ is determined by the $(\mathbf{p}, E)$-distribution alone, and this is valid for all the other points $\mathbf{r}', t'$. In other words, no space–time point influences the others, and this justifies in particular the probability interpretation for the wavefunctions $\Psi(\mathbf{r}, t)$ and $\Psi(\mathbf{p}, E)$.

The generalization of this situation is obvious and is almost required. If we have for example two interacting systems, say system 1 and system 2, there are $(\mathbf{p}, E)$-fluctuations between them which take place in $(\mathbf{p}, E)$-space. Then, generalization means that the information at a certain space–time point $\mathbf{r}_1, t_1$ of system 1 is not dependent on the space–time points $\mathbf{r}_2, t_2$ $(-\infty \leq \mathbf{r}_2, t_2 \leq \infty)$ of system 2 (and it is of course not dependent on all the other space–time points $\mathbf{r}_1', t_1'$ of system 1 with $\mathbf{r}_1' \neq \mathbf{r}_1, t_1' \neq t_1$). This must also be valid for system 2. The information at a certain space–time point $\mathbf{r}_2, t_2$ of system 2 is not dependent on the space–time points $\mathbf{r}_1, t_1$ $(-\infty \leq \mathbf{r}_1, t_1 \leq \infty)$ of system 1 (and it is of course not dependent on all the other space–time points $\mathbf{r}_2', t_2'$ of system 2 with $\mathbf{r}_2' \neq \mathbf{r}_2, t_2' \neq t_2$).

Let us express this situation in terms of the wavefunctions $\Psi(\mathbf{r}_1, t_1, \mathbf{r}_2, t_2)$ and $\Psi(\mathbf{p}_1, E_1, \mathbf{p}_2, E_2)$, where $\mathbf{p}_1, E_1$ are the $(\mathbf{p}, E)$-values for system 1 and $\mathbf{p}_2, E_2$ the corresponding value for system 2. Then, we have [12]

$$\begin{aligned}\Psi(\mathbf{r}_1, t_1, \mathbf{r}_2, t_2) = \frac{1}{(2\pi\hbar)^4} \int_{-\infty}^{\infty} & \Psi(\mathbf{p}_1, E_1, \mathbf{p}_2, E_2) \\ & \times \exp\left\{i\left[\frac{\mathbf{p}_1}{\hbar} \cdot \mathbf{r}_1 - \frac{E_1}{\hbar} t_1\right]\right\} \\ & \times \exp\left\{i\left[\frac{\mathbf{p}_2}{\hbar} \cdot \mathbf{r}_2 - \frac{E_2}{\hbar} t_2\right]\right\} \\ & \times dp_{1x} dp_{1y} dp_{1z} dE_1 \, dp_{2x} dp_{2y} dp_{2z} dE_2 .\end{aligned} \tag{3.47}$$

Using the ansatz

$$\Psi(\mathbf{p}_1, E_1, \mathbf{p}_2, E_2) = \Psi(\mathbf{p}_1, E_1)\Psi(\mathbf{p}_2, E_2) \tag{3.48}$$

we immediately get from Eq. (3.47)

$$\Psi(\mathbf{r}_1, t_1, \mathbf{r}_2, t_2) = \Psi(\mathbf{r}_1, t_1)\Psi(\mathbf{r}_2, t_2) \tag{3.49}$$

with

$$\begin{aligned}\Psi(\mathbf{r}_1, t_1) = \frac{1}{(2\pi\hbar)^2} \int_{-\infty}^{\infty} & \Psi(\mathbf{p}_1, E_1) \\ & \times \exp\left\{i\left[\frac{\mathbf{p}_1}{\hbar} \cdot \mathbf{r}_1 - \frac{E_1}{\hbar} t_1\right]\right\} dp_{1x} dp_{1y} dp_{1z} dE_1\end{aligned} \tag{3.50}$$

and

$$\begin{aligned}\Psi(\mathbf{r}_2, t_2) = \frac{1}{(2\pi\hbar)^2} \int_{-\infty}^{\infty} & \Psi(\mathbf{p}_2, E_2) \\ & \times \exp\left\{i\left[\frac{\mathbf{p}_2}{\hbar} \cdot \mathbf{r}_2 - \frac{E_2}{\hbar} t_2\right]\right\} dp_{2x} dp_{2y} dp_{2z} dE_2 .\end{aligned} \tag{3.51}$$

What does ansatz (3.48) mean? It does not mean that the $(\mathbf{p}, E)$-fluctuations of system 1 and system 2 are independent of each other. This is strictly forbidden because of the validity of the conservation laws for the momentum and the energy. At each time $\tau$, there is an exchange of momentum and energy between system 1 and system 2.

If system 1 takes at time $\tau$ the values $\mathbf{p}_1, E_1$ and system 2 the values $\mathbf{p}_2, E_2$ and if these fluctuations are $\Delta\mathbf{p}, \Delta E$, we have at time $\tau'$ (where $\tau'$ is the next value after $\tau$) for the momentums and the energies

$$\begin{aligned} \mathbf{p}_1' &= \mathbf{p}_1 \pm \Delta\mathbf{p}, \\ E_1' &= E_1 \pm \Delta E, \\ \mathbf{p}_2' &= \mathbf{p}_2 \mp \Delta\mathbf{p}, \\ E_2' &= E_2 \mp \Delta E. \end{aligned} \tag{3.52}$$

The equations expressed by (3.52) are dictated by the conservation laws for the momentum and the energy, which is valid at each time $\tau$.

There is a definite condition for the existence of $\Psi(\mathbf{p}_1, E_1)$ and $\Psi(\mathbf{p}_2, E_2)$. The probability densities $\Psi^*(\mathbf{p}_1, E_1)\Psi(\mathbf{p}_1, E_1)$ and $\Psi^*(\mathbf{p}_2, E_2)\Psi(\mathbf{p}_2, E_2)$ must have the same shape [12]. However, both functions may be shifted relative to each other in $(\mathbf{p}, E)$-space.

The meaning of Eq. (3.48) is more easily accessible through the properties that both systems have in $(\mathbf{r}, t)$-space. From (3.48) follows (3.49). Both functions $\Psi(\mathbf{r}_1, t_1)$ and $\Psi(\mathbf{r}_2, t_2)$ are not dependent on their absolute positions in $(\mathbf{r}, t)$-space and also not on their relative distance. This is also the case for the probability densities $\Psi^*(\mathbf{r}_1, t_1)\Psi(\mathbf{r}_1, t_1)$ and $\Psi^*(\mathbf{r}_2, t_2)\Psi(\mathbf{r}_2, t_2)$. What does that mean in connection with the interaction between system 1 and system 2?

Within projection theory, the interaction is defined by $(\mathbf{p}, E)$-fluctuations in $(\mathbf{p}, E)$-space. These interactions lead to correlations in $(\mathbf{r}, t)$-space. Thus, within projection theory, the notion "correlation" can be identified with the notion "interaction". The $(\mathbf{p}, E)$-fluctuations are directly described by the probability density $\Psi^*(\mathbf{p}, E)\Psi(\mathbf{p}, E)$. If $\Psi^*(\mathbf{p}, E)\Psi(\mathbf{p}, E)$ is changed, the interaction law between system 1 and system 2 is changed and simultaneously the correlations in $(\mathbf{r}, t)$-space. The interaction is distance-independent when the probability densities $\Psi^*(\mathbf{r}_1, t_1)\Psi(\mathbf{r}_1, t_1)$ and $\Psi^*(\mathbf{r}_2, t_2)\Psi(\mathbf{r}_2, t_2)$, which define system 1 and system 2 in $(\mathbf{r}, t)$-space, can be arbitrarily positioned in this space without changing the interaction laws $\Psi^*(\mathbf{p}_1, E_1)\Psi(\mathbf{p}_1, E_1)$ and $\Psi^*(\mathbf{p}_2, E_2)\Psi(\mathbf{p}_2, E_2)$. This is in fact possible in the case of ansatz (3.48) [which directly leads to (3.49)]. If we shift the function $\Psi(\mathbf{r}_1, t_1)$ arbitrarily by the constant values $\mathbf{r}_{1a}$ and $t_{1a}$ and the function $\Psi(\mathbf{r}_2, t_2)$

by $\mathbf{r}_{2a}$ and $t_{2a}$, we have for example

$$\begin{aligned} \mathbf{r}_1, t_1 &\rightarrow \mathbf{r}_1 - \mathbf{r}_{1a}, t_1 - t_{1a}, \\ \mathbf{r}_2, t_2 &\rightarrow \mathbf{r}_2 - \mathbf{r}_{2a}, t_2 - t_{2a}. \end{aligned} \tag{3.53}$$

Then, with (3.50) and (3.51) we get

$$\begin{aligned} \Psi(\mathbf{r}_1 - \mathbf{r}_{1a}, t_1 - t_{1a}) = \frac{1}{(2\pi\hbar)^2} \int_{-\infty}^{\infty} \Phi(\mathbf{p}_1, E_1) \\ \times \exp\left\{ i \left[ \frac{\mathbf{p}_1}{\hbar} \cdot \mathbf{r}_1 - \frac{E_1}{\hbar} t_1 \right] \right\} dp_{1x} dp_{1y} dp_{1z} dE_1 \end{aligned} \tag{3.54}$$

and

$$\begin{aligned} \Psi(\mathbf{r}_2 - \mathbf{r}_{2a}, t_2 - t_{2a}) = \frac{1}{(2\pi\hbar)^2} \int_{-\infty}^{\infty} \Phi(\mathbf{p}_2, E_2) \\ \times \exp\left\{ i \left[ \frac{\mathbf{p}_2}{\hbar} \cdot \mathbf{r}_2 - \frac{E_2}{\hbar} t_2 \right] \right\} dp_{2x} dp_{2y} dp_{2z}, dE_2 \end{aligned} \tag{3.55}$$

with

$$\Phi(\mathbf{p}_1, E_1) = \Psi(\mathbf{p}_1, E_1) \exp\left\{ -i \left[ \frac{\mathbf{p}_1}{\hbar} \cdot \mathbf{r}_{1a} - \frac{E_1}{\hbar} t_{1a} \right] \right\} \tag{3.56}$$

and

$$\Phi(\mathbf{p}_2, E_2) = \Psi(\mathbf{p}_2, E_2) \exp\left\{ -i \left[ \frac{\mathbf{p}_2}{\hbar} \cdot \mathbf{r}_{2a} - \frac{E_2}{\hbar} t_{2a} \right] \right\}. \tag{3.57}$$

For the probability densities, which directly describe the interaction, we obtain

$$\Phi^*(\mathbf{p}_1, E_1)\Phi(\mathbf{p}_1, E_1) = \Psi^*(\mathbf{p}_1, E_1)\Psi(\mathbf{p}_1, E_1) \tag{3.58}$$

and

$$\Phi^*(\mathbf{p}_2, E_2)\Phi(\mathbf{p}_2, E_2) = \Psi^*(\mathbf{p}_2, E_2)\Psi(\mathbf{p}_2, E_2). \tag{3.59}$$

That is, the interaction [$(\mathbf{p}, E)$-fluctuations] is not influenced when the functions $\Psi(\mathbf{r}_1, t_1)$ and $\Psi(\mathbf{r}_2, t_2)$ [$\Psi^*(\mathbf{r}_1, t_1)\Psi(\mathbf{r}_1, t_1)$ and $\Psi^*(\mathbf{r}_2, t_2)$ $\Psi(\mathbf{r}_2, t_2)$] are shifted arbitrarily relative to each other. There are no correlations between the system 1, described by $\Psi^*(\mathbf{r}_1, t_1)\Psi(\mathbf{r}_1, t_1)$, and system 2, which is described by $\Psi^*(\mathbf{r}_2, t_2)\, \Psi(\mathbf{r}_2, t_2)$. In other words, the interaction [$(\mathbf{p}, E)$-fluctuations] between both systems exclusively serves for the creation of the shape (form) of system 1 [defined by $\Psi^*(\mathbf{r}_1, t_1)\Psi(\mathbf{r}_1, t_1)$] and system 2 [defined by $\Psi^*(\mathbf{r}_2, t_2)\Psi(\mathbf{r}_2, t_2)$].

This analysis directly justifies the introduction of "distance-independent interaction." That is, there should also be interactions between systems possible that are not dependent on their mutual space–time distance. This is fulfilled if Eq. (3.48) is valid, i.e., if the function $\Psi(\mathbf{p}_1, E_1, \mathbf{p}_2, E_2)$ can be written as product $\Psi(\mathbf{p}_1, E_1)\Psi(\mathbf{p}_2, E_2)$. These distance-independent interactions describe the shapes (forms) of systems.

### 3.11.3. *Basic Equations in the Case of Distance-Independent Interactions*

In [12], we studied two systems $i$ and $j$ that interact by a distance-independent interaction. Then, the corresponding wavefunctions $\Psi_F(\mathbf{r}, t)_k, k = i, j$, may be expressed in terms of an interaction potential, say $V_F(x, y, z, t)_k$ with $k = i, j$, which must be independent on the space–time distance of both systems $i$ and $j$. Then, it is straightforward to show that the equation for $(\mathbf{r}, t)$-space is expressed by (see [12])

$$i\hbar\frac{\partial}{\partial t}\Psi_F(\mathbf{r}, t)_k = -\frac{\hbar^2}{2m_{0k}}\Delta\Psi_F(\mathbf{r}, t)_k + V_F(x, y, z, t)_k\Psi_F(\mathbf{r}, t)_k, \quad k = i, j. \tag{3.60}$$

The corresponding equation for $\Psi_F(\mathbf{p}, E)_k$ in $(\mathbf{p}, E)$-space has the form [12]

$$E\,\Psi_F(\mathbf{p}, E)_k = \frac{\mathbf{p}^2}{2m_{0k}}\Psi_F(\mathbf{p}, E)_k + V_F\left(i\hbar\frac{\partial}{\partial p_x}, i\hbar\frac{\partial}{\partial p_y}, i\hbar\frac{\partial}{\partial p_z}, -i\hbar\frac{\partial}{\partial E}\right)_k \Psi_F(\mathbf{p}, E)_k, \quad k = i, j. \tag{3.61}$$

Again, the potential $V_F(x, y, z, t)_k$ describes the form-interaction and is, at any time $\tau$, independent of the actual space–time distances

$\mathbf{r}_{12} = \mathbf{r}_1 - \mathbf{r}_2$ and $t_{12} = t_1 - t_2$ of both systems. In principle, both systems could additionally interact by a distance-dependent potential but this kind of interaction is assumed to be not effective in connection with our specific study given in this section.

If both systems are identical, none of them interact with any other system of the environment and, furthermore they must have the same mass and also the same potential, that is, we have

$$m_{0i} = m_{0j}, \tag{3.62}$$

$$V_F(x, y, z, t)_i = V_F(x, y, z, t)_j. \tag{3.63}$$

Then, from (3.60) and (3.61) it directly follows that the wavefunctions for both systems are identical:

$$\Psi_F(\mathbf{r}, t)_i = \Psi_F(\mathbf{r}, t)_j, \tag{3.64}$$

$$\Psi_F(\mathbf{p}, E)_i = \Psi_F(\mathbf{p}, E)_j. \tag{3.65}$$

The situation is symbolically expressed in Fig. 3.7.

### 3.11.4. *No Exchange of "Space–Time Pieces"*

Also in the case of distance-independent interactions, we may apply what we have outlined in Section 3.6.3, and we come to the following conception in connection with system $i$ and system $j$ (see in particular Fig. 3.7):

Since the $(\mathbf{p}, E)$-values fluctuate for both system $i$ and $j$ (in principal in the range $-\infty \leq \mathbf{p}, E \leq \infty$) there must be an interaction [$(\mathbf{p}, E)$-fluctuations] between system $i$ and systems $j$, i.e., the $(\mathbf{p}, E)$-values of the two interacting systems are strongly correlated at each time $\tau$. Only in this way the conservation laws for momentum and energy can be fulfilled at each time $\tau$. If system $i$ (system $j$) would perform $(\mathbf{p}, E)$-fluctuations independently from system $j$ (system $i$), these conservation laws would be violated.

Therefore, the systematic $(\mathbf{p}, E)$-fluctuations in $(\mathbf{p}, E)$-space reflect processes and we have to consider $(\mathbf{p}, E)$-space as reality. However, because the variables $\mathbf{p}$ and $E$ are auxiliary elements, this reality is a "fictitious reality". Therefore, also the $(\mathbf{p}, E)$-fluctuations (the

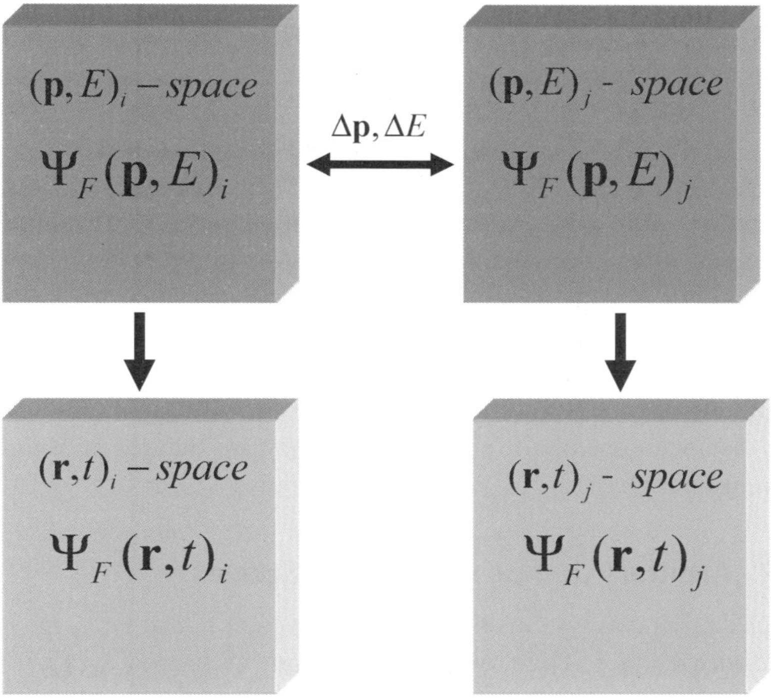


Fig. 3.7. Due to the conservation for momentum and energy, there are systematic $(\mathbf{p}, E)$-fluctuations $\Delta\mathbf{p}, \Delta E$ between the two $(\mathbf{p}, E)_k$-spaces, $k = i, j$. However, there are no fluctuations of certain "space–time elements" between the pictures for system $i$ and system $j$. Both systems are assumed to be identical. More details are given in the text.

interactions of the systems with other systems) are fictitious in character and merely simulate the real processes in basic reality.

Due to the conservation laws for the momentum $\mathbf{p}$ and the energy $E$, we must have systematic interaction processes $(\mathbf{p}, E)$-fluctuations between the system $i$ and system $j$ and, therefore, we have to consider $(\mathbf{p}, E)$-space as (fictitious) reality. However, there do not exist such conservation laws for the variables $\mathbf{r}$ and $t$, i.e., the realized $(\mathbf{r}, t)$-point at time $\tau$ of system $i$ is independent of the realized $(\mathbf{r}, t)$-points in connection with system $j$ at the same time $\tau$. There is no exchange of certain "space–time pieces" between the two pictures. In other words, there are no correlations between the $(\mathbf{r}, t)$-points at time $\tau$. This indicates that the variables $\mathbf{r}$ and $t$ are in fact elements of a picture. In other words, the information of fictitious reality [$(\mathbf{p}, E)$-space] is projected onto $(\mathbf{r}, t)$-space by a Fourier transform and we get the "picture of reality". The situation is summarized in Fig. 3.7.

Remark

It is important to note that within conventional physics, only distance-dependent interactions are known, that is, form-interactions are not defined here. This is a very principal point, in particular with respect to the notion "interaction". The source for the notion "interaction" is an invention of classical physics and goes back to Newton. We will briefly discuss below (Section 3.15) the main facts in connection with distance-dependent interactions as they are defined and used in conventional physics.

## 3.12. Arbitrary Jumps within $(\mathbf{r}, t)$-Space

Let us discuss two systems $i$ and $j$ that are positioned at $\mathbf{r}_k$, $t_k$ in $(\mathbf{r}, t)$-space with $k = i, j$. Then, we get on the basis of Section 3.11:

$$\Psi_F(\mathbf{p}, E)_k = \frac{1}{(2\pi\hbar)^2} \int_{-\infty}^{\infty} \Psi_F(\mathbf{r} - \mathbf{r}_k, t - t_k) \times \exp\left\{-i\left[\frac{\mathbf{p}}{\hbar} \cdot \mathbf{r} - \frac{E}{\hbar} t\right]\right\} dxdydzdt, \quad k = i, j. \tag{3.66}$$

We may shift the functions $\Psi_F(\mathbf{r} - \mathbf{r}_k, t - t_k)$, $k = i, j$ relative to $(\mathbf{r}, t)$-space without changing the interaction, i.e., the law for the $(\mathbf{p}, E)$-fluctuations, which are described by the probability densities $\Psi_F^*(\mathbf{p}, E)_k \Psi_F(\mathbf{p}, E)_k$, $k = i, j$, which are not changed when $\mathbf{r}$ and $t$ are shifted by the quantities $\Delta_{k,\mathbf{r}}$ and $\Delta_{k,t}$:

$$\begin{aligned} \mathbf{r} &\to \mathbf{r} - \Delta_{k,\mathbf{r}}, \quad k = i, j, \\ t &\to t - \Delta_{k,t}, \quad k = i, j, \end{aligned} \tag{3.67}$$

where the values $\Delta_{k,\mathbf{r}}$ and $\Delta_{k,t}$ are assumed not to be dependent on $\mathbf{r}$ and $t$.

### 3.12.1. *The* (p, $E$)*-Distributions*

The following is shown in [12] and in Section 3.11. The observer, who is resting relative to the system with $\mathbf{r}_k + \Delta_{k,\mathbf{r}}, t_k + \Delta_{k,t},\ k = i, j$, measures exactly the same $(\mathbf{p}, E)$-distribution

$$\Psi_F^*(\mathbf{p}, E)_k \Psi_F(\mathbf{p}, E)_k, \quad k = i, j \tag{3.68}$$

as the observer, which is resting with respect to the system with the space–time position $\mathbf{r}_k, t_k$ ($k = i$ or $j$). This scenario can be summarized schematically as follows:

$$\begin{gathered} \Psi_F^*(\mathbf{p}, E)_k \Psi_F(\mathbf{p}, E)_k \\ \downarrow \\ \Psi_F^*(\mathbf{r} - (\mathbf{r}_k + \Delta_{k,\mathbf{r}}), t - (t_k + \Delta_{k,t})) \\ \times \Psi_F(\mathbf{r} - (\mathbf{r}_k + \Delta_{k,\mathbf{r}}), t - (t_k + \Delta_{k,t})). \end{gathered} \tag{3.69}$$

In other words, the process in $(\mathbf{p}, E)$-space is not dependent on the position of the geometrical object (probability density)

$$\begin{aligned} \Psi_F^*(\mathbf{r} - (\mathbf{r}_k + \Delta_{k,\mathbf{r}}), t - (t_k + \Delta_{k,t})) \\ \times \Psi_F(\mathbf{r} - (\mathbf{r}_k + \Delta_{k,\mathbf{r}}), t - (t_k + \Delta_{k,t})) \end{aligned} \tag{3.70}$$

in $(\mathbf{r}, t)$-space (see in particular Section 3.11). In other words, the arbitrary jumps have no influence on the real properties of the systems. In particular, the function $\Psi_F(\mathbf{r} - \mathbf{r}_k, t - t_k)$ is shifted by the quantities $\Delta_{k,\mathbf{r}}$ and $\Delta_{k,t}$ but its form (shape) remains conserved. Then, we may conclude that it is a "motion relative to nothing". This point has been discussed in more detail in [12].

To each space-time position $\Delta_{k,\mathbf{r}}, \Delta_{k,t}$ belongs exactly the same interaction law $\Psi_F^*(\mathbf{p}, E)_k \Psi_F(\mathbf{p}, E)_k, k = i, j$. Therefore, all space-time points are equivalent and no one is preferred. Thus, the peaks with $\Delta_{k,\mathbf{r}}, \Delta_{k,t}$, [Eq. (3.70)] must occupy the space uniformly. This is only possible if each peak jumps statistically through $(\mathbf{r}, t)$-space; in this way it is possible to occupy space-time uniformly in the course of time $\tau$. It must be emphasized in this connection that the systems $i$ and $j$ are not static units. They are created new at each time $\tau$; at each time $\tau$

one of the possible values $\mathbf{p}$ and $E$ are created within the frame of the probability density $\Psi_F^*(\mathbf{p}, E)_k \Psi_F(\mathbf{p}, E)_k, k = i, j$. Therefore, the systems $i$ and $j$ are also created incessantly in $(\mathbf{r}, t)$-space together with the space-time positions $\Delta_{k,\mathbf{r}}$, $\Delta_{k,t}$, $k = i, j$.

From all these features, we have to conclude that the quantities $\Delta_{k,\mathbf{r}}$ and $\Delta_{k,t}$ must behave statistically (more details are given in [12]). That is, the values for $\Delta_{k,\mathbf{r}}$ and $\Delta_{k,t}$ are dependent on time $\tau$, and we have

$$\begin{aligned} \Delta_{k,\mathbf{r}} &= \Delta_{k,\mathbf{r}}(\tau), \quad k = i, j, \\ \Delta_{k,t} &= \Delta_{k,t}(\tau), \quad k = i, j. \end{aligned} \tag{3.71}$$

But both functions $\Delta_{k,\mathbf{r}}$ and $\Delta_{k,t}$ do not vary systematically with time $\tau$, and this is because such a law does not exist. If we have certain values $\Delta_{k,\mathbf{r}}(\tau_1)$ and $\Delta_{k,t}(\tau_1)$ at time $\tau_1$, we cannot say something about the values $\Delta_{k,\mathbf{r}}$ and $\Delta_{k,t}$ at time $\tau_2$, even when $\tau_2$ is the next time-value after $\tau_1$, i.e., the quantities $\Delta_{k,\mathbf{r}}(\tau_2)$ and $\Delta_{k,t}(\tau_2)$ cannot principally be predicted.

In conclusion, the quantities $\Delta_{k,\mathbf{r}}$ and $\Delta_{k,t}$ behave statistically. Thus, the space–time positions of the peaks (probability densities), defined by

$$\begin{aligned} &\Psi_F^*(\mathbf{r} - (\mathbf{r}_k + \Delta_{k,\mathbf{r}}(\tau)), t - (t_k + \Delta_{k,t}(\tau))) \\ &\quad \times \Psi_F(\mathbf{r} - (\mathbf{r}_k + \Delta_{k,\mathbf{r}}(\tau)), t - (t_k + \Delta_{k,t}(\tau))), \quad k = i, j \end{aligned} \tag{3.72}$$

must behave statistically, i.e., both peaks jump arbitrarily through space and time. These jumps are independent from each other because the interaction between both systems (system $i$ and system $j$) is not dependent on their distance but we assumed that this interaction is distance-independent and creates the form of the systems (Section 3.11).

### 3.12.2. *Effective Velocities*

The projection of $\Psi_F(\mathbf{p}, E)_k$, $k = i, j$ onto $(\mathbf{r}, t)$-space leads to the wavefunctions

$$\Psi_F(\mathbf{r} - (\mathbf{r}_k + \Delta_{k,\mathbf{r}}), t - (t_k + \Delta_{k,t})), \quad k = i, j$$

and the peaks (probability densities)

$$\Psi_F^*(\mathbf{r} - (\mathbf{r}_k + \Delta_{k,\mathbf{r}}), t - (t_k + \Delta_{k,t}))$$
$$\times \Psi_F(\mathbf{r} - (\mathbf{r}_k + \Delta_{k,\mathbf{r}}), t - (t_k + \Delta_{k,t})), \quad k = i, j$$

[Eq. (3.70)] should be considered as a definition of the form (shape) of the systems $i$ and $j$ and, as we have outlined above, these peaks (geometrical structures) jump arbitrarily relative to $(\mathbf{r}, t)$-space. In the following, let us briefly repeat and deepen the physical content of this effect.

We want to assume that system $i$ is at time $\tau_a$ at the space–time position $\mathbf{r}_{1a}, t_{1a}$ and that system $j$ is at the same time $\tau_a$ at the space–time position $\mathbf{r}_{2a}, t_{2a}$. Furthermore, let us assume that system $i$ is at time $\tau_b \neq \tau_a$ at the space–time position $\mathbf{r}_{1b}, t_{1b}$ and system $j$ is at the same time $\tau_b$ at $\mathbf{r}_{2b}, t_{2b}$. Then, the following is relevant. There is no law defined that would predict the values $\mathbf{r}_{1b}, t_{1b}$ on the basis of $\mathbf{r}_{1a}, t_{1a}$ and there is of course also no law that would predict the values $\mathbf{r}_{2b}, t_{2b}$ on the basis of $\mathbf{r}_{2a}, t_{2a}$. In other words, both systems take arbitrary space–time positions in the course of time $\tau$.

Because there is no distance-dependent interaction effective between system $i$ and system $j$, both peaks jump independently from each other through $(\mathbf{r}, t)$-space, i.e., there can also be no relationship between the space–time positions $\mathbf{r}_{1a}, t_{1a}$ and $\mathbf{r}_{2a}, t_{2a}$ and also not between $\mathbf{r}_{1b}, t_{1b}$ and $\mathbf{r}_{2b}, t_{2b}$. In other words, there are no relationships (physical laws) between the space–time positions $\mathbf{r}_{1a}, t_{1a}$, $\mathbf{r}_{1b}, t_{1b}$, $\mathbf{r}_{2a}, t_{2a}$ and $\mathbf{r}_{2b}, t_{2b}$, even when $\tau_b$ is the next time-value after $\tau_a$. Therefore, the values $\mathbf{r}_{1b}, t_{1b}$ and $\mathbf{r}_{2b}, t_{2b}$ can principally not be predicted. That is, all the quantities $\mathbf{r}_{1a}, t_{1a}$, $\mathbf{r}_{1b}, t_{1b}$, $\mathbf{r}_{2a}, t_{2a}$ and $\mathbf{r}_{2b}, t_{2b}$ are completely independent from each other and behave strictly statistically.

However, we have always to keep in mind that within projection theory the peaks (geometrical structures), defined by Eq. (3.70), do not move through space–time [$(\mathbf{r}, t)$-space] but are projected on it. Nevertheless, we may define "effective velocities" which we would like to call $v_{m\mathbf{r}}$ and $v_{mt}$; $v_{m\mathbf{r}}$ is the effective velocity with respect to the variable **r** and is given by

$$v_{m\mathbf{r}} = \frac{\mathbf{r}_{mb} - \mathbf{r}_{ma}}{\tau_b - \tau_a}, \quad m = 1, 2. \tag{3.73}$$

$v_{mt}$ is the effective velocity with respect to the variable $t$ and has the form

$$v_{mt} = \frac{t_{mb} - t_{ma}}{\tau_b - \tau_a}, \quad m = 1, 2. \tag{3.74}$$

The time interval $\tau_b - \tau_a$ can be close to zero and, on the other hand, $\mathbf{r}_{mb} - \mathbf{r}_{ma}$ and $t_{mb} - t_{ma}$ may be as large as the universe, where large really means with respect to the maximum space-extension as well as with respect to the maximum time-extension (that is, from the beginning to the end of time). With

$$\tau_b - \tau_a \to 0 \tag{3.75}$$

we obtain for the velocities

$$\begin{aligned} v_{m\mathbf{r}} &\to \infty, \quad m = 1, 2, \\ v_{mt} &\to \infty, \quad m = 1, 2. \end{aligned} \tag{3.76}$$

The situation is illustrated in Fig. 3.8.

Remark

We know from Special Theory of Relativity that the velocity of a body having the rest mass of $m_0$ cannot exceed the velocity of light. However, the property $v_{m\mathbf{r}} \to \infty$, $m = 1, 2$ [Eq. (3.76)] seems to be in contradiction to this relevant statement of the Special Theory of Relativity. But this is not the case. The reason is obvious: The observer, resting in system $i$, cannot say anything about the space–time positions of system $j$ and vice versa. Therefore, a relative motion of both systems is not definable. This is reason why the laws of Special Theory of Relativity cannot be applied here. In this connection, it should also be mentioned that the effective velocity $v_{mt}$ is not defined in Special Theory of Relativity because the system-specific time $t$ is not defined here. In [12], we have discussed possible relativistic effects within projection theory.

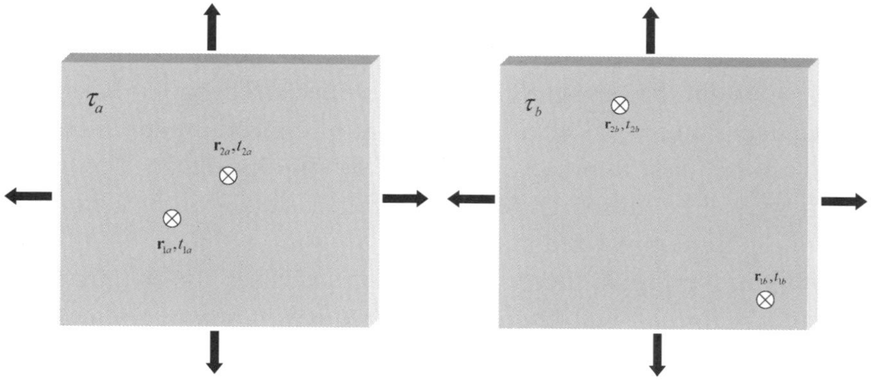


(a) (b)

Fig. 3.8. (a) At time $\tau_a$ system $i$ is at the space–time position $\mathbf{r}_{1a}, t_{1a}$ and system $j$ is at $\mathbf{r}_{2a}, t_{2a}$. (b) At time $\tau_b$ system $i$ is at the space–time position $\mathbf{r}_{1b}, t_{1b}$ and system $j$ is at $\mathbf{r}_{2b}, t_{2b}$. Because of $\tau_b - \tau_a \to 0$ and the fact that the distances $\mathbf{r}_{mb} - \mathbf{r}_{ma}$ and $t_{mb} - t_{ma}$ (with $m = 1, 2$) may be arbitrarily large (in principle, as large as the space–time extension of the universe), the velocities $v_{m\mathbf{r}}$ and $v_{mt}$ of the peaks may be close to infinity.

## 3.13. Mach's Principle: Preliminary Remarks

The arbitrary jumps of the geometrical structures in $(\mathbf{r}, t)$-space with arbitrary effective velocities $v_{m\mathbf{r}}$ and $v_{mt}$ (Section 3.12) does not lead to changes in the $(\mathbf{p}, E)$-states of the systems $i$ and $j$ (Section 3.11). In [12], we have characterized this situation by the statement "motion relative to nothing". This is in contrast to the facts in conventional physics. Let us briefly discuss this point in connection with Newton's theory.

### 3.13.1. *The Situation in Conventional Physics*

The physical picture within Newton's theory is quite different from that of projection theory. In classical mechanics the $(\mathbf{p}, E)$-states are definitely changed when a non-interacting body moves through space (relative to space) with varying velocity, and for the explanation of this scenario, space-effects (inertia) are essential in Newton's mechanics. In particular, here a non-interacting body may move through space with a certain constant velocity, and this property directly comes into play by unrealistic space-effects (inertia) in Newton's theory.

However, such a space-effect has to be considered as unphysical and has been strongly criticized by Ernst Mach and, in particular, also by Albert Einstein. Nevertheless, also within Special Theory of Relativity and General Theory of Relativity, the situation concerning space-effects could not be really improved. More details concerning the Theory of Relativity will be given in Chapter 5. The situation within Newton's theory can be summarized as follows:

Within Newton's theory the notion "inertia" is of particular relevance. Here all real bodies are embedded in space. Even when a body does not interact with other bodies, the effect of inertia is effective, i.e., the body moves through space with constant velocity $v$ where the velocity of zero is included, and the effect of inertia is entirely due the interaction of the body with space. The body moves relative to space, and its $(\mathbf{p}, E)$-state is dependent on its velocity relative to space, i.e., the variation of $v$ leads to changes of the bodies' $(\mathbf{p}, E)$-states. In particular, we have the following situation.

Let us consider an object having the mass $m_0$ that is resting relative to space, and we would like to assume that this body does not interact with any other body in the cosmos. Now we want to apply an accelerating force $\mathbf{K}$ on this object during the time interval $\Delta\tau = \tau_2 - \tau_1$, i.e., the effect of the force starts at $\tau_1$ and stops at $\tau_2$. During that time interval the work

$$A = \int_{\tau_1}^{\tau_2} \mathbf{K} \cdot d\mathbf{s} \tag{3.77}$$

is done on the body along the path $\mathbf{s}$, that is, due to the force $\mathbf{K}$ the object will be accelerated from the velocity $\mathbf{v}_1 = \mathbf{v}_1(\tau_1) = 0$ to the final velocity $\mathbf{v}_2 = \mathbf{v}_2(\tau_2) = \mathbf{v}$. While the body has no energy $E$ and no momentum $\mathbf{p}$ up to $\tau_1$, it has, as is well-known, at time $\tau_2$ the energy

$$E = m_0\mathbf{v}^2/2 \tag{3.78}$$

and the momentum

$$\mathbf{p} = m_0\mathbf{v}. \tag{3.79}$$

In other words, due to the motion relative to space there is a change of the $(\mathbf{p}, E)$-state of the object. This result is in distinct contrast to what we have found in connection within projection theory where the motion through $(\mathbf{r}, t)$-space has definitely no influence on the $(\mathbf{p}, E)$-states. This definitely means that Mach's principle is fulfilled within projection theory.

### 3.13.2. *The Situation within Projection Theory*

The fact that Mach's principle is fulfilled within projection theory is a quite general statement. This is because "only" geometrical structures (objects) are embedded in $(\mathbf{r}, t)$-space, and $(\mathbf{r}, t)$-space plays the role of an auxiliary element for the representation of the world, that is, to form pictures of it. In other words, since there can be no material objects embedded in $(\mathbf{r}, t)$-space, material objects cannot be influenced by space and time [$(\mathbf{r}, t)$-space]. Therefore, Mach's principle is from the beginning fulfilled and must be strictly valid for all physical situations.

In particular, we have recognized in Section 3.12 that in this case (Mach's principle is fulfilled) the geometrical objects (system $i$ and system $j$) may perform arbitrary jumps through space and time ($(\mathbf{r}, t)$-space).

Also within conventional classical physics (Newton's mechanics, Theory of Relativity) a lone material object, which is in conventional physics embedded in space (space–time), should perform statistical jumps if Mach's principle is fulfilled. This is however not the case. For example, de Sitter demonstrated within the framework of the General Theory of Relativity that the space has influence on the motion of a lone object though the universe. The lone body would be guided along a geodesic line as if it was made of inertial matter, i.e., the space would have within this theory an influence on this lone object because there is by definition no other matter in the universe. We will come back to Mach's principle in Chapter 5.

## 3.14. Can a Lone, Elementary Object Exist in the Cosmos?

We have outlined in Ref. 12 that static elementary systems (building blocks) with a definite, non-fluctuating momentum $\mathbf{p}_0$ and a definite, non-fluctuating energy $E_0$ cannot exist. The existence of certain elementary objects is always due to an interaction process of the object with other objects or systems. These interactions are *quantum processes* and are described by $(\mathbf{p}, E)$-fluctuations. In other words, quantum processes are a primary factor for the existence of the world.

Within projection theory, it is problematic to assume that matter is formed by putting certain basic units systematically together. In particular, within projection theory it is not possible to work with the concept that each of these elementary units can exist independently from the others and of other things in the cosmos. So, the concept of conventional physics, that matter is formed by putting certain static basic building blocks systematically together, is possibly not realizable within projection theory. Let us briefly explain the situation by means of a simple example.

For simplicity, let us consider a cosmos with only two subsystems. The two subsystems cannot exist independently from each other. In other words, this simple universe cannot be formed by putting two static units (the two subsystems) systematically together, and this is because the two subsystems cannot be thought as independent from each other. Each of these units needs the other unit in order to be able to exist [12], that is, both units come into existence in a self-creating process. They interact by $(\mathbf{p}, E)$-fluctuations and these interaction processes are a necessary condition for their existence. This situation is independent of the type of interaction. The two systems may interact by a form-interaction and/or by a distance-dependent interaction.

Again, the universe, defined here by two subsystems, which are under mutual influence, is not formed by putting the two independent subsystems systematically together but it emerges with two subsystems.

In other words, such a structure emerges as an ensemble and comes not into existence through basic building blocks, which are put together for the creation of the ensemble consisting of a certain number of subsystems. The reason is given by the following argument. There cannot exist independent (isolated) units in projection theory, but each unit is defined by specific interactions ($(\mathbf{p}, E)$-fluctuations). In other words, the subsystems of the ensemble can only be existent within the ensemble. Such units cannot be called "basic". Clearly, a certain subsystem (unit) may appear in connection with various ensembles that are different from each other. In contrast to conventional physics, we have no static units in projection theory but all comes into existence by fluctuations (interactions defined by $(\mathbf{p}, E)$-fluctuations).

The features discussed in connection with the simple universe consists of two units points on a critical and not trivial situation, and we have to analyze the situation for each case carefully. Here the following question is relevant. What parts of a system can be treated as a "unit construction system", and what parts cannot be analyzed in this way but simply emerge as a definite structure consisting of a certain number of subsystems (units)?

## 3.15. The Meaning of the Potential Functions

In Section 3.12, we have treated two systems $i$ and $j$ that interact by a distance-independent interaction. Both systems $i$ and $j$ jump arbitrarily in $(\mathbf{r}, t)$-space and their space–time distance at a certain time $\tau$ may be as large as the space-extension of the universe, where large really means with respect to the maximum space-extension as well as with respect to the maximum time-extension (that is, from the beginning to the end of time). Nevertheless, both systems $i$ and $j$ interact with a constant strength even when the space–time distance takes the largest possible value. In other words, both systems interact, but this interaction is independent on the actual space–time positions of both systems.

This property reflects the non-local character of projection theory. Because the interaction processes exclusively take place in $(\mathbf{p}, E)$-space and the structures in $(\mathbf{r}, t)$-space are projections from $(\mathbf{p}, E)$-space onto space–time, a position-independent interaction becomes possible in a quite natural way. Note that within $(\mathbf{r}, t)$-space no signals are exchanged between the two systems. Within $(\mathbf{r}, t)$-space there are "only" geometrical structures ($(\mathbf{r}, t)$-correlations) and nothing else.

In Section 3.11.3, we have formally defined a potential function that describes these geometrical structures (form of the systems) in $(\mathbf{r}, t)$-space, which reflect $(\mathbf{r}, t)$-correlations and correspond to

$(\mathbf{p}, E)$-fluctuations in $(\mathbf{p}, E)$-space, i.e., $(\mathbf{r}, t)$-correlations and $(\mathbf{p}, E)$-fluctuations take place simultaneously. In this section, we will discuss the physical meaning of the notion "interaction" within projection theory but also within conventional physics where the original thoughts by Newton will be of particular interest. The result can be anticipated as follows. Within projection theory, the interaction potential can be interpreted at best as auxiliary element without any imaginable background (there is no information or energy transfer through $(\mathbf{r}, t)$-space; see in particular the discussion in Section 3.13). On the other hand, we will recognize that within conventional physics the notion "interaction" takes in his original form a conception with an occult quality. [10] Before we discuss this point in more detail, let us summarize the main interaction characteristics in conventional physics as well as in projection theory.

### 3.15.1. *Interaction within Conventional Physics*

Let us consider two systems, say $i$ and $j$, that are embedded as real bodies in space. The variable for the system-specific time $t$ is not defined in conventional physics but only the reference time $\tau$. Also the real interactions processes exclusively take place in space, and we would like to assume that these two systems interact with each other, but none of them interact with other bodies.

The effect of an interaction between the systems, having at time $\tau$ the sharp space positions $\mathbf{r}_i$ and $\mathbf{r}_j$, is that the space positions are changed in the course of time $\tau$. (In contrast to projection theory, the form (shape) of the two systems $i$ and $j$ have to be assumed within conventional physics and have to be considered as pre-requisite here, and there is no conception to deduce the form of a system in conventional physics.)

That is, within conventional physics (Newton's mechanics) there is the following mechanism: The space positions $\mathbf{r}_i$ and $\mathbf{r}_j$, which the systems take at time $\tau$, are changed by the interaction process and we have at time $\tau + \Delta\tau$ the definite positions $\mathbf{r}_i + \Delta\mathbf{r}_i, \mathbf{r}_j + \Delta\mathbf{r}_j$, i.e., we have the transition

$$\mathbf{r}_i, \mathbf{r}_j \rightarrow \mathbf{r}_i + \Delta\mathbf{r}_i, \quad \mathbf{r}_j + \Delta\mathbf{r}_j. \tag{3.80}$$

In the case of point-like systems, the bodies take the forms $\delta(\mathbf{r}-\mathbf{r}_i)$ (system $i$) and $\delta(\mathbf{r}-\mathbf{r}_j)$ (system $j$) and we have after the interaction process [Eq. (3.94)] the following structural situation:

$$\text{system } i : \delta(\mathbf{r}-\{\mathbf{r}_i+\Delta\mathbf{r}_i\}), \tag{3.81}$$

$$\text{system } j : \delta(\mathbf{r}-\{\mathbf{r}_j+\Delta\mathbf{r}_j\}). \tag{3.82}$$

As is well known, in the case of conventional quantum systems, the variables $\mathbf{r}_i, \mathbf{r}_j$ are not changed by the deterministic laws of classical mechanics but by Schrödinger's equation of conventional quantum theory. That is, the variables $\mathbf{r}_i, \mathbf{r}_j$ for both systems behave statistically in the course of time $\tau$ and do not take definite values as in the classical case.

### 3.15.2. *Interaction within Projection Theory*

It is normally assumed within usual physics that the interaction strength decreases with increasing distance $\mathbf{r}_i - \mathbf{r}_j$ between the systems, and this is intuitively understandable because it is assumed in usual physics that the interaction processes take place in space and, therefore, the assumption that with decreasing potential function the influence of system $i$ on system $j$ and vice versa decreases becomes intuitively understandable.

However, we have to be careful. Within projection theory there cannot be such kind space–time connections since the interaction processes do not take place in $(\mathbf{r}, t)$-space. Within projection theory, we have "merely" $(\mathbf{r}, t)$-correlations in $(\mathbf{r}, t)$-space, and the real interaction processes are identified with $(\mathbf{p}, E)$-fluctuations in $(\mathbf{p}, E)$-space.

Just in the case of form (distance-independent) interactions the familiar potential picture is not applicable. Instead of one potential for both systems (distance-dependent interaction) we have here two potentials $V_F(x, y, z, t)_k$, $k = i, j$ [Eq. (3.60)]: one for system $i$ and another for system $j$. Thus, the usual interpretation in terms of more or less classical notions can no longer be applied in projection theory. In fact, the potential takes in $(\mathbf{p}, E)$-space the form of an operator

$$V_F\left(i\hbar\frac{\partial}{\partial p_x}, i\hbar\frac{\partial}{\partial p_y} i\hbar\frac{\partial}{\partial p_z}, -i\hbar\frac{\partial}{\partial E}\right)$$

[see Eq. (3.61)] and this expression is not accessible to descriptive pictures. Therefore, interaction potentials should be considered as abstract quantities for the determination of the $(\mathbf{p}, E)$-fluctuations in $(\mathbf{p}, E)$-space which appear in $(\mathbf{r}, t)$-space as correlations between the various space–time positions $\mathbf{r}$, $t$ [$\mathbf{r}, t$-correlations].

In other words, the potential is not the primary quantity but the $(\mathbf{p}, E)$-fluctuations and $(\mathbf{r}, t)$-correlations are. In $(\mathbf{r}, t)$-space, we have exclusively geometrical structures, and no real objects are embedded here. Therefore, it is wrong to assume that there is an exchange of energy $V_F(x, y, z, t)_k$, $k = i, j$ between the (geometrical) structures $i$ and $j$.

*Conclusion*: Within projection theory the interaction potential can be at best interpreted as an auxiliary element without any imaginable background. (There is no information or energy transfer through $(\mathbf{r}, t)$-space.) But also within conventional physics the notion "interaction" as an imaginable quality becomes questionable when we analyze this term in more detail.

### 3.15.3. *Further Remarks*

In contrast to conventional physics (Newton's mechanics, usual quantum theory), within projection theory we have a real extension of the formalism. A system is not only described by the variable $\mathbf{r} = (x, y, z)$ but in addition by the system-specific time $t$. In other words, when we go from usual physics to projection theory we have the transition $\mathbf{r} \rightarrow \mathbf{r}, t$ (see in particular Section 3.8). Then, the following mechanism is relevant within projection theory. The space–time positions $\mathbf{r}_i, t_i$ for system $i$ and $\mathbf{r}_j, t_j$ for system $j$ in $(\mathbf{r}, t)$-space, which the systems take at time $\tau$, are statistically changed due to the interaction process in $(\mathbf{p}, E)$-space and, therefore, we have at time $\tau + \Delta\tau$ the space–time positions $\mathbf{r}_i + \Delta\mathbf{r}_i, t_i + \Delta t_i, \mathbf{r}_j + \Delta\mathbf{r}_j, t_j + \Delta t_j$ with a certain probability, i.e., we have the transition

$$\mathbf{r}_i, t_i, \mathbf{r}_j, t_j \rightarrow \mathbf{r}_i + \Delta\mathbf{r}_i, t_i + \Delta t_i, \mathbf{r}_j + \Delta\mathbf{r}_j, t_j + \Delta t_j. \qquad (3.83)$$

The variables $\mathbf{r}_i, t_i, \mathbf{r}_j, t_j$ behave statistically in the course of time $\tau$; within conventional quantum theory only the quantities $\mathbf{r}_i, \mathbf{r}_j$ are random variables.

Within projection theory, no imaginable pictures can be designed for the potential $V(x, y, z, t)$ but this quantity has to be considered as an abstract function. Nevertheless, $V(x, y, z, t)$ describes interaction effects because in the case of $V(x, y, z, t) = 0$ we obtain the free, non-interacting case. The potential $V(x, y, z, t)$ describes interaction effects in $(\mathbf{r}, t)$-space as well as in $(\mathbf{p}, E)$-space; the interaction effects are expressed in $(\mathbf{r}, t)$-space by correlation effects between the various space–time points and in $(\mathbf{p}, E)$-space by $(\mathbf{p}, E)$-fluctuations between the system under investigation and its environment.

In contrast to projection theory, within conventional physics there exist some descriptive interpretations for the interaction which is here expressed by $U(x, y, z)$ and not by $V(x, y, z, t)$ (Section 3.7). Let us briefly quote some of these interpretations for $U(x, y, z)$ and their relevance in connection with the force laws as, for example, for the gravitational law. In the case of two bodies we have two masses $m_1$ and $m_2$, having the distance $r$, and the force is proportional to $m_1 m_2/r^2$ (gravitational law). However, this force law cannot be derived from one of the interpretations. This is a critical point and needs further consideration. In the next section we will give some general remarks in connection with classical interaction pictures where the mathematical description is based upon $U(x, y, z)$.

### 3.15.4. *Conventional Physics: What Mechanism is Behind the Interaction?*

We have already outlined several times that within conventional physics the real bodies are embedded in space and there is an information (energy) transfer through space. What does "information (energy) transfer through space" mean? How comes the mutual influence between two bodies (as, for example, between the earth and the sun) into existence? What mechanism can explain this mutual influence? In this connection two "pictures" turned out to be of particular relevance: the "proximity effect" and the "action-at-the-distance".

As is well known, Newton's equations of motion include the forces which act between the bodies. In the case of two bodies we have two masses $m_1$ and $m_2$, having the distance $r$, and the force is proportional to $m_1 m_2/r^2$ (gravitational law). On the other hand, in the case of two electrical point charges, say $q$ and $Q$, having also the distance $r$, the force between the charges is expressed by $qQ/r^2$ (Coulomb's law). The gravitational law and Coulomb's law of electrostatic force have exactly the same mathematical structure. Therefore, both laws have been interpreted by means of the same physical conceptions. As we have already remarked, here two interpretations are of particular relevance: "action-at-a-distance" and the so-called "proximity effect". Both interpretations can be characterized as follows:

1. "Action-at-a-distance"

According to Newton's mechanics, it appears to be the case that the forces between $N$ masses act across space instantly. The reason is straightforward: In the gravitational law there is a relationship between the spatially separated positions of the masses and no intermediate position appears in the law. This suggests the view that gravitational forces work at a distance, i.e., the interaction comes about through an "action-at-a-distance" as follows. The gravitational forces reside in body $i$, but come into effect at the location of the other $N - 1$ bodies. Within this view, the space between the bodies is free of gravitation. This "action-at-a-distance" picture can also be applied to electrical interactions. However, it must be emphasized that this effect ("action-at-a-distance") cannot be proved experimentally, i.e., it is a matter of belief, whether such an effect exists in reality or not.

2. Proximity effect

In contrast to the "action-at-a-distance" picture, the so-called "proximity effect" just teaches the opposite. While in the case of "action-at-a-distance", the space between the masses (or charges) is free of gravitation (or electricity), in the case of the "proximity effect" we have space-filling fields: gravitational fields and/or electrical fields. For example, let us consider again the interaction between two charges $q$ and $Q$. The charge $Q$ is associated with an electrical field $E(r)$ where $r$ is the position of charge $q$. The charge $q$ "feels" the field $E(r)$ and, therefore, the effect is a "proximity effect" since the force effect on charge $q$ is exactly determined by the electrical field at the position at which also the charge $q$ is positioned. However, this space-filling "medium" (field) also cannot be proved experimentally and it becomes entirely a matter of belief, whether an electrical field (gravitational field) exists in reality or not.

### 3.15.5. *Equivalent Conceptions*

The "proximity effect" and the "action-at-the-distance", as qualitatively different they may be, are nevertheless exactly equivalent. It is not possible to make a choice between these two conceptions, since there is no experimental way to distinguish between them, as they both have the same consequences. In conclusion, the "proximity effect" and the "action-at-the-distance" are equivalent conceptions which do not differ, neither mathematically nor experimentally.

It is of principal interest to note that the "proximity effect" and the "action-at-the-distance" are merely "expressions" or interpretations for the gravitational law $m_1 m_2/r^2$ and Coulomb's law $qQ/r^2$. These force laws cannot, however, be derived from these notions. Many people believe that a mechanism, which is composed of many familiar single processes (preferably from everyday life) can explain the mathematical structure of the force laws. What mechanism is, for example, responsible for the fact that the forces expressed by $m_1 m_2/r^2$ are inversely proportional to the square of the distance between the masses $m_1$ and $m_2$. As already mentioned, the ideas "proximity effect" and "action-at-the-distance" cannot give the answer to this question, since they interpret the relation $m_1 m_2/r^2$, but are not able to explain the mathematical structure of this force law.

It is therefore not surprising that there are further interpretations in connection with the notion "interaction". In classical mechanics the "principle of least action" is of particular interest. This principle is qualitatively different from Newton's original idea but leads to exactly the same result. Newton's equations of motion describe the trajectory of a body from point to point in space, and they are based on the principle of causality. The body feels a force and yields to it. To this idea there is an alternative. This alternative is the so-called "principle of least action". According to this the path of a body does not develop from place to place on the basis of cause and effect. Instead one path is selected from the totality of possible paths, and that is the one whose effect is most minimal. As already stated, although the "principle of least action" and Newton's place-to-place description are principally different from each other, the results obtained using both ideas are exactly equivalent, i.e., we can experimentally not distinguish between both ideas.

In conclusion, the interpretations for the notion "interaction" cannot help us further to understand the mutual influence in connection with two or more interacting bodies. In particular, all these interpretations are not able at all to deduce the force laws $m_1 m_2 / r^2$ and $qQ/r^2$. As in projection theory, we also come in conventional physics to the conclusion that the interaction laws described by $m_1 m_2 / r^2$ and $qQ/r^2$ are abstract laws.

Within projection theory, it is understandable why the interaction laws for $V(x, y, z, t)$ have to be considered as abstract laws that cannot be described on the basis of familiar models; this is mainly due to the fact that the interaction processes take place in $(\mathbf{p}, E)$-space and not in $(\mathbf{r}, t)$-space. Our intuitive concepts are simply not applicable to mechanisms in $(\mathbf{p}, E)$-space (Section 3.3). It is surprising that also the laws $m_1 m_2 / r^2$ and $qQ/r^2$ of conventional physics remain abstract laws, and this is because a mechanism (model) for the deduction of the laws $m_1 m_2 / r^2$ and $qQ/r^2$ could not be found although the whole scenario takes place in space for which our intuition is adapted.

### 3.15.6. *The Mechanical-Corpuscular World View*

Newton's scientific work was aligned to the mechanical-corpuscular world view. Also the effect of gravitation had to be attributed (explained) on mechanical-corpuscular effects. That is, the force law $m_1 m_2 / r^2$ had to be explained by models, which are entirely based on the principles of the mechanical-corpuscular world view. If that should not be possible, gravitation has to be considered as an occult (metaphysical) quality from the point of view of the mechanical-corpuscular world view. For example, the introduction of a "force acting-at-a-distance" has to be considered as a stranger within the mechanical-corpuscular world view. In other words, in the negative case we leave this world view for which, on the other hand, Newton developed his theory. The fact is that such a mechanical-corpuscular model could not be found for the description of gravity. Hence, strictly speaking, gravitation had to be considered as an occult quality [10]. This is an important and weighty statement. A really nice mechanical-corpuscular model has been discussed by Richard Feynman. Let us become acquainted with it using Feynman's own words [3]:

> "*Suppose that in the world everywhere there are a lot of particles, flying through us at very high speed. They come equally in all directions — just shooting by — and once in a while they hit us in a bombardment. We, and the sun, are practically transparent for them, practically but not completely, and some of them hit. Look, then, at what would happen* (*Fig. 3.9*).
>
> *S is the sun, and E the earth. If the sun were not there, particles would be bombarding the earth from all sides, giving little impulses by the rattle, bang, bang of the few that hit. This will not shake the earth in any particular direction, because there*

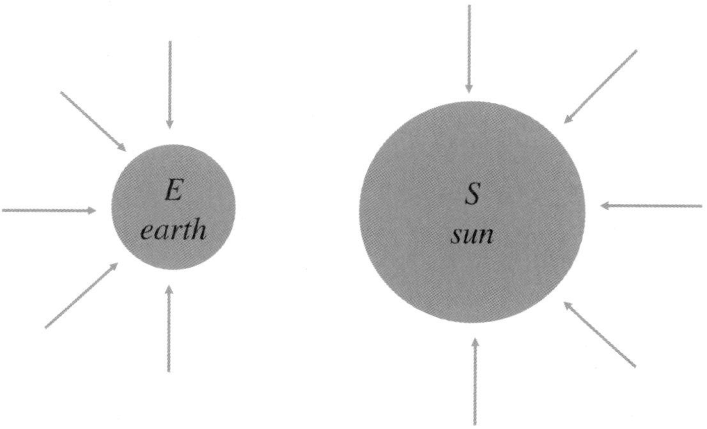


Fig. 3.9. Can the force of attraction between the earth and the sun be explained on the basis of impinging particles? The answer is given in the text by Richard Feynman.

*are as many coming from one side as from the other, from top as from bottom. However, when the sun is there the particles which are coming from that direction are partly absorbed by the sun, because some of them hit the sun and do not go through. Therefore the number coming from the sun's direction towards the earth is less than the number coming from the other sides, because they meet an obstacle, the sun. It is easy to see that the farther away the sun is, of all the possible directions in which the particles can come, a smaller proportion of the particles are being taken out. The sun will appear smaller — in fact inversely as the square of the distance. And this will be a result of large numbers of very simple operations, just hits, one after the other, from all directions. Therefore the strangeness of the mathematical relation will be very much reduced, because the fundamental operation is much simpler than calculating the inverse of the square of the distance. This design with the particles bouncing, does the calculation.*

*The only trouble with this scheme is that it does not work, for other reasons. Every theory that you make up has to be analyzed against all possible consequences, to see if it predicts anything else. This does predict something else. If the earth is moving, more particles will hit it from in front than from behind. (If you are running in the rain, more rain hits you in the front of the face than in the back of the head, because you are running into the rain.) Hence, if the earth is moving it is running into the particles coming towards it and away from the ones that are chasing it from behind. Hence, more particles will hit it from the front than from the back, and there will be a force opposing any motion. The force would slow the earth down in its orbit, and it certainly would not have lasted the three or four billion years (at least) that it has been going around the sun. So that is the end of that theory. 'Well', you say, 'it was a good one, and I got rid of the mathematics for a while. Maybe I could invent a better one'. Maybe you can, because nobody knows the ultimate. But up today, from the time of Newton, no one has invented another theoretical description of the mathematical machinery behind this law which does not either say the same thing over again, or make the mathematics harder, or predict some wrong phenomena. So there is no model for the theory of gravitation today, other than the mathematical form."*

This text by Feynman distinctly underlines the fact that there is no possibility to explain gravity by mechanical-corpuscular effects and, therefore, gravity had to be considered as a metaphysical element when the scientific standards are based on the mechanical-corpuscular world view. After Thomas Kuhn, the fact that there is no possibility to explain the interaction by mechanical-corpuscular effects has to be considered as problematic. He wrote the following [10]:

"*Yet, though Newton's work was directed to problems and embodied standards derived from the mechanical-corpuscular world view, the effect of the paradigm that resulted from his work was a further and partially destructive change in the problems and standards legitimate for science. Gravity, interpreted as an innate attraction between every pair of particles of matter, was an occult quality in the same sense as the scholastics' 'tendencies to fall' had been. Therefore, while the standards of corpuscularism remained in effect, the research for a mechanical explanation of gravity was one of the most challenging problems for those who accepted the Principia as paradigm. Newton devoted much attention to it and so did many of his eighteenth-century successors. The only apparent opinion was to reject Newton's theory for its failure to explain gravity, and that alternative, too, was widely adopted. Yet neither these views ultimately triumphed. Unable either to practice science without the principia or to make that work conform to the corpuscular standards of the seventeenth century, scientists gradually accepted the view that gravity was indeed innate. By the mid-eighteenth century, that interpretation had been almost universally accepted, and the result was a genuine revision (which is not the same as a retrogression) to a scholastic standard. Innate attractions and repulsions joined size, shape, position, and motion as physically irreducible primary 'properties of matter.'*"

In summary, there is no possibility to explain by a mechanism how the mutual influence between two bodies (as, for example, between the earth and the sun) comes into existence. The notion "interaction" has therefore been considered as an irreducible primary property of matter, more or less as an embarrassment solution. Instead of a certain mechanism (in analogy to the model of impinging particles discussed above) the gravity effect was declared afterwards to be a physically irreducible primary property of matter, i.e., it is "glued to", so to speak, onto matter.

### 3.15.7. *"Spring" between the Masses*

Then, we come to the following conception. There are static bodies (for example, earth and sun), which have the property to interact, and this interaction is often symbolized by a "spring" between the masses. This conception, born from an embarrassment solution, was more or less taken over from the physical disciplines as, for example, from modern elementary particle physics. Herwig Schopper remarked [17]:

> "*If we ask about the eternal, everlasting, indestructible elements in physics, then the answer would until recently have been: There are final impenetrable building blocks of matter. The forces permit the joining of these building blocks in various ways and thus cause the continual changes in nature. This image has endured in principle for several centuries, although certainly the question of what are to be considered the final building blocks is an issue which has been exposed to considerable change, as is also the case with our knowledge about the forces. Nevertheless — despite all progress brought about by the theory of relativity and quantum theory — the description of nature is in the last analysis based upon the materialistic idea of hard, tiny spheres with a kind of spring acting between them.*"

In other words, this conception (static bodies with a kind of spring (interaction) between them) is applied to the solutions in practically all disciplines of physics, i.e., from the planet system down to solids and to the area of elementary systems, whereby the range of dimensions here goes from $10^{15}$ cm (planet system) to $10^{-12}$ cm (atomic nuclei). Again, this concept emerged from classical mechanics, as an embarrassment solution so to speak, since a mechanism on the basis of the mechanical-corpuscular world view could not be found.

### 3.15.8. *Conclusion*

Our discussion in connection with the notion "interaction" showed that the situation in conventional physics, where the interaction processes are assumed to take place in space, is not convincing at all. Here it should be mentioned that Newton himself opposed this idea of "forces acting-at-a-distance", and he tried in vain to find a model for gravitation on the basis of a mechanical-corpuscular mechanism but without success. Newton's restraint in connection with "forces acting-at-a-distance" was soon given up by his successors, partly because of the success of Newtonian mechanics, but also because of the unsuccessful search for a mechanism. In fact, the description of the gravitational law $m_1 m_2/r^2$ on the basis of the familiar mechanisms of everyday life as, for example, by means of impinging particles (Section 3.15.6), was not possible.

The "proximity effect" and the "forces acting-at-the-distance" are, on the one hand, qualitatively quite different from each other but both notions are, on the other hand, exactly equivalent because it is not possible to distinguish between them, neither mathematically nor experimentally. In particular, none of these interpretations can be used for the derivation of the law $m_1 m_2/r^2$. Thus, the use of these interpretations is rather misleading. In both cases the objects are tightly linked with the space; the forces between the real objects (embedded in space) come into existence through space-effects. Therefore, such real space-effects have to be considered as misleading too. Within the original form of Newton's mechanics it seems to be the case that the forces between two masses act across space instantly ("forces acting-at-the-distance"). In other words, there is an information transfer through space. In the case of the "proximity effect" we have a space-filling field producing the force between the two objects. However, this space-filling "medium" (field) cannot be proved experimentally and it becomes entirely a matter of belief, whether a gravitational field (electrical field) exists in reality or not.

*Conclusion*

The above discussion showed that the conception of conventional physics (the material world is embedded in space) becomes basically doubtful. In contrast to this conception, projection theory offers a solution that is possibly more realistic.

### 3.15.9. *The Statements of Projection Theory*

The situation in projection theory is quite different. Here we have no material objects in space–time [$(\mathbf{r}, t)$-space] but only geometrical structures. These geometrical structures are not static elements. There is an exchange of momentum and energy in $(\mathbf{p}, E)$-space with other systems, and an object comes into existence by these $(\mathbf{p}, E)$-fluctuations, which are projected onto $(\mathbf{r}, t)$-space, leading to the shape of the systems and to relative structural configurations between them. Both, the variables $\mathbf{p}, E$ of $(\mathbf{p}, E)$-space and the variables $\mathbf{r}, t$ of $(\mathbf{r}, t)$-space are statistical quantities and fluctuate randomly in the course of time $\tau$, i.e., in contrast to conventional physics we have no static objects within projection theory.

There are $(\mathbf{r}, t)$-correlations in $(\mathbf{r}, t)$-space, but there is no exchange of information between the various geometrical structures, that is, nothing acts across $(\mathbf{r}, t)$-space as in the case of conventional physics ("proximity effect" and the "forces acting-at-the-distance"); the $(\mathbf{r}, t)$-correlations come exclusively into existence by the $(\mathbf{p}, E)$-fluctuations in $(\mathbf{p}, E)$-space; the $(\mathbf{r}, t)$-structures in $(\mathbf{r}, t)$-space are exclusively the result of projection processes and no geometrical part of $(\mathbf{r}, t)$-space is influenced by the others by actions through $(\mathbf{r}, t)$-space. From the point of view of the projection principle, such actions through space and time make no sense since $(\mathbf{r}, t)$-space merely contains pictures.

Furthermore, it is important to note that all effects (shape of the objects as well as their relative structural configuration) can be described by the elements of the theory, and these are $(\mathbf{p}, E)$-fluctuations. We do not need to graft the theory by strange elements. This is not possible within Newton's theory, as we have discussed above. Here we cannot describe the scenario by mechanical-corpuscular effects alone (for example, by impinging particles) but we need elements for the description of gravity which do not belong to the standards of the mechanical-corpuscular world view. Therefore, gravity can be qualified as an occult (metaphysical) element from the point of view of the mechanical-corpuscular world view. In this connection, Kuhn remarked [10]. "*Gravity, interpreted as an innate attraction between every pair of particles of matter, was an occult quality in the same sense as the scholastics' 'tendencies to fall' had been.*"

The situation in projection theory and conventional physics is also discussed in connection with Fig. 3.10.

### 3.15.10. *Interaction: Final Remarks*

The whole interaction scenario in physics (classical physics, usual quantum theory, quantum field theory, string theory) is based on processes between localized systems in space [$(\mathbf{r}, t)$-space]. Since the interactions in projection theory take place in $(\mathbf{p}, E)$-space, none of these specific processes can be applied here. Furthermore, in contrast to the conventional conceptions (here localized systems are used), within projection theory we are confronted with delocalized systems in $(\mathbf{p}, E)$-space, if the same systems are localized in $(\mathbf{r}, t)$-space [12].

The description of interacting delocalized systems in $(\mathbf{p}, E)$-space is not possible on the basis of interpretations and mechanisms, respectively, which we are using in classical physics or usual quantum theory. A mechanism is normally based on our intuition, and man's intuition is tailor-made to localized bodies that are embedded in $(\mathbf{r}, t)$-space and not to delocalized systems that are positioned in $(\mathbf{p}, E)$-space. Our intuition is adjusted to the structures in space which we experience in everyday life, and these space structures are composed of more or less localized bodies that are embedded in space [$(\mathbf{r}, t)$-space]. Our complete intuitive abilities have been developed in connection to situations which we experience in everyday life, and we call a

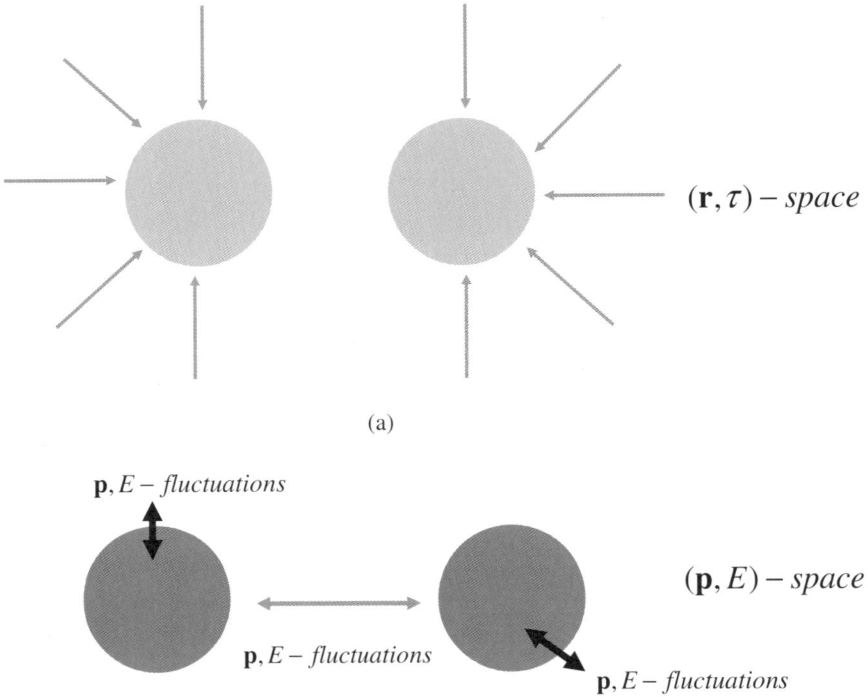


(a)

(b)

Fig. 3.10. (a) Two interacting bodies (see also Fig. 3.9) within the framework of the mechanical-corpuscular world view. It is an attempt to explain the interaction by means of impinging particles. However, this and similar models do not work as we have already outlined in Section 3.15.6. Therefore, the interaction (in this case gravity) has to be considered as an occult (metaphysical) quality when we judge from the point of view of the mechanical-corpuscular standards. The notion "force acting-at-a-distance" is a stranger within the mechanical-corpuscular world view. (b) The situation is different in projection theory. The scenario in (b) reflects the complete situation in $(\mathbf{p}, E)$-space, that is, both systems and their interactions are symbolically represented in $(\mathbf{p}, E)$-space. Here the basic mechanism is given by $\mathbf{p}, E$-fluctuations. This can be considered as the world view of projection theory, and it corresponds to the mechanical-corpuscular world view of Newton's mechanics. In $(\mathbf{r}, t)$ space, both, the form of the two bodies as well as their relative structural configuration, are equally explained by $\mathbf{p}, E$-fluctuations that take place in $(\mathbf{p}, E)$-space and which are projected onto $(\mathbf{r}, t)$-space); the form of the bodies are described by distance-independent interactions and their relative structural configuration by distance-dependent interactions. Due to the $(\mathbf{p}, E)$-fluctuations the bodies cannot be considered as static elements in $(\mathbf{p}, E)$ space and also not in $(\mathbf{r}, t)$ space).

certain process “imaginable” when it can be explained by the means of everyday life. A “mechanism” is normally defined on the basis of an imaginable process. However, interacting delocalized systems which are positioned in $(\mathbf{p}, E)$-space do not belong to the category of imaginable processes. Thus, it does not make much sense to explain these $(\mathbf{p}, E)$-fluctuations by means of such “mechanisms“ but we have to accept that they simply “emerge”.

## 3.16. Time

### 3.16.1. *Properties of Probability Distributions*

At time $\tau$ we have probability distributions $\{\mathbf{r}\}$, $\{t\}$, $\{\mathbf{p}\}$, $\{E\}$ for the variables $r$, $t$, $\mathbf{p}$ and $E$:

$$\tau : \{\mathbf{r}\}, \{t\}, \{\mathbf{p}\}, \{E\}. \tag{3.84}$$

But only *one* value of each distribution can be realized at time $\tau$, and there is a certain probability for the existence of these values which are expressed by $\Psi^*(\mathbf{r}, t)\Psi(\mathbf{r}, t)$ and $\Psi^*(\mathbf{p}, E)\Psi(\mathbf{p}, E)$. But how many values of each variable $\mathbf{r}$, $t$, $\mathbf{p}$ and $E$ could in principle be registered within a small time interval $\Delta\tau$ different from zero?

Let $\Delta_\mathbf{r}$, $\Delta_t$, $\Delta_\mathbf{p}$, $\Delta_E$ be the ranges of $\mathbf{r}$, $t$, $\mathbf{p}$ and $E$ for which the distributions $\Psi^*(\mathbf{r}, t)\Psi(\mathbf{r}, t)$ and $\Psi^*(\mathbf{p}, E)\Psi(\mathbf{p}, E)$ are not zero, and let us assume that the ranges $\Delta_\mathbf{r}$, $\Delta_t$, $\Delta_\mathbf{p}$, $\Delta_E$ are different from infinity. How many values $\mathbf{r}$, $t$, $\mathbf{p}$ and $E$ can come into existence in the time interval $\Delta\tau = \varepsilon$, where $\varepsilon$ is infinitesimal but different from zero? This number of values for each variable is identical with the number of $\tau$-values in the interval $\varepsilon$, and this number is given by the number of all real numbers within $\varepsilon$. We know from mathematics that the number of real numbers in $\varepsilon$ must be infinity, and in physics none of these real numbers are excluded. Therefore, within the infinitesimal interval $\varepsilon$, the number $N$ of values of a variable (as, for example $\mathbf{r}$) is also infinite in the interval $\Delta_\mathbf{r}$. Thus, the number density $\rho = N/\Delta_\mathbf{r}$ of $\mathbf{r}$-values is also infinite in the interval $\Delta_\mathbf{r}$, where the interval $\Delta_\mathbf{r}$ can take any value but it must be different from infinity.

In conclusion, the possible values defined by $\Psi^*(\mathbf{r},t)\Psi(\mathbf{r},t)$ and $\Psi^*(\mathbf{p},E)\Psi(\mathbf{p},E)$ are realized (with an infinite number density) already within an infinitesimal time-interval $\varepsilon$ of almost zero, although the ranges $\Delta_{\mathbf{r}}, \Delta_t, \Delta_{\mathbf{p}}, \Delta_E$ can take any value. (However, it must be different from infinity.) This is of particular importance for time $t$ and will be analyzed in more detail below.

This property (let us call it $\varepsilon$-property) is remarkable, but also in classical mechanics we work with an infinite number of space-points within an infinitesimal interval $\Delta\tau = \varepsilon$. Let us consider a classical particle, which moves with velocity $v_x$ from one space-position to another defining the space-distance of $\Delta_{\mathbf{r}}$. If this process takes place within an infinitesimal time interval $\Delta\tau = \varepsilon$, the space-interval $\Delta_{\mathbf{r}}$ must also be infinitesimal, otherwise we will not be able to define a reasonable classical velocity $v_x$. (Classical mechanics is a local theory). Nevertheless, the interval $\Delta_{\mathbf{r}}$ contains an infinite number of real numbers, that is, the particle runs monotonically over an infinite number of real numbers within the infinitesimal time interval $\Delta\tau = \varepsilon$. (In Fig. 3.11, only five points are drawn in).

However, in quantum theory, the world is non-local, and the laws are non-deterministic. In particular, the definition of the velocity $\mathbf{v}$ in the classical sense of the word is not possible. In conventional quantum theory only the space variable $\mathbf{r}$ behaves statistically. In contrast to classical mechanics and conventional quantum theory, within projection theory the space–time intervals $\Delta_{\mathbf{r}}$ and $\Delta_t$ can take any values (different from infinity) in the case of an infinitesimal time interval $\Delta\tau = \varepsilon$. Both intervals ($\Delta_{\mathbf{r}}$ and $\Delta_t$) are *statistically* occupied in the course of time $\tau$. Within the infinitesimal time interval $\Delta\tau = \varepsilon$, the variables of the system jumps statistically from one space–time point ($\mathbf{r}_i$, $t_i$) to another point ($\mathbf{r}_j$, $t_j$) where $\mathbf{r}_{ij} = \mathbf{r}_i - \mathbf{r}_j$ and $t_{ij} = t_i - t_j$ can take arbitrary values within the space intervals $\Delta_{\mathbf{r}}$, $\Delta_t$: $\mathbf{r}_{ij} \le \Delta_{\mathbf{r}}$, $t_{ij} \le \Delta_t$. Although the time interval $\Delta\tau = \varepsilon$ is infinitesimal, the number density in both intervals is infinity because the number of events within $\Delta\tau = \varepsilon$ is infinite. Note, that this property is independent on $\Delta_{\mathbf{r}}, \Delta_t, \Delta_{\mathbf{p}}, \Delta_E$; the ranges $\Delta_{\mathbf{r}}, \Delta_t, \Delta_{\mathbf{p}}, \Delta_E$ can take any value. (However, it must be different from infinity.) In Fig. 3.12, the relation between projection theory and usual quantum theory is given.

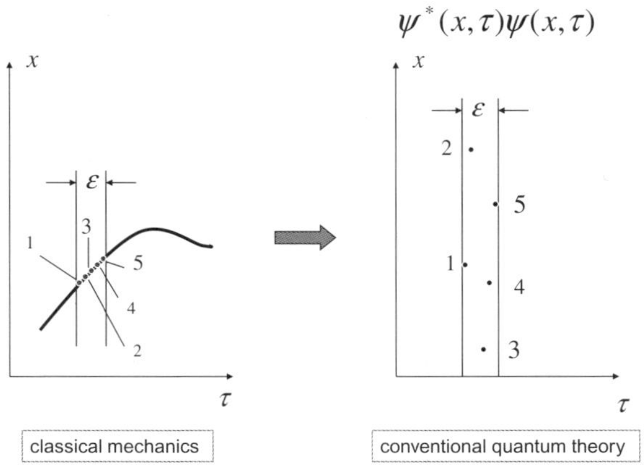


Fig. 3.11. Let us consider a classical particle, which moves in the direction of $x$ with velocity $v_x$ from one space-position to another defining the space-distance of $\Delta_{\mathbf{r}}$. If this process takes place within an infinitesimal time interval $\Delta\tau = \varepsilon$, the space-interval $\Delta_x$ must also be infinitesimal otherwise we will not be able to define a reasonable classical velocity $v_x$. Nevertheless, the interval $\Delta_{\mathbf{r}}$ contains an infinite number of real numbers, that is, the particle runs monotonically over an infinite number of real numbers within the infinitesimal time interval $\Delta\tau = \varepsilon$. (Here only five points are drawn in.) When we go from classical mechanics to conventional quantum theory the situation is different. Within the infinitesimal time interval $\Delta\tau = \varepsilon$, the variable $x$ of the system jumps statistically from one space–time point $x_i$ to another point $x_j$ where $x_{ij} = x_i - x_j$ can take arbitrary values within the space interval $\Delta_{\mathbf{r}} = \Delta_x$ with $x_{ij} \leq \Delta_x$, where $\Delta_x$ is the range of the probability density $\psi^*(x)\psi(x)$ of conventional quantum theory. Although the time interval $\Delta\tau = \varepsilon$ is infinitesimal the number density in the interval $\Delta_x$ is infinity because the number of events within $\Delta\tau = \varepsilon$ is infinite. (Here, only five points are drawn in also).

The situation in connection with the number of events ($\varepsilon$-property) within a small time interval $\Delta\tau = \varepsilon$ is close to Born's probability interpretation (Section 3.2.2) and, as we have outlined above, also in classical mechanics. While classical mechanics is a local theory, the theoretical picture given here led to an inherently non-local representation with respect to space and time.

There are essential differences between Born's theory and the investigation given here: (1) Within Born's probability interpretation point-like particles are needed; within the theoretical picture given here such point-like particles are not defined in this connection. (2) Within Born's theory, only the position $\mathbf{r}$ behaves statistically; within projection theory we have an extension due to the system-specific time $t$

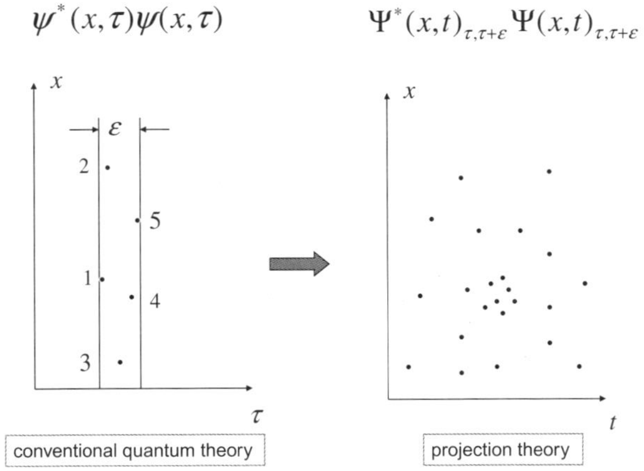


Fig. 3.12. The position of conventional quantum theory has been discussed in Fig. 3.11. When we go from conventional quantum theory to projection theory, the system-specific time $t$ appears, which is not defined in usual quantum theory. Not only the space coordinate $x$ behaves statistically but also the variable $t$. Therefore, we have the following situation: Within the infinitesimal time interval $\Delta\tau = \varepsilon$, the variables $x$ and $t$ of the system jump statistically from one space–time point $(x_i, t_i)$ to another point $(x_j, t_j)$ where $x_{ij} = x_i - x_j$ and $t_{ij} = t_i - t_j$ can take arbitrary values within the space intervals $\Delta_x$, $\Delta_t$: $x_{ij} \leq \Delta_x$, $t_{ij} \leq \Delta_t$, where $\Delta_x$ and $\Delta_t$ are the ranges of the probability density $\Psi^*(x,t)\Psi(x,t)$ of projection theory. Although the time interval $\Delta\tau = \varepsilon$ is infinitesimal, the number density in both intervals is infinity because the number of events within $\Delta\tau = \varepsilon$ is infinite (here only a few space–time points have been drawn in).

which is not defined in usual quantum theory. Not only the space positions **r** behave statistically (Born's theory) but each space–time point (**r**, $t$) jumps statistically.

## Does God play dice?

Whether or not the statistical behavior is also a property of basic reality can principally not be said from the observers' point of view. As we have discussed in detail in Chapter 2, a human cannot make any statement about basic (objective) reality. The only thing we know with certainty within projection theory is that basic (objective) reality must exist. In this connection Fig. 2.7 is instructive.

Albert Einstein asserted that "God does not play dice". But where does Einstein know what God is able or willing to do? What can people say about the abilities of God? Nothing! Such kind of questions are inappropriate, and this assessment is supported by the foundations of projection theory: From the point of view of projection theory we cannot state that "God plays dice" but also not that "God does not play dice". Both statements are not allowed within projection theory when we identify God with basic reality.

### 3.16.2. *The System-Specific Time*

Concerning time $t$ we have the following situation. At time $\tau$ only one value of the system-specific time $t$ is realized with a certain probability. However, if we consider an infinitesimal time interval $\Delta\tau = \varepsilon$ an infinite number of $t$-values are occupied, that is, the whole history (the complete past and future) of the system, described by the range $\Delta_t$ (lifetime) of the distribution $\Psi^*(\mathbf{r}, t)\Psi(\mathbf{r}, t)$, is given within the infinitesimal time interval $\Delta\tau = \varepsilon$ of our clocks (Section 3.16.1).

This is the case for any $\tau_i \pm \Delta\tau$, that is, the law $\Psi^*(\mathbf{r}, t)\Psi(\mathbf{r}, t)$ is independent of $\tau$. It is stationary with respect to time $\tau$. (See also Section 3.7.) Thus, we may say the following. Despite the statistical fluctuations, the whole of time — past, present and future — is existent as a frozen block at time $\tau$. In other words, there is no connection between $\tau$ and the system-specific time $t$, that is, the introduction of $\tau$ within projection theory, outlined so far, would make no sense. Therefore, we have to construct a connection between $\tau$ and the system-specific time $t$, because the existence of $\tau$ is a matter of fact. This connection between the reference time $\tau$ and the system-specific time $t$ will be constructed in the next section.

Let us state once more that the time $\tau$ has nothing to do with the system under investigation, but merely plays the role of a reference time and is defined by our clocks which we use in everyday life.

#### 3.16.2.1. *Reference Time and Selection Processes*

Principal remarks

So far there seems to be no connection between $\tau$ and the system-specific time $t$. Despite the statistical fluctuations, at time $\tau$ the whole of time $t$ (past, present and future) is existent as a frozen block:

$$\Psi^*(\mathbf{r},t)\Psi(\mathbf{r},t), \quad -\infty < \mathbf{r} < \infty, \quad -\infty < t < \infty. \tag{3.85}$$

This is valid for all times $\tau$. In other words, the reference time $\tau$ is not correlated to specific $t$-values for such a situation. From this point of view, it would make no sense to introduce $\tau$ within projection theory. The situation is summarized in Fig. 3.13.

For an observer without time-feeling ($\tau$ is not defined) such a process [defined by Eq. (3.85)] could not produce a certain time-feeling because $t$ jumps statistically between the various $t$-values without to give time $t$ a certain direction; there is no past, present and future definable.

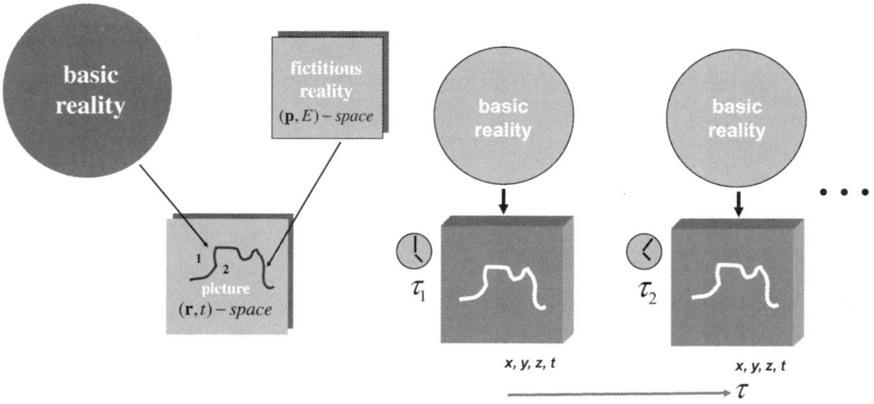


(a) (b)

Fig. 3.13. (a) In everyday life we have certain pictures spontaneously in front of us. In conventional physics, it is normally assumed that real matter is embedded in space and time. Within projection theory, there can only be geometrical structures in the picture and are exclusively elements (states) of the brain. All the real actions and processes, respectively, take place in absolute (basic) reality and the entire complex, which is involved, is projected onto $(\mathbf{r}, t)$-space (curve 1). We describe the geometrical structures by means of fictitious realties, that is, by $(\mathbf{p}, E)$-fluctuations (interactions) in $(\mathbf{p}, E)$-space. The $(\mathbf{p}, E)$-structures produced by these fluctuations are projected onto $(\mathbf{r}, t)$-space and we get curve 2. (b) If the system under investigation is stationary with respect to time $\tau$, the law $\Psi^*(\mathbf{r},t)\Psi(\mathbf{r},t)$ is independent on $\tau$, that is, we have at each time $\tau$ (for example, at $\tau_1$ and $\tau_2$) exactly the same $(\mathbf{r}, t)$-structure.

We know however from our observations in everyday life that we always observe only certain configurations of reality at time $\tau$:

$$\Psi^*(\mathbf{r}, t_0)\Psi(\mathbf{r}, t_0), \quad -\infty < \mathbf{r} < \infty \tag{3.86}$$

for $t_0 = \tau$. Each photography represents such a configuration in space, at a certain time $\tau = t_0$. Why does nature work in this way? Why such selection processes from $\Psi^*(\mathbf{r}, t)\Psi(\mathbf{r}, t)$ to $\Psi^*(\mathbf{r}, t_0)\Psi(\mathbf{r}, t_0)$ [see Eqs. (3.85) and (3.86)]? The answer is probably given by evolution. There is an important basic principle in connection with evolution [Chapter 2]. As little outside world as possible, this guarantees optimal chances for survival. This principle is clearly reflected in the transition from $\Psi^*(\mathbf{r}, t)\Psi(\mathbf{r}, t)$ to $\Psi^*(\mathbf{r}, t_0)\Psi(\mathbf{r}, t_0)$ and reflects a certain kind of selection. The occurrence of the reference time $\tau$ is obviously one of the features for that.

### Introduction of the reference system

How does nature organize that, i.e., the transition from the situation defined by Eq. (3.85) to that given by Eq. (3.86)? This cannot be due to an internal transformation within the system under investigation alone, that is, without the influence of another process. Besides the system under investigation, only the observers function appears within the frame of our analysis and, therefore, the transition from $\Psi^*(\mathbf{r}, t)\Psi(\mathbf{r}, t)$ to $\Psi^*(\mathbf{r}, t_0)\Psi(\mathbf{r}, t_0)$ [Eqs. (3.85) and (3.86)] must be due to an interplay between the system under investigation, described by $\Psi(\mathbf{r}, t)$, and the observer's function. However, the observer's function has been characterized so far by one parameter only. It is the reference time $\tau$ that is measured by our clocks in everyday life. Clearly, only one parameter ($\tau$) is not sufficient for the description of the interplay between the system characterized by $\Psi(\mathbf{r}, t)$ and the observer's function. How can we characterize the observation process more realistically?

For this purpose let us define a reference system that is produced inside the observer, and let us formally describe it by the wave function $\Psi_{ref}(t)$ and the probability distribution $\Psi^*_{ref}(t)\Psi_{ref}(t)$, respectively, and we would like to characterize the time variable for the reference system by $\gamma = t$. Clearly, also $\gamma = t$ is a system-specific quantity. (For simplicity we would like to assume that the reference system is not dependent on any position $\mathbf{r}$.)

The source for the existence of the wavefunction $\Psi_{ref}(t)$ are energy fluctuations in reality [$(\mathbf{p}, E)$-space] leading to $\Psi_{ref}(E)$ and $\Psi^*_{ref}(E)\Psi_{ref}(E)$, respectively. $\Psi_{ref}(E)$ and $\Psi_{ref}(t)$ are connected by (see also [12])

$$\Psi_{ref}(E) = \frac{1}{(2\pi\hbar)^{1/2}} \int_{-\infty}^{\infty} \Psi_{ref}(t) \exp\left\{i\frac{E}{\hbar}t\right\} dt. \tag{3.87}$$

In conclusion, at time $\tau$ we have two probability distributions, $\Psi^*_{ref}(t)\Psi_{ref}(t)$ and $\Psi^*(\mathbf{r}, t)\Psi(\mathbf{r}, t)$; one for the description of time $\gamma = t$ of the reference system and the other for the description of time $t$ of the system under investigation.

The reference system, characterized by $\Psi_{ref}(t)$ and $\Psi^*_{ref}(t)\Psi_{ref}(t)$, respectively, is located inside the brain of the observer. The system under investigation is positioned outside the observer and is characterized by $\Psi(\mathbf{p}, E)$ and $\Psi(\mathbf{r}, t)$, respectively. For the observation of the system $\Psi(\mathbf{r}, t)$ by $\Psi_{ref}(t)$, both systems must be coupled.

The reference system, described by $\Psi_{ref}(t)$, has two functions:

1. To describe the nature of the reference time more specifically, and
2. To select $\Psi^*(\mathbf{r}, t_0)\Psi(\mathbf{r}, t_0)$ from $\Psi^*(\mathbf{r}, t)\Psi(\mathbf{r}, t)$ [see Eqs. (3.85) and (3.86)].

We will recognize below that both functions are interconnected: selection is not possible without the existence of a systematically varying reference system. First we will discuss the reference time in more detail, and after that the selection process.

#### 3.16.2.2. *Structure of Reference Time*

So far, we have stated that time $\tau$ runs monotonically from the past to the future. However, this time-feeling must also be due to a process (inside the brain of the observer) and is therefore also a system-specific time.

This process might differ considerably from those treated within the frame of usual quantum theory. The reason is simple: within projection theory the equation for the determination of $\Psi(\mathbf{r}, t)$ [Eq. (3.22)] is more general than Schrödinger's equation [Eq. (3.21)] of usual quantum theory because the function $V(x, y, z, t)$ is in general more complex than the classical potential $U(x, y, z)$ which is used in usual quantum theory (Section 3.7.1). In principle, $V(x, y, z, t)$ could have an imaginary part. The appearance of the function $V(x, y, z, t)$ is a logical consequence within the theoretical structures of projection theory. (See in particular [12].)

We would like to describe the reference time by the distribution $\Psi^*_{\text{ref}}(t)\Psi_{\text{ref}}(t)$ introduced in the last section. The probability of finding a certain value $\gamma = t$ for the reference time in the interval $\Delta\gamma = \Delta t$ around $\gamma = t$ is given by $\Psi^*_{\text{ref}}(t)\Psi_{\text{ref}}(t)\Delta t$. In other words, the reference time $\gamma = t$ becomes uncertain because $\Psi^*_{\text{ref}}(t)\Psi_{\text{ref}}(t)$ has a certain width $\Delta_\tau$ and, therefore, no longer runs strictly from the past to the future as is suggested by our clocks used in everyday life. However, the probability distribution $\Psi^*_{\text{ref}}(t)\Psi_{\text{ref}}(t)$ for the reference time $\gamma = t$ should be a relatively sharp function and, furthermore, because the time $\tau$ of our clocks runs monotonically from the past to the future, the distribution $\Psi^*_{\text{ref}}(t)\Psi_{\text{ref}}(t)$ must run monotonically from the past to the future and we have

$$\Psi^*_{\text{ref}}(t)\Psi_{\text{ref}}(t) \rightarrow \Psi^*_{\text{ref}}(\tau - t)\Psi_{\text{ref}}(\tau - t). \tag{3.88}$$

We do not measure the time $\tau$ but the variable $\gamma = t$ of the distribution for the reference time, and this $\gamma = t$ is uncertain due to $\Psi^*_{\text{ref}}(\tau - t)$ $\Psi_{\text{ref}}(\tau - t)$. However, because $\Psi^*_{\text{ref}}(\tau - t)\Psi_{\text{ref}}(\tau - t)$ can be assumed to be a relatively sharp function, the variable $\gamma = t$ for the distribution of the reference time should be close to $\tau$. Due to $\tau$, the whole curve $\Psi^*_{\text{ref}}(\tau - t)\Psi_{\text{ref}}(\tau - t)$ moves strictly from the past to the future, but the values $\gamma = t$ for the reference time fluctuate around $\tau$.

It is easy to recognize that the expression $\Psi^*_{\text{ref}}(E)\Psi_{\text{ref}}(E)$ is independent on $\tau$. [In [12], we have outlined that not $\Psi_{\text{ref}}(E)$ is the relevant function but $\Psi^*_{\text{ref}}(E)\Psi_{\text{ref}}(E)$.] That is, in the case of the distribution $\Psi^*_{\text{ref}}(t)\Psi_{\text{ref}}(t)$ as well as for the "moving" distribution function $\Psi^*_{\text{ref}}(\tau - t)\Psi_{\text{ref}}(\tau - t)$, we have exactly the same energy distribution $\Psi^*_{\text{ref}}(E)\Psi_{\text{ref}}(E)$. Thus, all states $\tau$ are equivalent.

Since we experience, for example, a moving car with velocity $\mathbf{v}$ as a strict sequence of configurations $A(\mathbf{r} - \mathbf{v}\tau)$, we have for $\tau$ the strict sequence $\tau_1 < \tau_2 \cdots$. But this feature is based on our macroscopic observations and, therefore, in general we have to admit a certain small width $\Delta_\tau$ for $\Psi^*_{ref}(\tau - t)\Psi_{ref}(\tau - t)$.

### 3.16.2.3. *Selections*

We have outlined in Section 3.16.2.1 that the existence of the wavefunction $\Psi_{ref}(t)$ for the reference system should be responsible for selection processes. In fact, the transition from $\Psi^*(\mathbf{r}, t)\Psi(\mathbf{r}, t)$ to $\Psi^*(\mathbf{r}, t_0)\Psi(\mathbf{r}, t_0)$ can be explained on the basis of this kind of function, i.e., by $\Psi_{ref}(t)$. In other words, the interplay between the two systems (the reference system described by $[\Psi_{ref}(t), \Psi_{ref}(E)]$ and, on the other hand, the system under investigation described by $[\Psi(\mathbf{r}, t)$, $\Psi(\mathbf{r}, E)])$ should lead to the selection process. This process obviously filters the configuration $\Psi^*(\mathbf{r}, t_0)\Psi(\mathbf{r}, t_0)$ from $\Psi^*(\mathbf{r}, t)\Psi(\mathbf{r}, t)$ out.

However, the following has to be considered. The reference system, characterized by $\Psi_{ref}(t)$ and $\Psi^*_{ref}(t)\Psi_{ref}(t)$, respectively, is located inside the brain of the observer. The system under investigation is positioned outside the observer and is characterized by $\Psi(\mathbf{r}, t)$ and $\Psi^*(\mathbf{r}, t)\Psi(\mathbf{r}, t)$, respectively. For the observation of the system $\Psi(\mathbf{r}, t)$ by $\Psi_{ref}(t)$ both systems must be coupled. In [12], realistic models have been proposed, and we come to the following result.

The states $\Psi^*(\mathbf{r}, t)\Psi(\mathbf{r}, t)$ of the system under investigation will be systematically scanned by $\Psi^*_{ref}(\tau - t)\Psi_{ref}(\tau - t)$ and only those values of $t$ can be observed which correspond with the reference time $\tau$. (See also Figs. 3.14 and 3.15.) This leads to an effect of motion. The sense of time $\tau$ is to select a certain configuration $\Psi^*(\mathbf{r}, t_k)\Psi(\mathbf{r}, t_k)$ with $t_k = \tau$. Clearly, $\Psi^*(\mathbf{r}, t)\Psi(\mathbf{r}, t)$ is a static function (see Section 3.16.1) and does not change in the course of time $\tau$, and the effect of motion we experience in connection with $\Psi^*(\mathbf{r}, t)\Psi(\mathbf{r}, t)$ is entirely due to the "motion" of the reference time $\tau$.

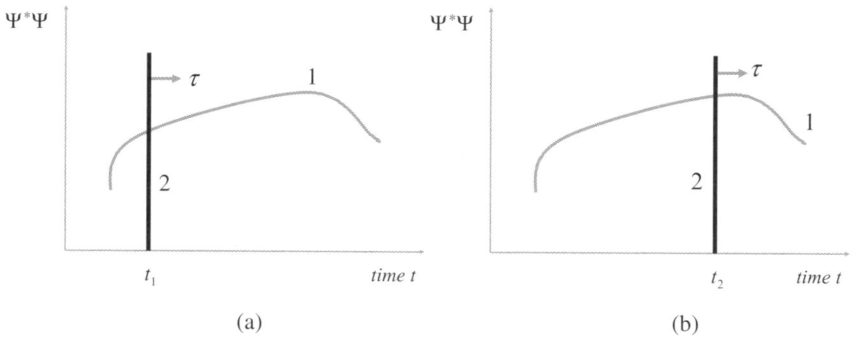


Fig. 3.14. The system under investigation, characterized by curve 1, fluctuates statistically between all possible configurations $t$ [defined by the range $\Delta_t$ of $\Psi^*(\mathbf{r},t)\Psi(\mathbf{r},t)$]. On the other hand, curve 2, described by $\Psi^*_{\text{ref}}(\tau - t)\Psi_{\text{ref}}(\tau - t)$, characterizes the observers time feeling and this must also be due to a process. $\Delta_\tau$ is the width of $\Psi^*_{\text{ref}}(\tau - t)\Psi_{\text{ref}}(\tau - t)$ and must be relatively small (in the case of a delta function we have $\Delta_\tau = 0$). Only those configurations can be observed which correspond with the reference time $t_B$ [with $\tau - \Delta_\tau/2 \le t_B \le \tau + \Delta_\tau/2$] since the human observer can only measure within the frame of curve 2 which is described by $\Psi^*_{\text{ref}}(\tau - t_B)\Psi_{\text{ref}}(\tau - t_B)$. In the case of $\Delta_\tau \cong 0$, the function $\Psi^*_{\text{ref}}(\tau - t)\Psi_{\text{ref}}(\tau - t)$ is an approximate delta function and we have $t_B = \tau$. Then, the selected $\mathbf{r}$-configuration of the system under investigation is given by $\Psi^*(\mathbf{r}, t = \tau)\Psi(\mathbf{r}, t = \tau)$ [12]. We measure one point of this $\mathbf{r}$-configuration at reference time $t_B = \tau$, and each point is given with the probability $\Psi^*(\mathbf{r}, t = \tau)\Psi(\mathbf{r}, t = \tau)\Delta t$. If the observation does not take place at a definite (sharp) time $\tau$ but within a certain time interval $\tau \pm \Delta\tau$ with $\Delta\tau = \varepsilon$, where $\varepsilon$ can be infinitesimal (Section 3.16.1), we measure an infinite number of values of the $\mathbf{r}$-configuration $\Psi^*(\mathbf{r}, t = \tau)\Psi(\mathbf{r}, t = \tau)$, that is, in this case the entire $\mathbf{r}$-configuration of the system under investigation is given at time $\tau$, and this is because $\varepsilon$ can be neglected. Within the frame of our observations in everyday life we actually have such a complex (entire) $\mathbf{r}$-configuration in front of us. Thus, we may conclude that we always make our observations at the macroscopic level within a certain time interval $\Delta\tau = \varepsilon$; however, $\varepsilon$ can be infinitesimal (Section 3.16.1). (a) is the situation for $\tau = \tau_1$ and $t = t_1$ leading to $t_{B1} = \tau_1$ and $\Psi^*(\mathbf{r}, t_1 = \tau_1)\Psi(\mathbf{r}, t_1 = \tau_1)$. (b) is the situation for $\tau = \tau_2$ and $t = t_2$ leading to $t_{B2} = \tau_2$ and $\Psi^*(\mathbf{r}, t_2 = \tau_2)\Psi(\mathbf{r}, t_2 = \tau_2)$.

### 3.16.3. *Information Inside, Information Outside*

The specific transition from $\Psi(\mathbf{r},t)[\Psi^*(\mathbf{r},t)\Psi(\mathbf{r},t)]$ to $\Psi(\mathbf{r},t = \tau)[\Psi^*(\mathbf{r},t = \tau)\Psi(\mathbf{r},t = \tau)]$ means that at time $\tau$ only a small fraction of the whole information $\Psi^*(\mathbf{r},t)\Psi(\mathbf{r},t)$ is accessible to the observer. The picture of reality at time $\tau$ is given by $\Psi^*(\mathbf{r},t = \tau)\Psi(\mathbf{r},t = \tau)$. All the other information $\Psi^*(\mathbf{r},t)\Psi(\mathbf{r},t)$, $t \neq \tau$, is ignored (filtered out) and is not *realized* in $(\mathbf{r},t)$–space but is *defined* for all values $\mathbf{r}$ and $t$;

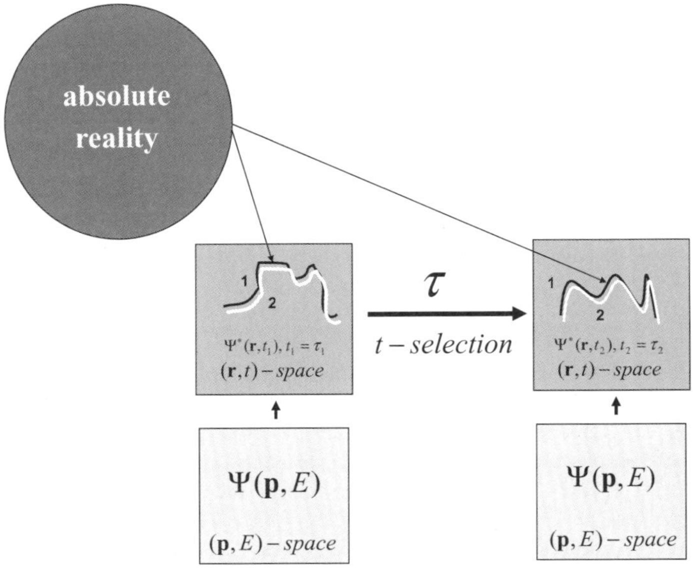


Fig. 3.15. The states $\Psi^*(\mathbf{r},t)\Psi(\mathbf{r},t)$ of the system under investigation are systematically scanned by $\Psi^*_{\text{ref}}(\tau - t)\Psi_{\text{ref}}(\tau - t)$ and only those values of $t$ of $\Psi^*(\mathbf{r},t)\Psi(\mathbf{r},t)$ can be observed which correspond with the reference time, which is given by $\tau$ if the $\Psi^*_{\text{ref}}(\tau - t)\Psi_{\text{ref}}(\tau - t)$ is very sharp (delta function). In other words, there is a $t$-selection with respect to $\Psi^*(\mathbf{r},t)\Psi(\mathbf{r},t)$. This leads to an effect of motion. The sense of time $\tau$ is to select a certain configuration of $\Psi^*(\mathbf{r},t_k)\Psi(\mathbf{r},t_k)$ with $t_k = \tau$. We experience the scanned $\mathbf{r}$-structures $\Psi^*(\mathbf{r},t_1)\Psi(\mathbf{r},t_1)$, $\Psi^*(\mathbf{r},t_2)\Psi(\mathbf{r},t_2)$, etc. (with $t_1 = \tau_1$ and $t_2 = \tau_2$, etc.) like a movie on a screen. Curves 1: projections from basic reality onto $(\mathbf{r},t)$-space. Curves 2: projections from fictitious reality [$(\mathbf{p},E)$-space] onto $(\mathbf{r},t)$-space. See also Fig. 3.13.

there are $(\mathbf{p},E)$-fluctuations described by $\Psi(\mathbf{p},E)$ [$\Psi^*(\mathbf{p},E)\Psi(\mathbf{p},E)$], and $\Psi(\mathbf{p},E)$ reflects the complete information leading to $\Psi(\mathbf{r},t)$ [$\Psi^*(\mathbf{r},t)\Psi(\mathbf{r},t)$] for all positions $\mathbf{r}$ and for all times $t$.

If the system with $\Psi(\mathbf{r},t)$ does not interact with the observer, characterized by $\Psi_{ref}(\tau - t)$, there can be no picture [$(\mathbf{r},t)$−structure] of the system in the observers brain, where the $(\mathbf{r},t)$−space is located. In other words, there can be no picture in $(\mathbf{r},t)$−space without an interaction between the system and the observer. Nevertheless, the system exists in the world outside, that is, there are $(\mathbf{p},E)$−fluctuations, described by $\Psi(\mathbf{p},E)$, independent from the observers existence. These $(\mathbf{p},E)$-fluctuations are complete at each time $\tau$ and, therefore, also the function $\Psi(\mathbf{r}, t)$ is defined at each time $\tau$ for all positions $\mathbf{r}$ and for all times $t$, and this is because we have an infinite number of events in the infinitesimal interval $\Delta\tau$ around $\tau$. In other words, if there is no interaction between the system and the observer, the information $\Psi(\mathbf{r}, t)$ $[\Psi^*(\mathbf{r}, t)\Psi(\mathbf{r}, t)]$ is *defined* at each time $\tau$ for $-\infty \leq \mathbf{r}, t \leq \infty$, but it is not observable; that is, it is not *realized* in the observers brain and is principally not accessible to the observers five senses.

However, an interaction between the observer and the system leads to the effect that only a selected part $\Psi^*(\mathbf{r}, t = \tau)\Psi(\mathbf{r}, t = \tau)$ is projected onto the observers' $(\mathbf{r}, t)$−space and not the entire information $\Psi(\mathbf{r}, t)$ $[\Psi^*(\mathbf{r}, t)\Psi(\mathbf{r}, t)]$ . Only this part is recognizable for the observer and only this part is accessible to the observers five senses.

In the case of $N$ observers, each of them is able to observe at each time $\tau$ the complete system with $\Psi(\mathbf{r}, t = \tau)$ if $N$ is not infinity. The reason is simple: In the infinitesimal interval $\Delta\tau$ we have an infinite number of events with respect to $\Psi(\mathbf{r}, t = \tau)$ $[\Psi^*(\mathbf{r}, t = \tau)\Psi(\mathbf{r}, t = \tau)]$ and, therefore, also each of the $N$ observers registers an infinite number of events within $\Delta\tau$ because we have $\infty/N = \infty$ for $N \neq \infty$.

Measurements and the experience of other physical effects, in particular, in connection with the human body can only come into existence if there is a relation to ($\mathbf{r}$, $t$)-space. (See in particular [12].) Real effects are not observable if they have no connection to $(\mathbf{r}, t)$-space. For a human observer only statements in connection with observations and measurements are possible within the frame of space and time.

A picture in space and time only exists for an observer at $t = \tau$ where we have of course to consider that the reference time $t$ in $\Psi_{ref}(\tau - t)$ is uncertain, and this because the function $\Psi^*_{ref}(\tau - t)\Psi_{ref}(\tau - t)$ has the width $\Delta_\tau$ which should however be much smaller than the width $\Delta_t$ of $\Psi^*(\mathbf{r}, t)\Psi(\mathbf{r}, t)$: $\Delta_\tau \ll \Delta_t$.

We have seen in Section 3.16.1 that within the interval $\Delta\tau \neq 0$ (where $\Delta\tau$ can be infinitesimal) an infinite number of $t$-values of $\Psi^*(\mathbf{r}, t)\Psi(\mathbf{r}, t)$ are occupied within the range $\Delta_t$ of $\Psi^*(\mathbf{r}, t)\Psi(\mathbf{r}, t)$ with $\Delta_t \neq \infty$ (Section 3.16.1). That is, the entire history (the complete past and future) of the system, described by $\Psi^*(\mathbf{r}, t)\Psi(\mathbf{r}, t)$ having the width $\Delta_t$, is given within the very small time interval $\Delta\tau \neq 0$ which can in principle be distinctly smaller than the width $\Delta_\tau$ of the function $\Psi^*_{ref}(\tau - t)\Psi_{ref}(\tau - t)$: $\Delta\tau \ll \Delta_\tau$. But this means that we have also an infinite number of $t$-values within the section described by

$$\Psi^*(\mathbf{r}, t)\Psi(\mathbf{r}, t)|_{\Delta\tau}, \quad \tau \leqslant t \leqslant \tau + \Delta\tau. \tag{3.89}$$

(If the interval $\Delta_t$ contains an infinite number of events, the fractions $\Delta_\tau/\Delta_t$ and $\Delta\tau/\Delta_t$ must also contain an infinite number of events.) In summary, we observe a mean **r**-configuration, defined by the **r**-configurations $\Psi^*(\mathbf{r}, t)\Psi(\mathbf{r}, t)$ within $\Delta\tau$ and $\Delta_\tau$, and each mean **r**-configuration is formed by an infinite number of values (events).

### 3.16.4. *Reality Outside*

The selected structure $\Psi(\mathbf{r}, t = \tau)$ $[\Psi^*(\mathbf{r}, t = \tau)\Psi(\mathbf{r}, t = \tau)]$ appears at time $\tau$ spontaneously (unconsciously) in the head of the observer. We principally do not know what is outside. Basic reality is not accessible to the observer. (See also the discussion in Chapter 2.) But we are able to construct a "fictitious reality" which is embedded in $(\mathbf{p}, E)$-space, that is, space outside is different from that in the head of the observer $[(\mathbf{r}, t)$-space$]$. This has drastic consequences. For example, geometrical optics describes how a certain geometrical structure (a tree etc.) is transferred from reality outside into the observers head. However, in this case it is assumed that reality outside is also embedded in $(\mathbf{r}, t)$-space, and this is not true from the point of view of projection theory since reality is embedded in $(\mathbf{p}, E)$-space and not in $(\mathbf{r}, t)$-space. Reality outside, described in $(\mathbf{p}, E)$-space, is projected onto $(\mathbf{r}, t)$-space. From this point of view, geometrical optics can merely be considered as an auxiliary method.

We have stated above that the time structure $\Psi(\mathbf{r}, t)$ $[\Psi^*(\mathbf{r}, t)\Psi(\mathbf{r}, t)]$ is systematically scanned by $\Psi(\mathbf{r}, t = \tau)$ $[\Psi^*(\mathbf{r}, t = \tau)\Psi(\mathbf{r}, t = \tau)$. This means that there must be a coupling between both time structures, and this is because both systems interact with each other.

### 3.16.5. *Constancy Phenomena*

The constancy of the shape of the function $\Psi^*_{\text{ref}}(\tau - t)\Psi_{\text{ref}}(\tau - t)$ is needed for a reliable observation of the world outside. This has to be managed by the physiological apparatus of the observer. It protects the perceptual functions against perturbations, in particular those from the outside world (environment). But this kind of protection is not a new phenomenon in connection with the perceptual functions of the observer's physiological apparatus. Various constancy phenomena are known in this respect; Konrad Lorenz state [18]:

> *"Of special interest to the scientist striving for objectivation is the study of those perceptual functions which convey to us the experience of qualities constantly inherent in certain things in our environment. If, of course, we perceive a certain object (say a sheet of paper) as "white", even when different coloured lights, reflecting different wavelengths, are thrown on it; this so-called constancy phenomenon is achieved by the function of a highly complex physiological apparatus which computes, from the colour of the illumination and the colour reflected, the object's constantly inherent property which we call its colour. ... Other neural mechanisms enable us to see that an object which we observe from various sides retains one and the same shape even though the image on our retina assumes a great variety of forms. Other mechanisms make it possible for us to apprehend that an object we observe from various distances remains the same size, although the size of the retina image decreases with distance."*

### 3.16.6. *Information in Connection with Space–Time Points*

Furthermore, we can state the following. (See also Section 3.16.1.) Each space–time point $(\mathbf{r}, t)$ contains, in a certain sense, the total information about the system. This is because we need, as a consequence of the Fourier transform (3.10), for the determination of $\Psi(\mathbf{r}_k, t_k)$ at space–time point $(\mathbf{r}_k, t_k)$ the whole information $\Psi(\mathbf{p}, E)$ about the system in $(\mathbf{p}, E)$-space (Section 3.3). In principle, we could measure at any space–time point $(\mathbf{r}_k, t_k)$ the complete distribution $\Psi^*(\mathbf{p}, E)\Psi(\mathbf{p}, E)$ within a certain time interval $\Delta\tau$, and $\Delta\tau$ can be infinitesimal but has to be different from zero. If we would be able to determine from these data, the wavefunction $\Psi(\mathbf{p}, E)$ we could, on the other hand, determine the complete wavefunction $\Psi(\mathbf{r}, t)$ for $-\infty \leq \mathbf{r} \leq \infty$, $-\infty \leq t \leq \infty$. That is, any space–time point $(\mathbf{r}_k, t_k)$ contains (within a certain infinitesimal time interval $\Delta\tau$) also the information of all the other space–time points for which $\Psi(\mathbf{r}, t)$ is defined. We come to the following scheme:

$$\begin{array}{r} \Psi^*(\mathbf{r}_k, t_k)\Psi(\mathbf{r}_k, t_k) \rightarrow \Psi^*(\mathbf{p}, E)\Psi(\mathbf{p}, E) \\ \downarrow \\ \Psi(\mathbf{p}, E) \rightarrow \Psi(\mathbf{r}, t), \\ -\infty \leq \mathbf{r} \leq \infty, \quad -\infty \leq t \leq \infty. \end{array} \quad (3.90)$$

Let us analyze this point in more detail. (See also Sections 3.2.2 and 3.16.1.)

1. At a certain space–time point, say $\mathbf{r}_k$ and $t_k$, the measurement of one of the possible values for $\mathbf{p}$ and for $E$ is done in the intervals $\mathbf{r}_k, \mathbf{r}_k + \Delta\mathbf{r}_k$ and $t_k, t_k + \Delta t_k$ with the probability density of $\Psi^*(\mathbf{r}_k, t_k)\Psi(\mathbf{r}_k, t_k)$, where $\Delta V_k = \Delta x_k \Delta y_k \Delta z_k \Delta t_k$ can be infinitesimal but different from zero.
2. Let $\Delta_\mathbf{r}$, $\Delta_t$, $\Delta_\mathbf{p}$, $\Delta_E$ be the ranges of $\mathbf{r}$, $t$, $\mathbf{p}$ and $E$ for which the distributions $\Psi^*(\mathbf{r}, t)\Psi(\mathbf{r}, t)$ and $\Psi^*(\mathbf{p}, E)\Psi(\mathbf{p}, E)$ are not zero, and let us assume that the ranges $\Delta_\mathbf{r}$, $\Delta_t$, $\Delta_\mathbf{p}$, $\Delta_E$ are different from infinity. How many values $\mathbf{r}$, $t$, $\mathbf{p}$ and $E$ can come into existence in the time interval $\Delta\tau = \varepsilon$, where $\varepsilon$ is infinitesimal but different from zero? (Here existence does not mean observation but definition.) This number of values for each variable is identical with the number of $\tau$-values in the interval $\varepsilon$, and this number is given by the number of all real numbers in $\varepsilon$. We already pointed out above that we know from mathematics that the number of real numbers within $\varepsilon$ must be infinity, and in physics none of these real numbers are excluded. Therefore, within the infinitesimal interval $\varepsilon$, the number $N$ of possible events with respect to a variable (as, for example $\mathbf{r}$) is

also infinite in the interval $\Delta_{\mathbf{r}}$. Thus, the number density $\rho = N/\Delta_{\mathbf{r}}$ of $\mathbf{r}$-values is also infinite in the interval $\Delta_{\mathbf{r}}$, where the interval $\Delta_{\mathbf{r}}$ can take any value but must be different from infinity.

3. Thus, within the interval $\Delta V_k = \Delta x_k \Delta y_k \Delta z_k \Delta t_k$ in connection with the function $\Psi^*(\mathbf{r}_k, t_k)\Psi(\mathbf{r}_k, t_k)$, we also have an infinite number of events because $\Delta_{\mathbf{r}}\Delta_t/\Delta V_k$ remains finite and is not infinity.
4. Conclusion: within the infinitesimal interval $\Delta V_k = \Delta x_k \Delta y_k \Delta z_k \Delta t_k$ around any space–time position, say $\mathbf{r}_k$ and $t_k$, the complete information about the system $[\Psi^*(\mathbf{p}, E)\Psi(\mathbf{p}, E)$ with $-\infty < \mathbf{p} < \infty, -\infty < E < \infty]$ is given on the basis of $\Psi^*(\mathbf{r}_k, t_k)\Psi(\mathbf{r}_k, t_k)$ to any degree of accuracy within the infinitesimal intervals $\Delta V_k = \Delta x_k \Delta y_k \Delta z_k \Delta t_k$ and $\Delta \tau = \varepsilon$. If we are able to determine the wavefunction $\Psi(\mathbf{p}, E)$ from $\Psi^*(\mathbf{p}, E)\Psi(\mathbf{p}, E)$ (eventually with the help of models) we could determine the complete wavefunction $\Psi(\mathbf{r}, t)$ for the complete region $-\infty \leq \mathbf{r} \leq \infty, -\infty \leq t \leq \infty$.

In summary, at any space–time point $(\mathbf{r}_k, t_k)$ the complete information about the system is determined. This is because projection theory is inherently non-local in character. (See also Section 3.4.) Within projection theory the entire information about a certain system (world) is enclosed in each space–time point $(\mathbf{r}, t)$. This has the following consequence. If the system with $\Psi^*(\mathbf{r}, t)\Psi(\mathbf{r}, t)$ is changed at a certain space–time point $(\mathbf{r}_k, t_k)$ with $\tau = t_k$, not only $\Psi^*(\mathbf{r}_k, t_k)\Psi(\mathbf{r}_k, t_k)$ is changed but also the entire function $\Psi^*(\mathbf{r}, t)\Psi(\mathbf{r}, t)$ within the whole region $-\infty \leq \mathbf{r} \leq \infty, -\infty \leq t \leq \infty$. That is, a local cause has a non-local effect.

### 3.16.7. *Time within Special Theory of Relativity*

Block universe

John Wheelers statement [19] "*Time is nature's way of keeping everything from happening at once*" is very close to that we obtained here ([12], Section 3.16.1) about the nature of time. Wheeler's statement is an interpretation of the situation given within Special Theory of Relativity. This is well summarized in [19]:

*"Minkowski's 4D space–time is often referred to as the block universe model. Once time is treated like a fourth dimension of space we can imagine the whole of space and time modeled as a four-dimensional block.... Here we have a view of the totality of existence in which the whole of time — past, present and future — is laid out frozen before us. Many physicists, including Einstein later in his life, pushed this model to its logical conclusion. In 4D space–time, nothing ever moves. All events which have ever happened or will ever happen exist together in the block universe and there is no distinction between past and future. This implies that nothing unexpected can ever happen. Not only is the future preordained but it is already out there and is as unalterably fixed as the past.*

*Is this picture really necessary? After all, we can just as easily imagine a Newtonian space–time modeled as a 4D block. The difference is that in that case space and time are independent of each other, whereas in relativity the two are linked. One of the consequences of relativity is that no two observers will be able to agree on when 'now' is. By abandoning absolute time we must also admit that the notion of a universal present moment does not exist. For one observer, all events in the Universe that appear to be simultaneous can be linked together to form a certain cross-sectional slice through space–time which that observer calls 'now'. But another observer, moving relative to the first, will have a different slice that will cross the first. Some events that lie on the first observer's 'now' slice will be in the second observer's past while others will be in his future. This mind-boggling result is known as the relativity of simultaneity, and is the reason why many physicists have argued that since there is no absolute division between past and future then there can be no passage of time, since we cannot agree on where the present should be.*

*Worse than that, if one observer sees an event A occur before an event B, then it is possible for another observer to witness B before A. If two observers cannot even agree on the order that things happen, how can we ever define an objective passage of time as a sequence of events?"* [19]

This behavior is also reflected in $\Psi^*(\mathbf{r},t)\Psi(\mathbf{r},t)$ because the range of the time interval $\Delta_t$ is *statistically* occupied in the course of time $\tau$ and we cannot agree on where the present should be, just as in the case of the block universe of Special Theory of Relativity. Without the reference time structure $\Psi^*_{\text{ref}}(\tau - t)\Psi_{\text{ref}}(\tau - t)$ (Section 3.16.2), which does not exist in Special Theory of Relativity, we would not be able to distinguish between past, present and future. Also the statement "in 4D space–time nothing ever moves" is in accordance with that what we have pointed out above in connection with projection theory: $\Psi^*(\mathbf{r},t)\Psi(\mathbf{r},t)$ is a static function and does not change; the general Eq. (3.22) just describes this situation. (However, in principle $\Psi^*(\mathbf{r},t)\Psi(\mathbf{r},t)$ could also vary with time $\tau$ [12], Section 3.7). Furthermore, the statement "Not only is the future preordained but is already out there and is as unalterably fixed in the past" is also in entire accordance with that we have outlined above about the nature of time.

There is however an essential difference between the result of Special Theory of Relativity and the investigation given here. While the block universe of Special Theory of Relativity is valid for the whole universe and cannot be influenced by the observer, $\Psi^*(\mathbf{r},t)\Psi(\mathbf{r},t)$ reflects only a certain system and can definitely be influenced by an observer (for example, to put the system at time $\tau_I$ into another environment). Such an influence means that the interaction between the system and its environment is changed and this is described by a new wavefunction, say $\Phi(\mathbf{r},t)$. In other words, instead of $\Psi^*(\mathbf{r},t)\Psi(\mathbf{r},t)$ we have now $\Phi^*(\mathbf{r},t)\Phi(\mathbf{r},t)$ for times $\tau > \tau_I$.

As in the case of $\Psi^*(\mathbf{r},t)\Psi(\mathbf{r},t)$, also in connection with $\Phi^*(\mathbf{r},t)\Phi(\mathbf{r},t)$ the following is valid. For certain times $\tau > \tau_I$ only one value of the system-specific time is realized with a certain probability. However, if we consider an infinitesimal time interval $\Delta\tau = \varepsilon$ an infinite number of $t$-values are occupied. In other words, the whole history (the complete past and future) of the system, described by the range $\Delta_t$ (life-time) of the distribution $\Phi^*(\mathbf{r},t)\Phi(\mathbf{r},t)$, is given within the infinitesimal time interval $\Delta\tau = \varepsilon$ of our clocks. This is the case for any $\tau_i \pm \Delta\tau$, that is, the law $\Phi^*(\mathbf{r},t)\Phi(\mathbf{r},t)$ is independent of $\tau > \tau_I$. It is stationary with respect to time $\tau > \tau_I$ (see also [12], Section

3.7). Thus, also here we may say the following. Despite the statistical fluctuations, the whole of time $t$-its past, present and future — is laid out frozen before us. But the past and the future of $\Phi^*(\mathbf{r}, t)\Phi(\mathbf{r}, t)$ have nothing to do with the past and the future of $\Psi^*(\mathbf{r}, t)\Psi(\mathbf{r}, t)$.

## Feynman diagrams

The need of a reference time also within the block universe of Special Theory of Relativity has been discussed very clearly by Kenneth Ford within the frame of elementary particle physics. Here certain particles can go backward in time and we obtain time-reversed trajectories. Ford wrote [20]:

> "*Unfortunately, no one knows how to test the theory of time-reversed trajectories experimentally. It must be accepted (if at all) for the symmetry it introduces into our picture of the world and for the simplicity it introduces into the description of antiparticles. To see that the direction of motion through time cannot be measured, imagine yourself a sub-microscopic observer of the scene depicted in Fig. 3.16. If you put a ruler horizontally across the bottom of the diagram and then push it slowly upward across the diagram, the intersections of the world line with the moving ruler edge give a rough history of your observations. The point is that it is the ruler edge of your observation which is moving in time, not the world lines themselves. The particle world lines may be regarded as perfectly static, simply there, painted in space–time like lines on a map. To a creature capable of comprehending the whole span of time as we comprehend the span of space, the activity of annihilation and creation represented in this diagram is not activity at all. It is a stationary display, a picture painted in space and time. It is the fact that man, the observer, can comprehend only one instant of time at each moment that converts the stationary display into motion and activity. At one time, the human observer sees an electron in one place. At a later time, he sees it at a different place. It is natural for him to believe that the electron, like the man, has moved forward in time to get from the first place to the second place. But there is no essential reason to*

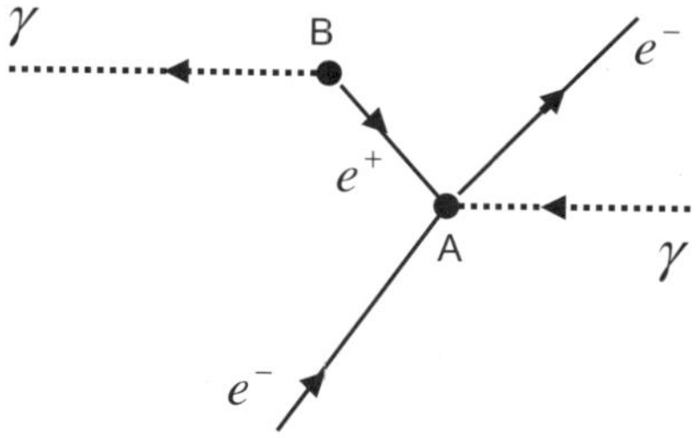


Fig. 3.16. Feynman diagram for photon–electron scattering. "*According to the normal view of time unrolling one direction, we start at the bottom of the diagram and read up. First, an electron and a photon are approaching each other. At vertex A, the photon creates an electron–positron pair. The new electron flies away, while the positron collides with the first electron at vertex B. There they undergo mutual annihilation and a new photon is born. The alternative view, which Feynman showed to be also consistent, is to picture the first electron proceeding to point B, where it emits a photon and reverses its path through time. It 'then' travels to point A where it absorbs the incoming photon and once again reverses its course through time, flying off in the 'right' direction. Either view is permissible and logically consistent.*" [20]

> *believe this. We know only that the electron world line has traced out a certain path in space–time, but we have no possible way of knowing in what order the points of that path were traced out, nor even whether it makes sense to speak of an order or a direction of tracing the world line.*" [20]

Here the role of the distribution $\Psi^*(\mathbf{r},t)\Psi(\mathbf{r},t)$ must be identified with the role of the Feynman diagram (Fig. 3.16), also the Feynman diagram should be considered as static situation in analogy to the distribution $\Psi^*(\mathbf{r},t)\Psi(\mathbf{r},t)$. On the other hand, the ruler edge exactly plays the role of the reference time $\tau$ described by $\Psi^*_{\text{ref}}(\tau - t)\Psi_{\text{ref}}(\tau - t)$.

### $\tau$-dependent systems

In the case of $E = \mathbf{p}^2/(2m_0) + U(x, y, z, \tau)$, our basic eq. (3.22), which we have used so far, becomes non-stationary with respect to time $\tau$. If the reference time structure $\Psi^*_{\text{ref}}(\tau - t)\Psi_{\text{ref}}(\tau - t)$ is a sufficiently sharp function (see in particular Section 3.16.2), we get Eq. (3.26) for the determination of $\Psi(\mathbf{r},t)_\tau$. In other words, if $\Psi(\mathbf{r},t)$ varies with time $\tau$ we have $\Psi(\mathbf{r},t)_\tau = \Psi(\mathbf{r},t,\tau)$ and the whole system (reality) changes with respect to the variables $\mathbf{r}$ and $t$. Let $\Psi^*(\mathbf{r},t,\tau_1)\Psi(\mathbf{r},t,\tau_1)$ be the

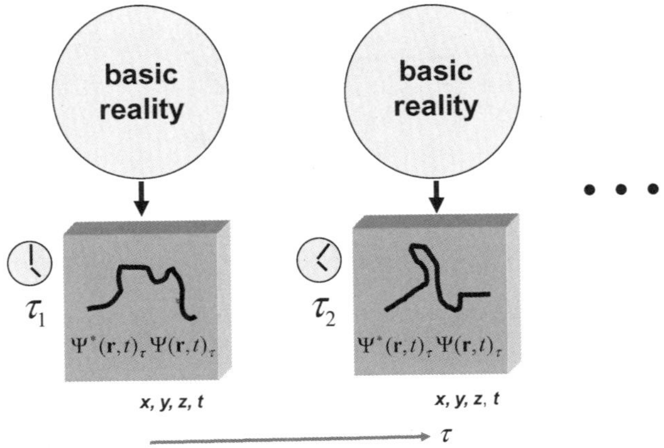


Fig. 3.17. In the case of $\tau$-dependent systems (systems, which are not stationary with respect to $\tau$) the wavefunction varies in the course of time $\tau$. That is, instead of $\Psi(\mathbf{r},t)$ we get $\Psi(\mathbf{r},t)_\tau$ and $\Psi^*(\mathbf{r},t)_\tau\Psi(\mathbf{r},t)_\tau$, respectively. In contrast to Fig. 3.13(b), the space–time structure at $\tau_1$ is different from that at time $\tau_2$.

configuration at time $\tau_1$ and $\Psi^*(\mathbf{r},t,\tau_2)\Psi(\mathbf{r},t,\tau_2)$ be the configuration at time $\tau_2$, then the whole space–time structure $(-\infty \le \mathbf{r} \le \infty,\ -\infty \le t \le \infty)$ is changed by the transition from $\tau_1$ to $\tau_2$. This is shown in Fig. 3.17.

## 3.17. Time Travel in Physics

In this section we would like to discuss the possibility for time travel within projection theory. Within projection theory, there is the theoretical possibility for time travel without using the elements of the Theory of Relativity. Due to the relationship between reference time $\tau$ and system-specific time $t$, the principal potentiality for time travel is given, i.e., for time travel to the future as well as to the past.

In Newton's cosmos, time travel was not conceivable and was exclusively a topic of science fiction. However, by the development of the Theory of Relativity time travel becomes principally possible on the basis of realistic physical laws. Before we discuss time travel within projection theory, let us first quote some basic features in connection with time travel, which are typical in the Theory of Relativity.

### 3.17.1. *Time Travel to the Past within the Theory of Relativity*

Scientists already know that time travel to the future is permitted in the Theory of Relativity, but time travel to the past is still under investigation and is obviously a problematic topic because certain contradictions can appear, which has not yet been solved satisfactorily within the Theory of Relativity.

Within the *Special Theory of Relativity*, only time travel to the future is possible because we cannot move faster than the velocity of light. With increasing speed the clocks are slowing down and time is stopped if the velocity of light is reached. For velocities larger than that of light, time (clocks) could go back but, unfortunately, we cannot move faster than the velocity of light and, therefore, time travel to the past is not possible within the frame of the Special Theory of Relativity. However, in the case of the *General Theory of Relativity*, time travel to the past become principally possible under certain conditions. Within this theory we have curved space–time structures, and certain curvatures permit short-cuts through space–time, allowing the observer to overhaul in a certain sense a light beam, and time travel to the past becomes possible. Here we only want to discuss time travel to the past because within the description of this type of time travel contradictions can appear, and the challenge is to overcome the problems.

#### 3.17.1.1. *Paradoxes*

A travel to the past can lead to paradoxes, and this seems to be a serious problem. For example, a time traveller to the past can kill his grandmother before she gave birth to his mother. However, this action cannot be possible for the following reason. If the time traveller would kill his grandmother, then his mother would have never been born and, therefore, the time traveller could not come into existence. Thus, if he was never born, he would never been able to travel to the past in order to kill his grandmother. In other words, we are confronted with a paradox, the so-called grandmother paradox. This paradox is often considered as sufficient to rule out that time travel to the past is possible within the frame work of Theory of Relativity.

What does that mean? The ruling out of time travel to the past would be against the laws of Theory of Relativity because this theoretical possibility is without doubt anchored in this theory. Thus, we might take the position that Theory of Relativity can lead to unrealistic results. Why? The answer could be relatively simple. As we will outline in Chapter 5, even in General Theory of Relativity, Mach's principle is not fulfilled, that is, within this theory material objects can be influenced by the elements of space and time, and this is strictly forbidden when we take Mach's principle as a relevant condition. Furthermore, within General Theory of Relativity, the notion "absolute space" is not completed eliminated. However, the principal possibility for the existence of an absolute space reflects without any doubt an unphysical situation, as we will still investigate closely in Chapter 5. In conclusion, all these points indicate that the Theory of Relativity should not be considered as a complete framework but it should be assumed that it is open for improvements. It is however questionable whether such improvements can really solve the grandmother paradox.

The grandmother paradox is a typical example for certain contradictions within the Theory of Relativity. It particularly shows that within this theory, time travel to the past can lead to contradictions. But "physical solutions" for such paradoxes and contradictions, respectively, exist but scientists cannot say what approach (physical solution for the grandmother paradox) can be considered as correct.

However, the tools for these "physical solutions" do not belong to the Theory of Relativity, i.e., these paradoxes cannot be solved by the means of the theory (Theory of Relativity) itself, which predicts such time travel to the past. In other words, the Theory of Relativity itself cannot deliver the physical elements for the elimination of the paradoxes that come into existence by the theory. On the other hand, the theoretical framework would certainly be an ill construction if certain paradoxes are admitted by parts of the theory, but have to be eliminated by other parts of the same theory. Instead, the complete situation has to be based on a theory of a piece.

The physical solutions discussed in literature for the elimination of the contradictions are additional tools and physical elements, respectively, that is, they do not belong to the Theory of Relativity. This cannot be considered as a satisfactory situation, and most of the members of the scientific community know that.

### 3.17.1.2. *Shaking Hands with the Younger Self*

Let us discuss another example. A time traveller may also go back to the past and may shake hands with himself. That is, the older man meets his younger self. In this case the world line of the observer loops back in time, where it intersects itself (Fig. 3.18). But what if the older man kills his younger self? Also here a paradox appears similar to the grandmother paradox.

J. Richard Gott wrote in connection with this example [21]: "*A time traveller who visits the past is just someone whose world line somehow loops back, where it could even intersect itself. This would allow the time traveller to shake hands with himself. The older man could meet up with his younger self and say, 'Hi! I'm your future self! I've travelled back in time to say hello!' … The surprised younger man would reply, 'Really?' He would then continue his life, becoming old and eventually looping back to that same event — where he would recognize his younger self, shake hands, and say, 'Hi! I'm your future self! I've travelled back in time to say hello!' … But what, as an older man, the time traveller refuses to say hello and instead simply kills his younger self? Time travel to the past suggests such a paradox.*"

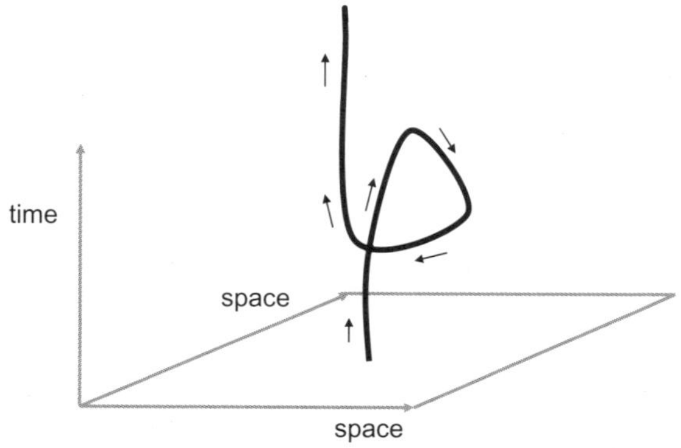


Fig. 3.18. An older man meets its younger self, and the world line intersects itself.

### 3.17.2. *Physical Solutions*

In connection with the grandmother paradox, there are certain physical solutions discussed in literature for overcoming the contradictions. Two of them are of particular interest. The "many-worlds theory" and the "argument of self-consistency". Let us briefly discuss both proposals.

Since the many-worlds theory seems to be of particular relevance in connection with time travel to the past, let us first give some general remarks on the interpretations of conventional quantum theory.

#### 3.17.2.1. *Two Relevant Interpretations of Conventional Quantum Theory*

Within conventional quantum theory there is an immense diversity of opinion and a huge variety of interpretations, and the many-worlds theory is one of them. Most interpretations lead to completely different conceptions of the world. For example, the *Copenhagen Interpretation* is quite different from the *many-worlds theory*.

##### Brief remarks on the copenhagen interpretation

According to the standard Copenhagen interpretation, the world (or any system) consists of options which are equally unreal. By the act of observation, a system is forced to select one of its options and this becomes real, that is, with the Copenhagen interpretation reality is produced by the act of observation so that any system (for example, an electron) cannot be thought of as having an independent existence. We know nothing about what it is doing when we are not looking at it. Within the Copenhagen interpretation, nothing is real unless we look at it. As soon as we stop looking, it ceases to be real.

Before an observation (measurement), the properties of a quantum system are not independent realities, and it is meaningless to talk about physical reality. They are not "either/or" alternative worlds, but there is a superposition of all possible properties (worlds). The effect of observation (measurement) is to chop the overlapping worlds apart into disconnected alternative worlds. This chopping process is called the "collapse of the wavefunction". The collapse of the wavefunction and the assignment of the statistical weight cannot be described by Schrödinger's equation of conventional quantum theory, although it belongs to the basics of this theory. That is, the collapse of the wave function and the assignment of the statistical weight are not explained within this picture "*they are consequences of an external a priori metaphysics, which is allowed to intervene at this point and suspend the Schrödinger equation.*" [16]

Remark

Usual quantum theory is based on the superposition principle which has to be assumed as correct because this theory is basically linear. The collapse of the wavefunction is the weak point within the Copenhagen interpretation. In order to avoid the collapse of the wavefunction, the many-worlds theory has been introduced. In fact, it is possible to assume within the many-worlds interpretation that the superposition never collapses and, therefore, we have no problems with the description of the observation process, that is, Schrödinger's equation of usual quantum theory is not suspended during the observation process. In the next section some relevant details for the many-worlds interpretation are given, which are in particular useful for the analysis of time travel to the past.

Brief remarks on the many-worlds theory

Within the many-worlds theory, there is not just one world, but as many alternative real worlds as options exist within the Copenhagen interpretation. "*What happens when we make a measurement at the quantum level that we are forced by the process of observation to select one of these alternatives, which becomes part of what we see as the real world; the act of observation cuts the ties that bind alternative realities together, and allows them to go on their own separate ways through superspace, each alternative reality containing its own (real) observer who has made the same observation but has a different quantum answer.*" [22]

In other words, within the many-worlds theory there is a real splitting of the world into many real worlds. For example, within the many-worlds theory both cats of Schrödinger's famous thought experiment are real: There is a real world with an observer and a box with a live cat in it, and there is another real world with an observer and a dead cat in the box.

How convincing is the many-worlds theory? It must be emphasized once more that the many-worlds theory does not follow directly from the basics of usual quantum theory but is merely one of the possible interpretations. So far, the many-worlds theory is not a compelling outcome of usual quantum theory. Thus, this concept is to be used with caution.

### Critical remarks

The reason for the problems in connection with the Copenhagen interpretation is the linearity of conventional quantum theory and not the kind of interpretation. Therefore, the goal should not be to find an alternative interpretation to the Copenhagen interpretation (e.g., the many-worlds theory), but to find a non-linear theory. (The projection theory is a nonlinear theory without superposition principle.) Thus, from this point of view the many-world theory should be considered as an embarrassment solution.

### Using the many-worlds-theory for the analysis of time travel to the past

In conclusion, within the many-worlds theory there is not only one real world but many, which exist simultaneously. This scenario is obviously suitable for solving the grandmother paradox. That is, in principle it could be possible that a time traveller goes back in time in order to kill his grandmother when she was a young girl. Contradictions are avoided because in this case the universe is caused to branch onto a different track (world), and this additional world contains the dead grandmother and the time traveller. The world with the grandmother and her daughter who gave birth to the time traveller still exist. There are parallel worlds that exist simultaneously.

It is outlined in [21] that only on the basis of the many-worlds theory we may admit that a time traveller can perform changes in the past, which would also be possible for him without time travel to the past (for example, to kill his grandmother). In this connection J. Richard Gott wrote [21]:

> "*To someone hoping to find a time machine in order to go to the past to save a lost loved one, the most comforting thing I can say is that as far as we understand today, this can only be accomplished if the many-world theory of quantum mechanics is true. If that is true, then there is already a parallel universe in which your loved one is okay now. That's because all the possible universes exist. Unfortunately, you are just in the wrong one.*"

However, the solution of the grandmother paradox is in this case not based on the theoretical elements of the theory, which predicts time travel, i.e., General Theory of Relativity is not able to solve this problem and needs a foreign tool (many-worlds theory) that does not follow from it. Strictly speaking, from the point of view of the General Theory of Relativity, as independent and closed physical frame, the many-worlds theory represents a metaphysical element. Both theories (Theory of Relativity and conventional quantum theory) have been developed independently from each other.

Furthermore, the many-worlds interpretation is a quantum picture and the following principal question arise. Can we connect without difficulties a classical framework (General Theory of Relativity) with a quantum-theoretical picture (many-worlds interpretation)? No, we cannot. We will still discuss in Chapter 5 that these two theories, as they are currently formulated, cannot both be right; they are mutually incompatible. Then, to consider the many-worlds theory as the solution for the grandmother paradox becomes rather questionable.

#### 3.17.2.2. *Argument of Self-Consistency*

The alternative argument for the solution of the grandmother paradox can be expressed as follows: "*Time travellers do not change the past because they were always part of it.*" [21] We have outlined above that a time traveller may shake hands with an earlier version of himself, but only those actions between the older man and his younger self are allowed that lead to a self-consistent solution. Therefore, all actions are forbidden that would lead to contradictions.

In connection with the grandmother paradox, the principle of self-consistency is not violated if the following rule is fulfilled. The time traveller is able to drink a cup of coffee with his grandmother, but he is not able to kill her — otherwise he would not exist, and we know that he is.

In conclusion, it is not allowed that a time traveller may perform all the changes in the past, which are possible for him without time travel to the past. That means the time traveller has lost the ability for killing his grandmother before she gave birth to his mother. This is not a matter of consciousness because the time traveller can meet his grandmother accidentally and he must not know that it is his grandmother. Therefore, the transition from the present to the past is somehow accompanied with a specific biological change of the human observer. In other words, the time travel to the past is coupled to the biological structure, i.e., the time traveller changes his biological structure during the time travel. If the General Theory of Relativity would be a self-consistent framework in this respect it would be able to describe such a relationship between time travel and biological structure, but this is not the case. There is no physical law in the General Theory of Relativity that tells us what is allowed and what not. The above mentioned rule "The time traveller is able to drink a cup of coffee with his grandmother, but he is not able to kill her" follows from the logical point of view (principle of self-consistency) but the theory does not deliver this rule (law).

### 3.17.2.3. *Chronology Protection Conjecture*

Without any doubt, within the General Theory of Relativity there are problems in connection with time travel to the past. Stephen Hawking proposed the so-called "chronology protection conjecture". The contents of this conjecture is that the basic laws of physics always work towards situations in which time travel to the past are prevented. In fact, such a hypothesis would solve all the problems in connection with time travel to the past, i.e., a human would be never able to go back in time and nobody could kill his grandmother, i.e., the grandmother paradox and similar situations would be eliminated if the "chronology protection conjecture" is valid.

However, the "chronology protection conjecture" can obviously not be considered as a general law, and it is therefore disproved. In the case of the so-called Rindler vacuum, the "chronology protection conjecture" is not fulfilled. J. Richard Gott wrote [21]: "*This wrapped Rindler vacuum had zero energy density and pressure throughout the entire Misner space — in both time travel and non-time travel regions — and therefore it solved Einstein's equations exactly. This was a self-consistent solution: the geometry which includes time travel, gave rise to the quantum vacuum properly, and that quantum vacuum state, through Einstein's equations, in turn produced the geometry one started with. This solution provids an important counterexample to the chronology protection conjecture, for it concerns the very example that had helped motivate the conjecture in the first place.*" In conclusion, this counterexample demonstrates that the chronology protection conjecture may be considered as disproved.

### 3.17.3. *Time Travel within Projection Theory*

The paradoxes that that we have discussed in connection with time travel to the past within General Theory of Relativity do not appear in projection theory. Again, within projection theory, time travel is principally possible, at least theoretically, and for the description of this effect we do not need the elements of the Theory of Relativity.

#### 3.17.3.1. *Reference Time Structures for N Observers*

The reference time $\tau$ is an external parameter and is measured with our clocks. What does "external time" mean? $\tau$ has nothing to do with the system under investigation that is characterized by $\Psi(\mathbf{r}, t)$, where $t$ is the system-specific time. That is why $\tau$ can be considered as an external quantity.

However, it turned out that $\tau$ is also connected to a system-specific time, which is characterized by the function $\Psi_{\text{ref}}(\tau - t)$ (Section 3.16.2), and the effects due to $\Psi_{\text{ref}}(\tau - t)$ come into existence through specific brain functions of the observer. The purpose of $\Psi_{\text{ref}}(\tau - t)$ has been discussed in Section 3.16.2. At time $\tau$ the observer selects from $\Psi(\mathbf{r}, t)$ specific configurations $\Psi(\mathbf{r}, t_0)$ with $t_0 = \tau$.

In principle, we could have as many functions $\Psi_{\text{ref}}(\tau - t)$ as there are observers, i.e., each observer could in principle define (produce) its own reference system. Then, in the case of $N$ observers we have

$$\Psi_{\text{ref}}(\tau_i - t)_i, \quad i = 1,\ 2, \ldots, N. \tag{3.91}$$

In principle, each reference time structure could be independent of the others

$$\Psi_{\text{ref}}(\tau_1 - t)_1 \neq \Psi_{\text{ref}}(\tau_2 - t)_2 \neq \cdots \neq \Psi_{\text{ref}}(\tau_N - t)_N. \tag{3.92}$$

However, this would make not much sense because a communication between the various observers would not be possible in the case of non-correlated reference time structures. Therefore, it is reasonable to require that all reference times $\tau_k$, $k = 1, 2, \ldots, N$ are the same:

$$\tau_1 = \tau_2 = \cdots = \tau_N = \tau. \tag{3.93}$$

Then, the form (shape) of the functions $\Psi_{\text{ref}}(\tau_k - t)_k, k = 1, 2, \ldots, N$ should also be the same and, instead of (3.92) and with (3.93) we get

$$\Psi_{\text{ref}}(\tau - t)_1 = \Psi_{\text{ref}}(\tau - t)_2 = \cdots = \Psi_{\text{ref}}(\tau - t)_N = \Psi_{\text{ref}}(\tau - t). \tag{3.94}$$

In conclusion, the reference time $\tau$ and also the reference time structure $\Psi_{\text{ref}}(\tau - t)$ should be the same for all $N$ observers. Otherwise, communication between the observers would not be possible. In particular, co-ordinated actions between the observers would be impossible. The values $\tau_k$, $k = 1, 2, \ldots, N$ must be collectively adjustable leading to Eq. (3.93). We have to search in this direction in order to find an adequate mechanism that explains the property $\tau_1 = \tau_2 = \cdots = \tau_N = \tau$ [Eq. (3.93)]. More details and arguments will be given below.

As we know, the reference time $\tau$ is also used in conventional physics as, for example, in Newton's mechanics. In this connection, it is interesting to outline the qualitative difference between $\tau$ of projection theory and that of Newton's mechanics (conventional physics). In the next section we will briefly discuss this point.

### 3.17.3.2. *Situation in Conventional Physics* (*Newton's Mechanics*)

Within usual physics, everything, sun, moon and stars as well as stones, humans and animals are embedded and arranged, respectively, in space. These spatial structures change in the course of time $\tau$. In other words, we have an abundance of trajectories and curves $s = f(\tau)$, which characterize each of the various objects and which all come into existence by the external time $\tau$, i.e., they all are carried forward by the external time $\tau$. Sun, moon and stars and everything else that exists swim in the stream of time.

This is the "world of the external time $\tau$" as it appears to a human observer, and this concept has been taken over in this form by conventional physics, in particular, by Newton in his formulation of classical mechanics. We cannot make a theory about time $\tau$ within the frame of classical mechanics (conventional physics) because $\tau$ is used here as an input parameter and, this is the reason why it is treated as an external quantity, and we come to the simple and imaginative idea that "everything swims in the stream of time". Concerning external time, in [23] another metaphor that is more detailed is given and it allows the following beautiful comparison [23]:

> "*The time is like the cable of the cable railway of San Francisco. The cable is driven by a far and invisible machine, is however hidden. We know that it moves, because the carriages are connected and are carried with it. Completely similarly we usually see the time in everyday life as a universal, cosmic movement of the present, of the now; they are propelled by natural or divine forces and matter, life, humans and society are connected with them and are propelled and moved along for a while.*"

Thus, as the cable of this train remains invisible and inaudible and which is moved by unseen and inescapable forces, we have to imagine the effect of the external time $\tau$ of conventional physics. "Cable train" or "stream of time" are simple auxiliary elements to make that, which we call "external time", plausible. However, such metaphors cannot explain the physics behind the reference time $\tau$.

Such a time definition is almost metaphysical. We have no other concepts for the external time $\tau$ in classical mechanics and usual physics, respectively, i.e., we have no deeper ideas about that what we call time in conventional physics. Einstein was asked: "What is time?" His answer was: "Time is just that what we measure with our clocks." But this inevitably leads to the conception that "sun, moon and stars and everything else that exists swim in the stream of time" or that "the time is like the cable of the cable railway of San Francisco". However, all these auxiliary constructions have to be considered as rather unphysical.

This kind of external time means somehow that there is a privileged (absolute) process, because none of the admitted physical processes, as they are defined in the context of conventional physics, is able to describe such a time-producing machine. Within projection theory, the situation is different (Section 3.16.2). Here we principally have as many functions $\Psi_{\text{ref}}(\tau_i - t)_i$, $i = 1, 2, \ldots, N$ as there are observers, and the reference times $\tau_i$ and the reference time structure, expressed by $\Psi_{\text{ref}}(\tau_i - t)_i$, come into existence by the brain functions of the observer. However, as we have stated above, all values $\tau_i$ and all functions $\Psi_{\text{ref}}(\tau_i - t)_i$ should be the same and we come to *one* value $\tau$ and to *one* function $\Psi_{\text{ref}}(\tau - t)$. The reference time appears as an external time because it has nothing to do with the system under investigation, which is characterized by $\Psi(\mathbf{r}, t)$ and the system-specific time $t$. In conclusion, $\tau$ and $\Psi_{\text{ref}}(\tau - t)$ are explained within the frame of the theory (projection theory) for which these notions are used. More details (models etc.) are given in [12]. In contrast to conventional physics, within projection theory we do not have to assume that there is a privileged process (an external time machine) for the production of $\tau$. Such a privileged process has to be classified as metaphysical element, that is, it has to be considered as unphysical.

Also, the Theory of Relativity works with the time of Newton's mechanics. However, there is no longer an absolute external parameter but it transforms in gravitational fields, and the time in a moving frame of reference is different from that in a resting frame of reference. In other words, Newton's time is still an element in the theory of relativity but it is transformed depending on the physical situation.

### 3.17.3.3. *Observations and Descriptions*

The world outside becomes conscious to an observer at time $\tau$. This is an experimental fact. We know nothing about the world for times $\tau' > \tau$ and also not for times $\tau' < \tau$. (What we know for $\tau' < \tau$ is merely a memory effect.) The picture of reality in $(\mathbf{r}, t)$-space is always a selected picture. At time $\tau$ the observer selects from $\Psi(\mathbf{r}, t)$ specific configurations $\Psi(\mathbf{r}, t_0)$ with $t_0 = \tau$. This selected information, expressed by $\Psi(\mathbf{r}, t_0)$, is produced by the observer's brain in an "unconscious" way on the basis of $\Psi_{\text{ref}}(\tau - t)$.

"Conscious" states, also inside of an observer, come into play as follows. Conscious actions are only possible on the basis of this unconscious picture of reality, which are expressed at time $\tau$ by $\Psi(\mathbf{r}, t_0)$ and is positioned in $(\mathbf{r}, t)$-space, where the selected information $\Psi(\mathbf{r}, t_0)$ comes into existence by $(\mathbf{p}, E)$-fluctuations in $(\mathbf{p}, E)$-space that are projected on $(\mathbf{r}, t)$-space.

Furthermore, in connection with the term "consciousness", the feature "thinking" comes into play and, on the other hand, thinking can take place on various levels (levels of reality, see Chapter 4 and in particular [8]). Concerning the notion "levels of reality", let us briefly summarize the following.

We cannot make statements about basic reality, but we can describe certain aspects of it within the framework of levels (levels of reality) which are vertically arranged in accordance with the degree of generality. In [8], we called this procedure "principle of level-analysis". In Fig. 3.19, this principle is used for the classification of an observer with and without the possibility of $t$-selections (level 1 and level 2 in Fig. 3.19) but also with the possibility for the theoretical description of the picture of reality (level 3 and level 4 in Fig. 3.19). Here "theoretical description" involves the notions consciousness and thinking. In connection with Fig. 3.19, the system under investigation is the human observer and, instead of $\Psi(\mathbf{r}, t)$, we would like to use the designation $\Psi_{\text{human}}(\mathbf{r}, t)$.

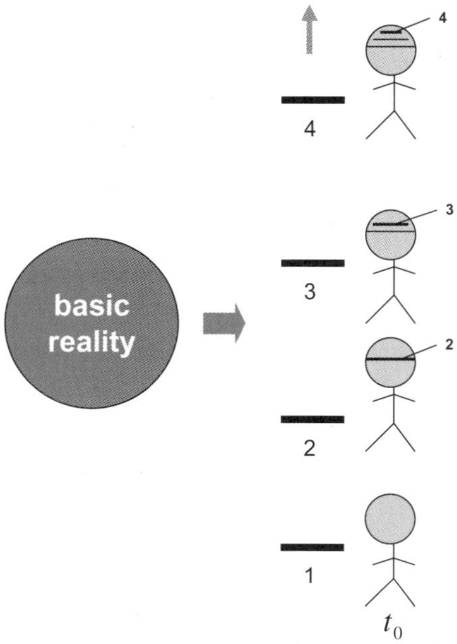


Fig. 3.19. We may describe certain aspects in connection with a human observer on the basis of levels. Here four levels are shown. Level 1: No $t$-selection takes place and the observer has no picture of reality [expressed by $\Psi_{\text{human}}(\mathbf{r},t)$] in his head. Level 2: There are $t$-selections ($\Psi_{\text{ref}}(\tau-t)$ is effective), i.e., at time $\tau$ the observer selects from $\Psi_{\text{human}}(\mathbf{r},t)$ a specific configuration, say $\Psi_{\text{human}}(\mathbf{r},t_0)$ with $\tau \leq t_0 \leq \tau+\Delta\tau$. Level 3: A theory, say $a$, is involved for the description of the picture of reality [again expressed by $\Psi_{\text{human}}(\mathbf{r},t)$] that is unconsciously observed at level 2. Level 4: Here an improved (refined) theory, say $b$, is involved for the description of the same picture of reality. Level 4 is positioned above level 3 because it is assumed that the theory $b$ is more general than theory $a$.

Conscious actions in space and time [$(\mathbf{r},t)$-space] need the process of thinking but also the unconsciously observed picture of reality (Fig. 3.20).

In conclusion, the appearance of "consciousness" and "thinking" is obviously tightly-linked with the appearance of the unconsciously observed "picture of realty" on the basis of $\Psi_{\text{ref}}(\tau-t)$. In other words, consciousness and thinking are activated by $\Psi_{\text{ref}}(\tau-t)$. If the function $\Psi_{\text{ref}}(\tau-t)$ [or better the probability density $\Psi^*_{\text{ref}}(\tau-t)\Psi_{\text{ref}}(\tau-t)$] is between $\tau$ and $\tau+\Delta\tau$ different from zero, consciousness and thinking are not activated for $t<\tau$ and $t>\tau+\Delta\tau$. If we characterize a

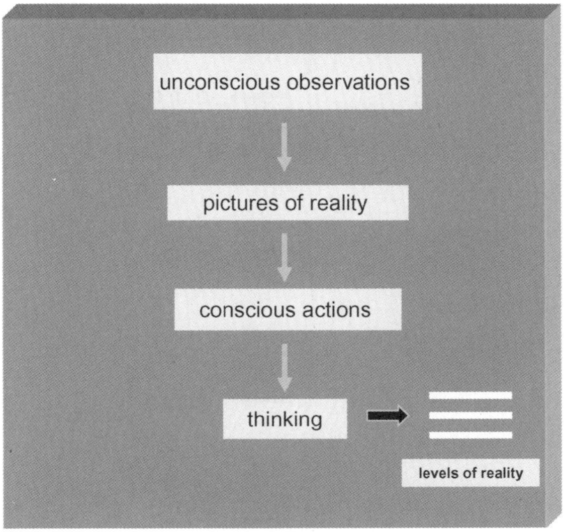


Fig. 3.20. For conscious actions, an observer needs an unconsciously observed picture of reality and he must have the ability for thinking where thinking is a multilayered procedure, that is, we have certain "levels of reality". More details in connection with the notion "levels of reality" are given in Chapter 4.

human by the wavefunction $\Psi_{\text{human}}(\mathbf{r}, t)$, then only the unselected function $\Psi_{\text{human}}(\mathbf{r}, t)$ $[\Psi^*_{\text{human}}(\mathbf{r}, t)\Psi_{\text{human}}(\mathbf{r}, t)]$ of him exits for $t < \tau$ and $t > \tau + \Delta\tau$, i.e., only the material part of the human (Fig. 3.21).

### 3.17.3.4. *Only the Material Part of an Observer is Depictable in $(\mathbf{r}, t)$-Space*

The "elements of thinking" are not depictable in $(\mathbf{r}, t)$-space but only the so-called material objects which come into existence by $(\mathbf{p}, E)$-fluctuations and are projected from $(\mathbf{p}, E)$-space onto $(\mathbf{r}, t)$-space. Clearly, the "products of thinking" can lead to constructions (houses, cars, etc.) that are depictable in $(\mathbf{r}, t)$-space but not the elements of thinking (formulas etc.) which have led to the construction in $(\mathbf{r}, t)$-space. In Chapter 4, we will study this further.

From the principles of evolution we have to conclude that basic reality contains much more information than an observer can depict in space and time (see in particular Chapter 2, Fig. 2.5). This space–time

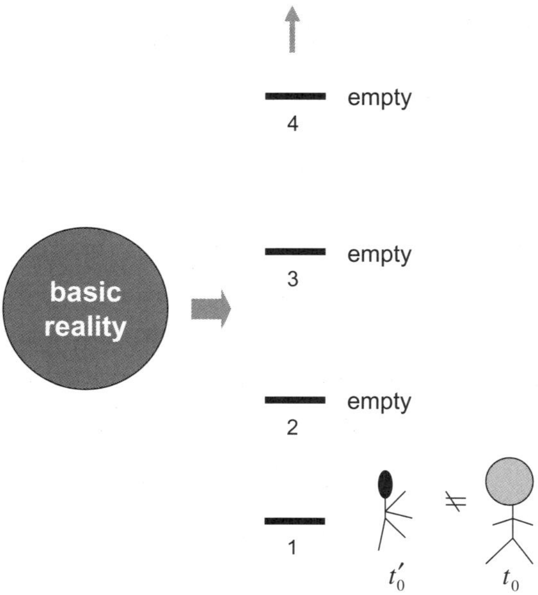


Fig. 3.21. It is assumed that the function $\Psi_{\rm ref}(\tau - t)$ $[\Psi^*_{\rm ref}(\tau - t)\Psi_{\rm ref}(\tau - t)]$ is between $\tau$ and $\tau + \Delta\tau$ different from zero. In the case of $t < \tau$ and $t > \tau + \Delta\tau$, the reference time mechanism cannot be effective, and we obtain for the human observer the following levels: In contrast to Fig. 3.19, levels 2, 3 and 4 must be empty and only the unselected function $\Psi_{\rm human}(\mathbf{r}, t)$ $[\Psi^*_{\rm human}(\mathbf{r}, t)\Psi_{\rm human}(\mathbf{r}, t)]$ of him can exist for $t < \tau$ and $t > \tau + \Delta\tau$, which takes at $t = t_0'$ the form $\Psi_{\rm human}(\mathbf{r}, t_0')$ $[\Psi^*_{\rm human}(\mathbf{r}, t_0')\Psi_{\rm human}(\mathbf{r}, t_0')]$. Since $t_0'$ is different from $t_0$, the structures $\Psi^*_{\rm human}(\mathbf{r}, t_0')\Psi_{\rm human}(\mathbf{r}, t_0')$ and $\Psi^*_{\rm human}(\mathbf{r}, t_0)\Psi_{\rm human}(\mathbf{r}, t_0)$ must be different from each other, that is, the shape of the human observer at $t_0'$ is different from that at $t_0$.

information is mainly that part of the outside world which a human observer needs for survival.

In Chapter 2 we quoted the following: "*The contents of the space–time information, selected from basic reality, is dictated by the principles of evolution. The unselected information, on which the selected picture of reality is based, remains principally unknown. For example, we know the observer's shape in space and time but not his structure in basic reality (reality outside). However, the projection principle also says that the observer must be more than what is pictured in space and time which exclusively shows the material part of the observer. But this material part is only the selected information about the observer, and the unselected information must necessarily be "richer" than the material part which is pictured in space and time (see also Figs. 2.6 and 2.7), that is, we may not conclude that the space–time pictures contain the complete information about the reality outside. In fact, we know that a human observer has mind and feelings; his intellectual abilities are accompanied by logical laws that have to be considered as real extensions of the material world. However, these logical laws etc. are not depictable in space and time."*

### 3.17.3.5. *Two Kinds of Selection*

We have to distinguish between two kinds of selection: 1. A selection from basic reality to the space–time information $\Psi(\mathbf{r}, t)$. 2. At time $\tau$ the observer selects from $\Psi(\mathbf{r}, t)$ specific configurations $\Psi(\mathbf{r}, t_0)$ with $t_0 = \tau$.

Remarks concerning 1

We have a selection with respect to basic reality, i.e., only a part of basic reality is accessible to the observer (Chapter 2), and this selected part is described by $\Psi(\mathbf{r}, t)$ $[\Psi^*(\mathbf{r}, t)\Psi(\mathbf{r}, t)]$. This kind of selection is due to the biological evolution. How this kind of selection is managed remains principally unknown because a human observer can nothing say about basic reality.

Remarks concerning 2

We have a second selection, and this selection refers to the system-specific time $t$, i.e., only a certain part of $\Psi(\mathbf{r}, t)$ $[\Psi^*(\mathbf{r}, t)\Psi(\mathbf{r}, t)]$ becomes observable at time $\tau$, and this kind of selection is managed by the reference time mechanism, which we have expressed by $\Psi_{ref}(\tau - t)$. It is outlined above that consciousness, thinking etc. are obviously activated by $\Psi_{ref}(\tau - t)$.

### 3.17.3.6. *Local Changes*

If the $\mathbf{r}$-structure is changed at $t_0 = \tau$, the picture in $(\mathbf{r}, t)$-space, which is characterized by the probability density $\Psi^*(\mathbf{r}, t_0)\Psi(\mathbf{r}, t_0)$, gets the form $\Phi^*(\mathbf{r}, t_0)\Phi(\mathbf{r}, t_0)$, that is, we have at time $\tau$ the transition

$$\Psi^*(\mathbf{r}, t_0)\Psi(\mathbf{r}, t_0) \rightarrow \Phi^*(\mathbf{r}, t_0)\Phi(\mathbf{r}, t_0), \quad -\infty \leq \mathbf{r} \leq \infty. \tag{3.95}$$

This is of course fulfilled if we replace $\Psi(\mathbf{r}, t_0)$ by $\Phi(\mathbf{r}, t_0)$:

$$\Psi(\mathbf{r}, t_0) \rightarrow \Phi(\mathbf{r}, t_0), \quad -\infty \leq \mathbf{r} \leq \infty. \tag{3.96}$$

Then, instead of

$$\begin{aligned}\Psi(\mathbf{r}, t_0) = \frac{1}{(2\pi\hbar)^2} \int_{-\infty}^{\infty} & \Psi(\mathbf{p}, E) \\ & \times \exp\left\{i\left[\frac{\mathbf{p}}{\hbar} \cdot \mathbf{r} - \frac{E}{\hbar} t_0\right]\right\} dp_x dp_y dp_z dE\end{aligned} \tag{3.97}$$

[see Eq. (3.10)] we obtain

$$\begin{aligned}\Phi(\mathbf{r}, t_0) = \frac{1}{(2\pi\hbar)^2} \int_{-\infty}^{\infty} & \Phi(\mathbf{p}, E) \\ & \times \exp\left\{i\left[\frac{\mathbf{p}}{\hbar} \cdot \mathbf{r} - \frac{E}{\hbar} t_0\right]\right\} dp_x dp_y dp_z dE,\end{aligned} \tag{3.98}$$

that is, we have the transition

$$\Psi(\mathbf{p}, E) \rightarrow \Phi(\mathbf{p}, E), \quad -\infty \leq \mathbf{p}, E \leq \infty. \tag{3.99}$$

Then, we come to the following statement. While in $(\mathbf{r}, t)$-space the $\mathbf{r}$-structure is only changed at $t_0 = \tau$ [transition (3.96)], in $(\mathbf{p}, E)$-space the entire $(\mathbf{p}, E)$-structure is concerned [transition (3.99)].

In other words, the information $\Phi(\mathbf{r}, t_0)$ at $t_0 = \tau$ contains the complete information of the system under investigation, also those for $t_1 \neq t_0$, $t_2 \neq t_0, \ldots$. The reason is simple, because we may apply the same function $\Phi(\mathbf{p}, E)$ [Eq. (3.98)], which we used for the determination of $\Phi(\mathbf{r}, t_0)$ and also for the calculation of the wavefunction $\Phi(\mathbf{r}, t)$ at times $t_1 \neq t_0$, $t_2 \neq t_0, \ldots$ In analogy to

(3.98), we obtain the following set of equations:

$$\Phi(\mathbf{r}, t_1) = \frac{1}{(2\pi\hbar)^2} \int_{-\infty}^{\infty} \Phi(\mathbf{p}, E) \times \exp\left\{i\left[\frac{\mathbf{p}}{\hbar}\cdot\mathbf{r} - \frac{E}{\hbar}t_1\right]\right\} dp_x dp_y dp_z dE, \quad (3.100)$$

$$\Phi(\mathbf{r}, t_2) = \frac{1}{(2\pi\hbar)^2} \int_{-\infty}^{\infty} \Phi(\mathbf{p}, E) \times \exp\left\{i\left[\frac{\mathbf{p}}{\hbar}\cdot\mathbf{r} - \frac{E}{\hbar}t_2\right]\right\} dp_x dp_y dp_z dE, \quad (3.101)$$

$$\vdots$$

with

$$\Phi(\mathbf{r}, t_1) \neq \Psi(\mathbf{r}, t_1), \quad (3.102)$$

$$\Phi(\mathbf{r}, t_2) \neq \Psi(\mathbf{r}, t_2), \quad (3.103)$$

$$\vdots$$

and

$$\Phi^*(\mathbf{r}, t_1)\Phi(\mathbf{r}, t_1) \neq \Psi^*(\mathbf{r}, t_1)\Psi(\mathbf{r}, t_1), \quad (3.104)$$

$$\Phi^*(\mathbf{r}, t_2)\Phi(\mathbf{r}, t_2) \neq \Psi^*(\mathbf{r}, t_2)\Psi(\mathbf{r}, t_2), \quad (3.105)$$

$$\vdots$$

In other words, when we change the **r**-structure at $t_0 = \tau$, all the other possible **r**-structures at $t_1, t_2, \ldots, t_i, \ldots$ are changed simultaneously. This is independent on the position of $t_i$ relative to $t_0 = \tau$, i.e., the **r**-structures are changed for $t_i < t_0 = \tau$ as well as for $t_i > t_0 = \tau$. This must have serious consequences. Let us first briefly outline the situation in conventional physics and after that we will discuss the past, present and future within the framework of projection theory.

Conventional physics

We know from conventional physics that changes in the present influence the future but not the past. We know certain things from the past but this is classified as memory effect within conventional physics.

Projection theory

Within projection theory, we have the following situation: the changes made at $t_0 = \tau$ in the present changes the whole $t$-structure, that is, the past is changed as well as the future. The function $\Phi^*(\mathbf{r}, t)\Phi(\mathbf{r}, t)$ exists for $t < t_0 = \tau$ and for $t < t_0 = \tau$. In fact, from the $\varepsilon$-property (Section 3.16.1) it follows that within an infinitesimal $\tau$-interval $\varepsilon$ the number density $\rho = N/\Delta_t$ of $t$-values is infinite in the interval $\Delta_t$, that is, all $t$-intervals are occupied with an infinite density, if the function $\Phi^*(\mathbf{r}, t)\Phi(\mathbf{r}, t)$ has the finite range $\Delta_t$ with respect to the $t$-variable (and also a finite range $\Delta_\mathbf{r}$ with respect to the $\mathbf{r}$-variable). In other words, the function $\Phi^*(\mathbf{r}, t)\Phi(\mathbf{r}, t)$ is determined within the entire $(\mathbf{r}, t)$-region between $\tau$ and $\tau + \varepsilon$, where the interval $\varepsilon$ can be infinitesimal, i.e., the past, present and future are equally occupied at time $\tau$ (or more precisely, within an infinitesimal interval around time $\tau$).

This is interesting and is a new phenomenon. For $t < t_0 = \tau$, there can be memory effects with respect to $\Psi^*(\mathbf{r}, t)\Psi(\mathbf{r}, t)$ but not in connection with the new structure $\Phi^*(\mathbf{r}, t)\Phi(\mathbf{r}, t)$ which exists for times $\tau' > \tau$ for $-\infty \leq \mathbf{r}, t \leq \infty$. Clearly, the structure $\Psi^*(\mathbf{r}, t)\Psi(\mathbf{r}, t)$ no longer exists for times $\tau' > \tau$ for $-\infty \leq \mathbf{r}, t \leq \infty$.

### 3.17.3.7. *Time Shifts*

We have outlined above that all reference times $\tau_k$, $k = 1, 2, \ldots, N$ for the $N$ observers have to be the identical: $\tau_1 = \tau_2 = \cdots = \tau_N = \tau$ [Eq. (3.93)]. If an individual, say $i$, would be able to shift his personal reference time $\tau_i = \tau_{\text{ind}}$ by a constant term $\alpha$, we have, instead of $\tau_i = \tau$

$$\tau_{\text{ind}} = \tau + \alpha \tag{3.106}$$

with

$$\tau_{\text{ind}} \neq \tau_1, \quad \tau_{\text{ind}} \neq \tau_2, \ldots, \tau_{\text{ind}} \neq \tau_N; \; \tau_{\text{ind}} = \tau_i. \tag{3.107}$$

Consequently, we have

$$\begin{aligned} &\Psi_{\text{ref}}(\tau_{\text{ind}} - t)_{\text{ind}} \neq \Psi_{\text{ref}}(\tau_1 - t)_1, \\ &\Psi_{\text{ref}}(\tau_{\text{ind}} - t)_{\text{ind}} \neq \Psi_{\text{ref}}(\tau_2 - t)_2, \ldots, \\ &\Psi_{\text{ref}}(\tau_{\text{ind}} - t)_{\text{ind}} \neq \Psi_{\text{ref}}(\tau_N - t)_N \end{aligned} \tag{3.108}$$

and

$$\Psi_{\text{ref}}(\tau_{\text{ind}} - t)_{\text{ind}} = \Psi_{\text{ref}}(\tau_i - t)_i. \tag{3.109}$$

The shift of $\tau_i = \tau_{\text{ind}}$ by the constant term $\alpha$ means that there is a time travel with respect to observer $i$ where both, $\tau$ and $\tau_{\text{ind}}$, go strictly from the past to the future. In the case of a positive value for $\alpha$, we have time travel to the future and in the case of negative $\alpha$ there is time travel to the past, that is, we have

$$\alpha \begin{cases} + & \text{time travel to the future,} \\ - & \text{time travel to the past.} \end{cases} \tag{3.110}$$

Let us only discuss time travels to the past. This case is of particular relevance in connection with Theory of Relativity, as we have discussed in Section 3.17.2. The situation with respect to projection theory is represented in Fig. 3.22.

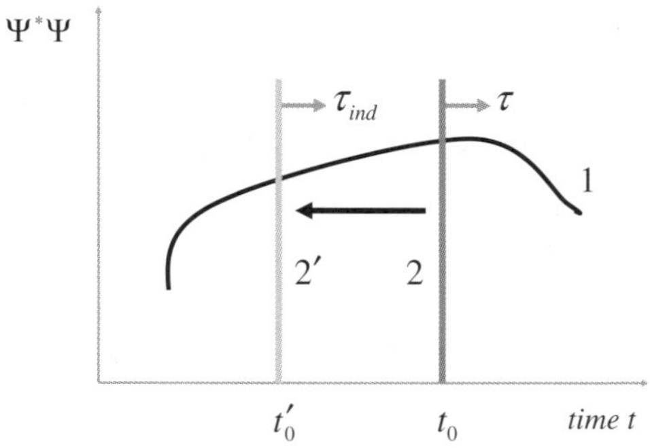


Fig. 3.22.

All reference times $\tau_k,\ k = 1, 2, \ldots, N$ for $N$ human observers are strictly coupled to the biological structure of these observers. Then, the definition ranges for $\tau_k,\ k = 1, 2, \ldots, N$ are given by the birth and the death of the $N$ observers. If $\tau_k^{\text{birth}}$ is the date of birth of the $k$th observer and $\tau_k^{\text{death}}$ his date of death, we must have

$$\tau_k^{\text{birth}} \leq \tau_k \leq \tau_k^{\text{death}}, \quad k = 1, 2, \ldots, N. \tag{3.111}$$

In the case of observer $i$ who travels to the past ($\alpha$ is negative), we have the following strict condition:

$$\tau_i^{\text{birth}} \leq \tau_{\text{ind}} \tag{3.112}$$

with

$$\tau_{\text{ind}} = \tau - |\alpha|. \tag{3.113}$$

If we have

$$\tau_i^{\text{birth}} = \tau = \tau_{\text{ind}} \tag{3.114}$$

the value for $\alpha$ must be zero, i.e., time travel for observer $i$ to the past would not be allowed. This is an interesting and relevant point because this property solves the grandmother paradox which we have already discussed in connection with General Theory of Relativity in Section 3.17.1. Observer $i$ is principally able to kill his grandmother but only after his birth, because $\tau_i^{\text{birth}} > \tau_{\text{ind}}$ is not defined. This situation means that observer $i$ cannot go back in time to kill his grandmother before his mother has been born. This is strictly forbidden because condition (3.111) has to be fulfilled.

In other words, within projection theory there does not exist a grandmother paradox. We have always taken into account that the reference time $\tau_{\text{ind}}$ and of course also the reference time structure, expressed by $\Psi_{\text{ref}}(\tau_{\text{ind}} - t)_{\text{ind}}$, are coupled to the biological system (observer $i$). This property is not formulated within the Theory of Relativity and, therefore, we can have paradoxical situations as, for example, in connection with the grandmother paradox.

The time travel of observer $i$ does not take place in connection with the $\mathbf{r}$-structure $\Psi(\mathbf{r}, t_0)$, he has at time $t_0 = \tau$, but it is his $\mathbf{r}$-structure

$\Psi(\mathbf{r}, t_0')$ which he has at time $t_0' = \tau_{\text{ind}}$, where $\Psi(\mathbf{r}, t_0') \neq \Psi(\mathbf{r}, t_0)$. Since it is time travel to the past, observer $i$ is at $\tau_{\text{ind}}$ younger than at time $\tau$. For example, if he is a man of thirty at time $\tau$, he is a boy of ten years at time $\tau_{\text{ind}}$ in the case of $|\alpha| = 20$, i.e., it is the boy of ten years who is principally able to kill his grandmother at $\tau_{\text{ind}}$ but it is not the man of thirty (see in particular Figs. 3.19 and 3.21).

Without time travel, the grandmother could still be alive at time $\tau$ [which we would like to characterize by $\Psi_{\text{grandmother}}(\mathbf{r}, t)$], but not in the other case: If the boy kills his grandmother at time $\tau_{\text{ind}}$ (in the case of a time travel), the grandmother is definitely dead at time $\tau$. This is consistent with the fact that any change at time $t_0' = \tau_{\text{ind}}$ leads to a global change of the space–time structure (Section 3.17.3.6), i.e., instead of $\Psi_{\text{grandmother}}(\mathbf{r}, t)$ we have the new characteristic function $\Phi_{\text{grandmother}}(\mathbf{r}, t)$ for times larger than $\tau_{\text{ind}}$, where the functions $\Psi_{\text{grandmother}}(\mathbf{r}, t)$ and $\Phi_{\text{grandmother}}(\mathbf{r}, t)$ are in general different for all values $\mathbf{r}$ and $t$

$$\Psi_{\text{grandmother}}(\mathbf{r}, t) \neq \Phi_{\text{grandmother}}(\mathbf{r}, t), \quad -\infty \leq \mathbf{r}, t \leq \infty. \tag{3.115}$$

That is, we have not only changes with respect to the future relative to $\tau_{\text{ind}}$ but also changes with respect to the past relative to $\tau_{\text{ind}}$ (Section 3.17.3.6).

### 3.17.3.8. *Decoupling of Subsystems*

Let us take up at time $\tau$ a certain subsystem with $\psi(\mathbf{r}, t)$ from a system that is characterized by $\phi(\mathbf{r}, t)$. Then, only the parts $\phi(\mathbf{r}, t_0)$ and $\psi(\mathbf{r}, t_0)$ are accessible to an observer (Section 3.16.2). However, we cannot only take up the part $\psi(\mathbf{r}, t_0)$ with $-\infty \leq \mathbf{r} \leq \infty$ but only the whole subsystem $\psi(\mathbf{r}, t)$ with $-\infty \leq \mathbf{r}, t \leq \infty$. The part $\psi(\mathbf{r}, t_0)$ cannot be detached from the subsystem since this subsystem is only definable over the whole $(\mathbf{r}, t)$-range $(-\infty \leq \mathbf{r}, t \leq \infty)$. We may also argue as follows. The system with $\psi(\mathbf{r}, t)$ comes into existence by $(\mathbf{p}, E)$-fluctuations which are described by the probability density $\psi^*(\mathbf{p}, E)\psi(\mathbf{p}, E)$ in $(\mathbf{p}, E)$-space, i.e., the function $\psi(\mathbf{p}, E)$ is the primary information about the subsystem under investigation. Then, we obtain the function $\psi(\mathbf{r}, t)$ through a projection from $(\mathbf{p}, E)$-space onto $(\mathbf{r}, t)$-space. As we have outlined several times, this is managed by a Fourier transform [12]. On the other hand, when we want to reproduce the primary information $\psi(\mathbf{p}, E)$, we need the entire function $\psi(\mathbf{r}, t)$ with $-\infty \leq \mathbf{r}, t \leq \infty$ since $\psi(\mathbf{p}, E)$ is obtained from $\psi(\mathbf{r}, t)$ by the inverse Fourier transform [12]

$$\psi(\mathbf{p}, E) = \frac{1}{(2\pi\hbar)^2} \int_{-\infty}^{\infty} \psi(\mathbf{r}, t) \exp\left\{-i\left[\frac{\mathbf{p}}{\hbar} \cdot \mathbf{r} - \frac{E}{\hbar}t\right]\right\} dxdydzdt. \tag{3.116}$$

In conclusion, we only observe the selected part $\psi(\mathbf{r}, t_0)$ of the subsystem but the subsystem itself is indivisible and is defined by $\psi(\mathbf{r}, t)$ with $-\infty \leq \mathbf{r}, t \leq \infty$, which has to be considered as the minimal information of the subsystem because it is indivisible.

Thus, if we want to take up at time $\tau$ a certain subsystem from the system with $\phi(\mathbf{r}, t)$ only the information $\psi(\mathbf{r}, t)$ with $-\infty \leq \mathbf{r}, t \leq \infty$ can be detached from the system, that is, we get

$$\phi(\mathbf{r}, t) \rightarrow \phi_{\text{empty}}(\mathbf{r}, t), \psi(\mathbf{r}, t), \tag{3.117}$$

where the systems $\phi_{\text{empty}}(\mathbf{r}, t)$ and $\psi(\mathbf{r}, t)$ are decoupled, and $\phi_{\text{empty}}(\mathbf{r}, t)$ is the system without subsystem $\psi(\mathbf{r}, t)$.

If subsystem $\psi(\mathbf{r}, t)$ is identical with the grandmother of observer $i$, we have

$$\psi(\mathbf{r}, t) = \Psi_{\text{grandmother}}(\mathbf{r}, t) \tag{3.118}$$

and, instead of (3.95), we obtain

$$\phi(\mathbf{r}, t) \rightarrow \phi_{\text{empty}}(\mathbf{r}, t), \Psi_{\text{grandmother}}(\mathbf{r}, t) \tag{3.119}$$

and, furthermore, if an external observer, different from observer $i$, takes up the grandmother from system $\phi(\mathbf{r}, t)$ at $\tau' < \tau_{\text{birth, mother}}$, where $\tau_{\text{birth, mother}}$ is the birth of the mother of observer $i$, then nothing can be left from the mother and observer $i$, that is, for times $\tau'' > \tau'$, the function $\phi_{\text{empty}}(\mathbf{r}, t)$ can no longer contain the mother, observer $i$ and of course all the other things which are connected to the grandmother's actions for times $\tau'' > \tau'$. This is a condition which follows directly from the principle of self-consistency. Any model for $\phi_{\text{empty}}(\mathbf{r}, t)$ may only contain self-consistent solutions [21], and the grandmother, etc. may not appear in $\phi_{\text{empty}}(\mathbf{r},t)$. On the other hand, for times $\tau'' < \tau'$ the function $\phi(\mathbf{r},t)$ must be valid, that is, this solution must contain the grandmother.

### 3.17.3.9. *Configurations at Various Times*

We have $N$ observers, and only observer $i$ is a time traveller, which is characterized by $\Psi^i_{\text{human}}(\mathbf{r},t)$. The remaining $(N-1)$ observers do not go back in time and are characterized by the functions $\Psi^k_{\text{human}}(\mathbf{r},t)$, $k = 1, 2, \dots, N$, $k \neq i$. That is, at $\tau_{\text{ind}}$ for all these $(N-1)$ observers Fig. 3.21 has to be applied since they are in the state $t'_0$ with $\Psi^k_{\text{human}}(\mathbf{r},t'_0)$, $k = 1, 2, \dots, N$, $k \neq i$, where no reference time structure can be effective for the $(N-1)$ observers because we have in this case

$$\begin{aligned}\Psi_{\text{ref}}(\tau - t)_1 &= \Psi_{\text{ref}}(\tau - t)_2 = \cdots = \Psi_{\text{ref}}(\tau - t)_N \\ &\neq \Psi_{\text{ref}}(\tau_i - t)_i = \Psi_{\text{ref}}(\tau_{\text{ind}} - t)_{\text{ind}}. \end{aligned} \tag{3.120}$$

However, a reference time structure for the $(N-1)$ observers is effective at time $\tau$, where all these observers are in the state $t_0$ with $\Psi^k_{\text{human}}(\mathbf{r},t_0)$, $k = 1, 2, \dots, N$, $k \neq i$. That is, for each of the $(N-1)$ observers the scheme that is represented in Fig. 3.19 has to be applied.

In the case of observer $i$, who travels back in time, the situation is characterized as follows. At $\tau_{\text{ind}}$, a reference time structure is defined and is given by

$$\Psi_{\text{ref}}(\tau_k - t)_k = \Psi_{\text{ref}}(\tau_{\text{ind}} - t)_{\text{ind}}, \quad k = i, \ k \neq 1, 2, \dots, N \tag{3.121}$$

and observer $i$ is at $\tau_{\text{ind}}$ in the state $t'_0$ with $\Psi^k_{\text{human}}(\mathbf{r},t'_0)$, $k = i$, $k \neq 1, 2, \dots, N$. In this situation the scheme has to be applied that is represented in Fig. 3.19. On the other hand, a reference time structure for observer $i$ is not defined at time $\tau$, where it is in the state $t_0$ with $\Psi^k_{\text{human}}(\mathbf{r},t_0)$, $k = i$, $k \neq 1, 2, \dots, N$ and, therefore, Fig. 3.21 is valid.

In summary, at time $\tau$ we have the following reference time structures:

$$\Psi_{\text{ref}}(\tau - t)_k, \quad k = 1, 2, \dots, N, \ k \neq i \tag{3.122}$$

and the configurations

$$\Psi^k_{\text{human}}(\mathbf{r}, t_0), \quad k = 1, 2, \ldots, N, \;\; k \neq i. \tag{3.123}$$

On the other hand, at time $\tau_{\text{ind}}$ we have

$$\Psi_{\text{ref}}(\tau_k - t)_k = \Psi_{\text{ref}}(\tau_{\text{ind}} - t)_{\text{ind}}, \quad k = i, \;\; k \neq 1, 2, \ldots, N \tag{3.124}$$

with

$$\Psi^k_{\text{human}}(\mathbf{r}, t_0'), \quad k = i, \;\; k \neq 1, 2, \ldots, N. \tag{3.125}$$

Only for these structures [Eqs. (3.123) and (3.125)] conscious actions are possible for the observers because unconscious pictures are selectable for these configurations (Section 3.17.3.3).

On the other hand, at time $\tau_{\text{ind}}$ no conscious actions are possible for the $(N - 1)$ observers. In particular, since there are no reference time structures

$$\Psi_{\text{ref}}(\tau_{\text{ind}} - t)_k, \quad k = 1, 2, \ldots, N, \;\; k \neq i \tag{3.126}$$

defined, no $t$-selections can take place and there is no "picture of reality" defined for the $(N - 1)$ observers at time $\tau_{\text{ind}}$, but only at $\tau$. At time $\tau_{\text{ind}}$ the configurations are given in the form

$$\Psi^k_{\text{human}}(\mathbf{r}, t_0'), \quad k = 1, 2, \ldots, N, \;\; k \neq i. \tag{3.127}$$

Furthermore, at time $\tau$ conscious actions are not possible for observer $i$. In particular, since there is no reference time structures

$$\Psi_{\text{ref}}(\tau_k - t)_k = \Psi_{\text{ref}}(\tau - t)_{\text{ind}}, \quad k = i, \;\; k \neq 1, 2, \ldots, N \tag{3.128}$$

defined, no $t$-selections can take place and there is no "picture of reality" defined for observer $i$ at time $\tau$, but only at $\tau_{\text{ind}}$. At time $\tau$, the configuration is given in the form

$$\Psi^k_{\text{human}}(\mathbf{r}, t_0), \quad k = 1, 2, \ldots, N, \;\; k \neq i. \tag{3.129}$$

In summary, pictures of reality, i.e., $t$-selected configurations, are only given for all $(N - 1)$ observers with the structure (3.129), and for observer $i$ with

$$\Psi^{k}_{\text{human}}(\mathbf{r}, t'_0), \quad k = i, \ k \neq 1, 2, \ldots, N. \qquad (3.130)$$

Only in these cases a human observer is able to perform conscious actions.

### 3.17.4. *Final Remarks*

Within General Theory of Relativity, paradoxes (contradictions) can appear in connection with time travel to the past. It turned out that the proposed physical solutions (many-world theory, argument of self-consistency, chronology protection conjecture) for eliminating the contradictions are not convincing.

On the other hand, also within projection theory, time travel is principally possible without using relativistic elements. Due to the relationship between reference time $\tau$ and system-specific time $t$ the principal potentiality for time travel is given, i.e., for time travel to the future as well as to the past. The grandmother paradox and similar situations do not appear in projection theory. Here time travel becomes possible if we are able to decouple an individual reference time structure from the others. Shifts with respect to the individual reference time are connected to specific changes of the individual brain functions.

The reason why the General Theory of Relativity leads to paradoxes in connection with time travel to the past is possibly due to the fact that this theory is not able to connect the biological structure of the time traveller with the basic elements of the relativistic framework. We have briefly addressed this point in Section 3.17.1.2.

Within projection theory, the time traveller (observer $i$) cannot shake hands with his younger self as is principally possible within General Theory of Relativity (Section 3.17.1.2). Such a relativistic scenario is not possible in projection theory. Here the situation is expressed by the following picture. The time travel to the past means that the human observer no longer occupies consciously the state $t_0$, but now the state $t'_0$ is experienced consciously, in which he is younger than at time $t_0$. Then, we have at $t'_0$ a situation in analogy to Fig. 3.19. After time travel the state at $t_0$ becomes a pure material state without consciousness, and we have now a situation in analogy to that which we have said in connection with Fig. 3.21, that is, at time $t_0$ only the material level is occupied and all the other levels are empty.

## 3.18. Summary

In this chapter, we have discussed the consequences which follow from the projection principle. There are some essential results, and here is a summary of the most relevant features.

1. The outside world (basic reality) is not accessible to the human observer. Therefore, the pictures in front of us are the most direct and essential event. Thus, we may state the following. Within the memory of man all essential things are represented within the frame of pictures that appear spontaneously in front of us. This is the most basic statement within projection theory. These pictures of reality are represented in $(\mathbf{r}, t)$-space and are described on the basis of processes which take place in fictitious reality. Each process is characterized by $(\mathbf{p}, E)$-fluctuations in $(\mathbf{p}, E)$-space (fictitious reality). The processes in $(\mathbf{p}, E)$-space are projected onto $(\mathbf{r}, t)$-space.
2. Each process and system, respectively, defines its own time-structure where the letter $t$ is the variable for describing these system-specific time structures. The quantity $t$ does not replace the time $\tau$, but the external parameter $\tau$ also exists in projection theory as reference time. In other words, projection theory extends the frame of description by $t$. Instead of $\mathbf{r}, \tau$ (conventional physics), within projection theory we have $\mathbf{r}, t, \tau$.
3. Not only does $\mathbf{r}$ act as a statistical variable but also $t$. Thus, the wavefunction of a system is defined in terms of the variables $\mathbf{r}, t$, i.e., we have $\Psi = \Psi(\mathbf{r}, t)$, and $\Psi(\mathbf{r}, t)$ can be dependent on $\tau$ or not. The probability distributions for $\mathbf{r}$ and $t$ can only be determined in connection with $\mathbf{p}$ and $E$ [the variables of $(\mathbf{p}, E)$-space], and we come to the following statement: "The measurement of one of the possible values for $\mathbf{p}$ and for $E$ at time $\tau$ is done in the space–time intervals $\mathbf{r}$, $\mathbf{r} + d\mathbf{r}$ and $t$, $t + dt$ with

the probability density of $\Psi^*(\mathbf{r}, t)\Psi(\mathbf{r}, t)$." Clearly, this instruction is more general than Born's probability description.

4. The system-specific time $t$ behaves statistically. In general, $t$ jumps arbitrarily from one $t$-position to another, and we cannot know when we observe an event in the past, present or in the future. In contrast to that, the reference time $\tau$, measured with our clocks, goes by definition always strictly from the past to the future. This feature of time $t$ is completely new. Even the notion "system-specific time" $t$ is not known in conventional quantum theory. Here only the reference time $\tau$ appears which however only plays the role of an external parameter.
5. Point-like particles with definite energy and definite momentum are not definable within the frame of projection theory. This "point-model", used in usual quantum theory, is obviously a hold-over from classical mechanics and is mainly based on our observations in everyday life. Furthermore, there are non-local effects, that is, we observe in $(\mathbf{r}, t)$-space global instantaneous changes in the structure, and it is a feature of the projection that all space–time points $\mathbf{r}, t$ $(-\infty \leq \mathbf{r} \leq \infty, \ -\infty \leq t \leq \infty)$ are instantaneously involved. This effect is comparable with a flashlight on a screen.
6. The systematic $(\mathbf{p}, E)$-fluctuations in $(\mathbf{p}, E)$-space reflect processes and we have to consider $(\mathbf{p}, E)$-space as reality. However, because the variables $\mathbf{p}$ and $E$ are auxiliary elements this reality is a "fictitious reality". Therefore, also the $(\mathbf{p}, E)$-fluctuations (the interactions of the systems with other systems) are fictitious in character and merely simulate the real, but unknown processes in basic reality.

   Due to the conservation laws for the momentum $\mathbf{p}$ and the energy $E$ we must have systematic interaction processes ($(\mathbf{p}, E)$-fluctuations) of the system under investigation with other systems of the environment and, therefore, we have in fact to consider $(\mathbf{p}, E)$-space as (fictitious) reality. However, there do not exist such conservation laws for the variables $\mathbf{r}$ and $t$, i.e., the realized $(\mathbf{r}, t)$-point at time $\tau$ of the system under investigation is independent of the realized $(\mathbf{r}, t)$-points at the same time $\tau$ of the other systems of the environment. There is no exchange of certain "space–time pieces" between the pictures. In other words, there are no correlations between the $(\mathbf{r}, t)$-points at time $\tau$. This indicates that $\mathbf{r}$ and $t$ are in fact elements of a picture. In other words, the information of fictitious reality $[(\mathbf{p}, E)$-space$]$ is projected onto $(\mathbf{r}, t)$-space by a Fourier transform and we get the "picture of reality".

7. Projection theory that non-interacting systems cannot exist in nature. Why does nature not admit such systems, i.e., free elementary systems without internal structure? Such free systems with constant momentum $\mathbf{p}_0$ and constant energy $E_0$ are in a certain sense "dead systems" because they are not involved in the processes in nature; there are no $(\mathbf{p}, E)$-fluctuations between a free system and its surroundings, and such systems would be useless.

   Only interacting systems with non-vanishing $(\mathbf{p}, E)$-fluctuations $(\Delta\mathbf{p} \neq 0,\ \Delta E \neq 0)$ are relevant when we base our considerations on the "principle of usefulness". That is also a requirement of projection theory because within this theory, free (non-interacting) systems without substructure cannot exist. Thus, projection theory obviously contains the "principle of usefulness". Only processes are relevant because static systems are dead systems and cannot be involved in the scenario in nature. Like Whitehead and Bergson, we argue for the primacy of process.

8. The features of projection theory directly justify the introduction of "distance-independent interaction". That is, there should also be interactions between systems possible that are not dependent on their mutual space–time distance. Such kind of interaction describe the form of objects (systems).

   Objects that do not communicate by distance-dependent interactions but only by distance-independent interaction behave statistically. We showed that the space–time positions of the probability densities, which define the systems in $(\mathbf{r}, t)$-space, must behave statistically, i.e., these peaks jump arbitrarily through space and time. These jumps are independent from each other because the interaction between the systems is not dependent on their distance but we assumed that this interaction is only distance-independent and creates the form of the systems.

9. The physical picture within Newton's theory is quite different from that of projection theory. In classical mechanics, the $(\mathbf{p}, E)$-states are definitely changed when a body moves through space (relative to space) with varying velocity, and for the explanation of this scenario, space-effects (inertia) are essential in Newton's mechanics. In particular, here a non-interacting body may move through space with a certain constant velocity, and this property directly comes into play by unrealistic space-effects (inertia) in Newton's theory. However, such kind of space-effect has to be considered as unphysical and has been strongly criticized by Ernst Mach and, in particular, also by Albert Einstein. Nevertheless, also within the Special Theory of Relativity and General Theory of Relativity the situation concerning space-effects could not really be improved. More details concerning the Theory of Relativity will be given in Chapter 5.

   The situation within projection theory is quite different. Here Mach's principle is fulfilled and this is a quite general statement. The reason is obvious because "only" geometrical structures (objects) are embedded in $(\mathbf{r}, t)$-space, and $(\mathbf{r}, t)$-space plays the role of an auxiliary element for the representation of the world, that is, to form pictures of it. In other words, since there can be no material objects embedded in $(\mathbf{r}, t)$-space, material objects cannot be influenced by space and time [$(\mathbf{r}, t)$-space]. Therefore, Mach's principle is from the beginning fulfilled and must be strictly valid for all physical situations. In particular, we have recognized that in this case (Mach's principle is fulfilled) the geometrical objects (system $i$ and system $j$) may perform arbitrary jumps through space and time [$(\mathbf{r}, t)$-space].

10. Within projection theory, the interaction potential can be at best interpreted as auxiliary element without any imaginable background. (There is no information or energy transfer through $(\mathbf{r}, t)$-space.) But also within conventional physics the notion "interaction" as an imaginable quality becomes questionable when we analze this term in more detail.

In contrast to projection theory, within conventional physics there exist some descriptive interpretations for the interaction (action-at-a-distance, proximity effect, principle of least action), which is here expressed by $U(x, y, z)$ and not by $V(x, y, z, t)$. We have briefly discussed these interpretations for $U(x, y, z)$ and their relevance in connection with the force laws as, for example, for the gravitational law $m_1 m_2/r^2$. However, this force law cannot be derived from one of the interpretations. This is a critical point.

We came to the conclusion that these interpretations for the notion "interaction" cannot help us further to understand the mutual influence in connection with two or more interacting bodies. In particular, all these interpretations (action-at-a-distance, proximity effect, principle of least action) are not able at all to deduce the force law $m_1 m_2/r^2$. As in projection theory, we have to conclude that also in conventional physics, the interaction law described by $m_1 m_2/r^2$ is an abstract law.

In particular, there is no possibility to explain gravity by mechanical-corpuscular effects and, therefore, gravity had to be considered as a metaphysical element when the scientific standards are based on the mechanical-corpuscular world view. Gravity can be qualified as an occult element from this point of view. In this connection Kuhn remarked [10]. "*Gravity, interpreted as an innate attraction between every pair of particles of matter, was an occult quality in the same sense as the scholastics' 'tendencies to fall' had been.*"

The situation in projection theory is quite different. Here we have no material objects in space–time [$(\mathbf{r}, t)$-space] but only geometrical structures. These geometrical structures are not static elements. There is an exchange of momentum and energy in $(\mathbf{p}, E)$-space with other systems, and an object comes into existence by these $(\mathbf{p}, E)$-fluctuations, which are projected onto $(\mathbf{r}, t)$-space leading to the shape of the systems and to relative structural configurations between them.

There are $(\mathbf{r}, t)$-correlations in $(\mathbf{r}, t)$-space, but there is no exchange of information between the various geometrical structures, that is, nothing acts across $(\mathbf{r}, t)$-space as in the case of conventional physics ("proximity effect" and the "forces acting-at-the-distance"); the $(\mathbf{r}, t)$-correlations come exclusively into existence by the $(\mathbf{p}, E)$-fluctuations in $(\mathbf{p}, E)$-space. The $(\mathbf{r}, t)$-structures in $(\mathbf{r}, t)$-space are exclusively the result of projection processes and no geometrical part of the $(\mathbf{r}, t)$-space is influenced by the others by actions through $(\mathbf{r}, t)$-space. From the point of view of the projection principle such actions through space and time make no sense since $(\mathbf{r}, t)$-space merely contains pictures.

11. In contrast to usual physics, within projection theory a system-specific time $t$ is defined, that is, each system produces its own time scale. The time $\tau$ (measured with our clocks in everyday life) serves as reference time and has nothing to do with the system under investigation. Despite the statistical fluctuations, at time $\tau$ the whole of time $t$-past, present and future — is laid out frozen before us: $\Psi^*(\mathbf{r}, t)\Psi(\mathbf{r}, t)$, $-\infty < \mathbf{r}, t < \infty$. However, we do not observe the whole scenario $\Psi^*(\mathbf{r}, t)\Psi(\mathbf{r}, t)$ at time $\tau$, but only selected configurations of reality: $\Psi^*(\mathbf{r}, t_0)\Psi(\mathbf{r}, t_0)$, $-\infty < \mathbf{r} < \infty$, for $t_0 = \tau$. Each photography represents such a configuration in space, at a certain time $\tau = t_0$. The transition from $\Psi^*(\mathbf{r}, t)\Psi(\mathbf{r}, t)$ to $\Psi^*(\mathbf{r}, t_0)\Psi(\mathbf{r}, t_0)$ must be due to an interplay between the system under investigation, described by $\Psi(\mathbf{r}, t)$, and the observer's function that is characterized by the reference time structure $\Psi_{ref}(t)$.
12. Within General Theory of Relativity, paradoxes (contradictions) can appear in connection with time travel to the past. It turned out that the proposed physical solutions (many-world theory, argument of self-consistency, chronology protection conjecture) for eliminating the contradictions are not convincing.

    On the other hand, also within projection theory, time travel is principally possible without using relativistic elements. Due to the relationship between reference time $\tau$ and system-specific time $t$, the principal potentiality for time travel is given, i.e., for time travel to the future as well as to the past. The grandmother paradox and similar situations do not appear in projection theory. Here time travel becomes possible if we are able to decouple an individual reference time structure from the others. Time shifts with respect to the individual reference time is connected to specific changes of the individual brain functions.

The reason why the General Theory of Relativity leads to paradoxes in connection with time travel to the past is possibly due to the fact that this theory is not able to connect the biological structure of the time traveller with the basic elements of the relativistic framework.

# 4

# Basic Reality and Levels of Reality

In this chapter we will discuss "levels of reality". This notion has already been used in Chapter 3 in connection with Figs. 3.18–3.20. Basic reality (outside world) is not accessible to human observers and, therefore, we have to find other principles for the description of the outside world. This can be done on the basis of various standards, which express the possibility for the description of the world in terms of "levels" and we come to the notion "levels of reality". In this chapter we will justify this method. In particular, we will discuss the difference between "hard objects" (cars, trees, houses, etc.) and "physical laws". What is the difference between the states of matter and the states of mind? Let us start our considerations with the notion "hard objects".

## 4.1. Hard Objects

The world in front of us is given in terms of so-called "hard objects" (cars, trees, houses, etc.). It is the world we observe in everyday life which is represented as pictures in $(\mathbf{r}, t)$-space. These hard objects are described within projection theory by $(\mathbf{p}, E)$-fluctuations reflecting the interactions between the objects in $(\mathbf{p}, E)$-space. As we have already outlined several times, we get the pictures of the hard objects by transmitting the information from $(\mathbf{p}, E)$-space to $(\mathbf{r}, t)$-space.

### 4.1.1. *Outside World and Basic Reality*

The pictures that appear spontaneously in front of us come from the basic (objective) reality which is principally not accessible to the human observer (Chapter 2). We know nothing about the process of how this information is transmitted and transformed from basic reality to the human brain where it appears as the "picture of reality". All these processes are principally unknown. These pictures with the hard objects (trees, cars, etc.) are exclusively located in the brain of the observer and nowhere else, but we have the impression that all these objects are in front of us (outside of us). The statement by Brückner explains this fact very clearly (Section 2.5). Let us repeat the main facts here.

> "*It is characteristic of the sight-process that our awareness does not register the picture of a candle on the retina inside the eye, but we have the impression that we are standing opposite the candle-light which is located in the space outside, not standing on his head but upright. All observations are 'projected' by our senses outwardly into space.*"

That is, the hard objects appear to be embedded in an outside world. This is our impression. Fact is however that the hard objects (cars, trees, houses, etc.) are located in the brain as pictures. Most people believe that this observed outside world is identical with basic (objective) reality. This is obviously a fallacy as we have recognized in Chapter 2. In the following, the situation will be further investigated.

### 4.1.2. *The "Hard Objects" are Elements of the Brain*

Again, from the point of view of projection theory basic reality remains hidden. However, we can construct "fictitious realities" in $(\mathbf{p}, E)$-space producing pictures of reality in $(\mathbf{r}, t)$-space (see in particular Fig. 3.1, Chapter 3). These models for the pictures of reality are, if the models are realistic, more or less identical with those pictures produced spontaneously, i.e. without conscious actions, by the brain on the basis of the processes taking place in basic reality, which are however not directly recognizable to us, but only in the form of species-dependent pictures that appear spontaneously in $(\mathbf{r}, t)$-space.

From the viewpoint of projection theory, there is no other possibility for the modeling of the geometrical pictures for the hard objects that are located in our brain but actually appear in front of us (Chapter 2).

In connection with the analysis given in this chapter, it is relevant to repeat that also space and time do not belong to basic reality but have to be considered as auxiliary elements for the representation of the pictures. Thus, space and time are exclusively elements of the brain. Already the philosopher Immanuel Kant recognized that. The main features of space and time have been discussed in Chapter 2.

## 4.2. General Physical Laws

Because the pictures in $(\mathbf{r}, t)$-space, produced on the basis of $(\mathbf{p}, E)$-fluctuations in $(\mathbf{p}, E)$-space (fictitious reality), agree more or less with those produced by the principally unknown basic reality, the $(\mathbf{p}, E)$-fluctuations in $(\mathbf{p}, E)$-space (fictitious reality) can be considered as *real processes* although they are merely intellectual (theoretical) constructions of the human mind. [In this connection, we should not forget that the hard objects (trees, cars, etc.) in front of us are also nothing else than "states of the brain". An outside world is feigned which is obviously more useful for a human observer (see in particular Section 4.1)].

### 4.2.1. *The Hard Objects in Front of us and the Physical Laws are Equally States of the Observer's Brain*

These specific theoretical constructions (models for $(\mathbf{p}, E)$-fluctuations in $(\mathbf{p}, E)$-space) cannot be found without a general physical law. On the other hand, the general physical law [in this case $i\hbar\partial/\partial t\Psi(\mathbf{r}, t) = -\hbar^2/(2m_0)\Delta\Psi(\mathbf{r}, t) + V(x, y, z, t)\Psi(\mathbf{r}, t)$; see Eq. (3.22), Section 3.7.1] contains the features of all systems and, therefore, of all possible pictures $\Psi^*(\mathbf{r}, t)\Psi(\mathbf{r}, t)$ in $(\mathbf{r}, t)$-space and all possible models for $(\mathbf{p}, E)$-fluctuations in $(\mathbf{p}, E)$-space, respectively. This is because the characteristics of each specific model for $(\mathbf{p}, E)$-fluctuations [picture $\Psi^*(\mathbf{r}, t)\Psi(\mathbf{r}, t)$] is dictated by the features of the physical law.

Consequently, like the hard objects, the physical laws [as, for example, $i\hbar\partial/\partial t\Psi(\mathbf{r},t) = -\hbar^2/(2m_0)\Delta\Psi(\mathbf{r},t) + V(x,y,z,t)\Psi(\mathbf{r},t)$] have to be considered as *real objects* within projection theory. Physical laws are *objects* in generalized form, more general than the hard objects. As we have already remarked above, both, the hard objects as well as the physical laws are "states of the observer's brain" within projection theory.

All these points need further consideration, and we want to do that step-by-step in this chapter. We will particularly talk about the following topics: (1) What is the difference between a physical law and the specific models? (2) Are the brain functions describable by the physical brain activities? (3) Outside world and basic reality. (4) The principle of deduction.

Let us first talk about the difference between a concrete picture expressed by a specific pattern $\Psi^*(\mathbf{r},t)\Psi(\mathbf{r},t)$ (model) and the physical law on which $\Psi(\mathbf{r},t)$ is based. We have already discussed this point above. Let us further discuss this topic in the next few section.

### 4.2.2. *Specific Models*

The picture $\Psi^*(\mathbf{r},t)\Psi(\mathbf{r},t)$ is always the solution for a specific model (for any natural or artificial system) that is characterized by a specific function

$$V(x,y,z,t) = V(x.y.z,t)^{\text{specific}} \tag{4.1}$$

leading to

$$\Psi^*(\mathbf{r},t)\Psi(\mathbf{r},t) = [\Psi^*(\mathbf{r},t)\Psi(\mathbf{r},t)]^{\text{specific}}. \tag{4.2}$$

Without a specific model (system), we cannot have a picture (pattern) in $(\mathbf{r},t)$-space. The physical law $i\hbar\partial/\partial t\Psi(\mathbf{r},t) = -\hbar^2/(2m_0)\Delta\Psi(\mathbf{r},t)+ V(x,y,z,t)\Psi(\mathbf{r},t)$ describes general features and cannot be represented as a picture in $(\mathbf{r},t)$-space.

Equation $i\hbar\partial/\partial t\Psi(\mathbf{r},t) = -\hbar^2/(2m_0)\Delta\Psi(\mathbf{r},t) + V(x,y,z,t)\Psi(\mathbf{r},t)$ is more general than any specific model because it contains the characteristics of all possible physical systems, that is, it is valid for all possible specific functions $V(x, y, z, t)$:

$$
\begin{aligned}
V(x, y, z, t)_1^{\text{specific}} &\rightarrow [\Psi^*(\mathbf{r}, t)\Psi(\mathbf{r}, t)]_1^{\text{specific}} \\
V(x, y, z, t)_2^{\text{specific}} &\rightarrow [\Psi^*(\mathbf{r}, t)\Psi(\mathbf{r}, t)]_2^{\text{specific}} \\
&\vdots \\
V(x, y, z, t)_n^{\text{specific}} &\rightarrow [\Psi^*(\mathbf{r}, t)\Psi(\mathbf{r}, t)]_n^{\text{specific}} \\
&\vdots
\end{aligned}
\tag{4.3}
$$

Remark

Only the specific models $[\Psi^*(\mathbf{r}, t)\Psi(\mathbf{r}, t)]^{\text{specific}}$ are depictable in $(\mathbf{r}, t)$-space and have to be determined on the basis of the general equation

$$
i\hbar\frac{\partial}{\partial t}\Psi(\mathbf{r}, t) = -\frac{\hbar^2}{2m_0}\Delta\Psi(\mathbf{r}, t) + V(x, y, z, t)\Psi(\mathbf{r}, t) \tag{3.22}
$$

(Section 3.7.1) leading to the specific wave function $[\Psi(\mathbf{r}, t)]^{\text{specific}}$ in $(\mathbf{r}, t)$-space. Equation (3.22) is not depictable in $(\mathbf{r}, t)$-space. The same situation holds for $(\mathbf{p}, E)$-space. Here, instead of Eq. (3.22), we have

$$
\begin{aligned}
E\Psi(\mathbf{p}, E) = {}& \frac{\mathbf{p}^2}{2m_0}\Psi(\mathbf{p}, E) \\
& + V\left(i\hbar\frac{\partial}{\partial p_x}, i\hbar\frac{\partial}{\partial p_y}, i\hbar\frac{\partial}{\partial p_z} - i\hbar\frac{\partial}{\partial E}\right)\Psi(\mathbf{p}, E)
\end{aligned}
\tag{3.23}
$$

(Section 3.7.1) which contains — as we know from Chapter 3 — exactly the same physical information as Eq. (3.22). From Eq. (3.23), we can calculate specific $(\mathbf{p}, E)$-fluctuations described by the pattern $[\Psi^*(\mathbf{p}, E)\Psi(\mathbf{p}, E)]^{\text{specific}}$ that are depictable in $(\mathbf{p}, E)$-space. Clearly, as in the case of $(\mathbf{r}, t)$-space, also the general Eq. (3.23) is not depictable in $(\mathbf{p}, E)$-space. $\Psi(\mathbf{p}, E)$ and $\Psi(\mathbf{r}, t)$ are related to each other by a Fourier transform (Chapter 3)

$$\Psi(\mathbf{p}, E)^{\text{specific}} = \frac{1}{(2\pi\hbar)^2} \int_{-\infty}^{\infty} \Psi(\mathbf{r}, t)^{\text{specific}} \times \exp\left\{-i\left[\frac{\mathbf{p}}{\hbar}\cdot\mathbf{r} - \frac{E}{\hbar}t\right]\right\} dtdxdydz. \quad (4.4)$$

More details in connection with Eqs. (3.22), (3.23) and (4.4) are given in [12].

### 4.2.3. *Can a General Physical Law be Deduced from its Specific Solutions?*

We can determine the specific pictures (4.3) on the basis of the physical law but, on the other hand, we cannot deduce the scientific characteristics and rules, on which the physical law is based, on the basis of a specific model (picture). Again, Eq. (3.22) is more general than one of its specific solutions. It contains the characteristics of all possible physical systems.

In order to deduce the specific solution $[\Psi^*(\mathbf{r}, t)\Psi(\mathbf{r}, t)]_n^{\text{specific}}$ from the potential $V(x, y, z, t)_n^{\text{specific}}$ [see Eq. (4.3)] we need the basic Eq. (3.22). However, it is not possible to conclude from $\Psi(\mathbf{r}, t)_n^{\text{specific}}$ alone on Eq. (3.22). The structure of Eq. (3.22) cannot be deduced on the basis of a specific solution $\Psi(\mathbf{r}, t)_n^{\text{specific}}$ alone.

This in particular also means that we cannot deduce from the features of a "specific picture" the features of another "specific picture". There is a certain analogy to the following case. We cannot explain the features of an aircraft on the basis of the features of a car. Another example is that we cannot draw the features of a cat from those of a cow.

In other words, we are not able to construct a general physical law [as, for example, the law given by Eq. (3.22)] on the basis of a given specific pattern (given in the form of a mathematical expression or by a set of numbers) for $\Psi(\mathbf{r}, t)^{\text{specific}}$ and $\Psi(\mathbf{p}, E)^{\text{specific}}$, respectively. All specific functions $\Psi(\mathbf{r}, t)^{\text{specific}}, n = 1, 2, \ldots$ $[\Psi(\mathbf{p}, E)^{\text{specific}}, n = 1, 2, \ldots]$ must obey the physical law (3.23), but the physical law itself cannot be recognized by one of its specific solutions (models).

In particular, the physical law is not depictable in $(\mathbf{r}, t)$-space [$(\mathbf{p}, E)$-space)]. Only specific solutions can be pictured in $(\mathbf{r}, t)$-space [$(\mathbf{p}, E)$-space)]. Another kind of representation should be considered for the abstract physical law (3.22), which contains potentially the characteristics of all possible physical systems, i.e., all possible functions $V(x, y, z, t)$, and this feature can be expressed symbolically by $\Psi = \Psi(V)$ in the case of the physical law $i\hbar\partial/\partial t\Psi(\mathbf{r}, t) = -\hbar^2/(2m_0)\Delta\Psi(\mathbf{r}, t) + V(x, y, z, t)\Psi(\mathbf{r}, t)$.

Furthermore, we know that $(\mathbf{r}, t)$-space is tightly linked to $(\mathbf{p}, E)$-space (and vice versa), and the statements formulated above particularly mean that the "object" physical law cannot be the outcome of $(\mathbf{p}, E)$-fluctuations as in the case of the hard objects in front of us. As we have stated several times, within projection theory there is the following strict law: If there are no $(\mathbf{p}, E)$-fluctuations in $(\mathbf{p}, E)$-space there cannot be a physical picture in $(\mathbf{r}, t)$-space. A picture for a physically real process is not defined without $(\mathbf{p}, E)$-fluctuations and vice versa! Then, by definition, systems that are not depictable in $(\mathbf{r}, t)$-space are not physical systems (are not material objects).

### 4.2.4. *States of Matter, States of Mind*

All objects that are based on specific $(\mathbf{p}, E)$-processes in $(\mathbf{p}, E)$-space always produce specific pictures in $(\mathbf{r}, t)$-space (Section 4.2.3) and define *material objects* (hard objects like trees, cars, atoms, molecules, etc.) and have to be considered as "states of matter". On the other hand, the general physical law $i\hbar\partial/\partial t\Psi(\mathbf{r}, t) = -\hbar^2/2m_0\Delta\Psi(\mathbf{r}, t) + V(x, y, z, t)\Psi(\mathbf{r}, t)$ is not depictable in $(\mathbf{r}, t)$-space. Therefore, such a law cannot be produced by $(\mathbf{p}, E)$-processes in $(\mathbf{p}, E)$-space. We would like to define these physical laws as "states of mind".

#### 4.2.4.1. *What Can a Specific System "see"?*

From the point of view of a specific system with a specific function for $V(x, y, z, t)$, as, for example, $V(x, y, z, t)_1^{\text{specific}}$ with the transition $V(x, y, z, t)_1^{\text{specific}} \rightarrow [\Psi^*(\mathbf{r}, t)\Psi(\mathbf{r}, t)]_1^{\text{specific}}$, there exists only one $(\mathbf{r}, t)$-structure [and of course only one kind of $(\mathbf{p}, E)$-fluctuation in

$(\mathbf{p}, E)$-space)], and this is given by its own, i.e., by $[\Psi^*(\mathbf{r}, t)\Psi(\mathbf{r}, t)]_1^{\text{specific}}$ and $[\Psi^*(\mathbf{p}, E)\Psi(\mathbf{p}, E)]_1^{\text{specific}}$, respectively. For such a system, there does not exist a $(\mathbf{r}, t)$-structure in the world different from its own. This system is *completely* described by $[\Psi^*(\mathbf{r}, t)\Psi(\mathbf{r}, t)]_1^{\text{specific}}$ or by the equivalent information $[\Psi^*(\mathbf{p}, E)\Psi(\mathbf{p}, E)]_1^{\text{specific}}$ in $(\mathbf{p}, E)$-space, i.e., for this specific system there is by definition nothing else, and nothing remains hidden. In contrast, in connection with the human observer we have often the impression that not everything can be recognized; certain things seem to be hidden.

For example, our specific system with $[\Psi^*(\mathbf{r}, t)\Psi(\mathbf{r}, t)]_1^{\text{specific}}$ could be a hydrogen atom: $[\Psi^*(\mathbf{r}, t)\Psi(\mathbf{r}, t)]_1^{\text{specific}} = [\Psi^*(\mathbf{r}, t)\Psi(\mathbf{r}, t)]^{\text{hydrogen}}$, and we would like to assume that the hydrogen atom is completely described by $[\Psi^*(\mathbf{r}, t)\Psi(\mathbf{r}, t)]^{\text{hydrogen}}$ and, on the other hand, by the specific $[(\mathbf{p}, E)\text{-fluctuations}]^{\text{hydrogen}}$ which are expressed in $(\mathbf{p}, E)$-space by $[\Psi^*(\mathbf{p}, E)\Psi(\mathbf{p}, E)]^{\text{hydrogen}}$. Then, *all* the properties of the hydrogen atom are given in $(\mathbf{r}, t)$-space (Section 4.2.3). The hydrogen atom is captured in $(\mathbf{r}, t)$-space, and, therefore, it only defines a "state of matter". For example, it cannot have brain functions in connection with $[\Psi^*(\mathbf{r}, t)\Psi(\mathbf{r}, t)]^{\text{specific}}$ for the production of certain "states of mind", i.e., the ability of thinking has to be excluded, and there can be no possibility for the hydrogen atom to go beyond $(\mathbf{r}, t)$-space. In other words, the hydrogen atom cannot "observe" objects outside $(\mathbf{r}, t)$-space. Therefore, the hydrogen atom (or any other specific system) is in principle not able to "see" or to make "statements" about a general physical law as, for example, that which is expressed by Eq. (3.22) on which the function $\Psi(\mathbf{r}, t)^{\text{hydrogen}}$ is based, and this is because the physical law [Eq. (3.22)] is not depictable in $(\mathbf{r}, t)$-space and, on the other hand, because the hydrogen atom is captured in $(\mathbf{r}, t)$-space.

#### 4.2.4.2. *Conclusion*

Clearly, the properties of a system with $[\Psi^*(\mathbf{r}, t)\Psi(\mathbf{r}, t)]_1^{\text{specific}}$ as, for example, a hydrogen atom with $[\Psi^*(\mathbf{r}, t)\Psi(\mathbf{r}, t)]^{\text{hydrogen}}$, can in nature be different from those described by $[\Psi^*(\mathbf{r}, t)\Psi(\mathbf{r}, t)]^{\text{hydrogen}}$.

This could in principle be the case but here only the description in $(\mathbf{r}, t)$-space is concerned. However, also an improved representation for $[\Psi^*(\mathbf{r}, t)\Psi(\mathbf{r}, t)]^{\text{hydrogen}}$ cannot change the situation because it remains a description in $(\mathbf{r}, t)$-space $[(\mathbf{p}, E)$-space$)]$. On the other hand, since the general physical law $i\hbar\partial/\partial t\Psi(\mathbf{r}, t) = -\hbar^2/(2m_0)\Delta\Psi(\mathbf{r}, t) + V(x, y, z, t)\Psi(\mathbf{r}, t)$ is not depictable in $(\mathbf{r}, t)$-space, also a general physical law that is different from this law would also not be depictable in $(\mathbf{r}, t)$-space. In other words, there is a clear line between the "states of matter" and the "states of mind".

### 4.2.4.3. *Atoms with Mind?*

Even when the hydrogen atom, characterized by $[\Psi^*(\mathbf{r}, t)\Psi(\mathbf{r}, t)]^{\text{hydrogen}}$ (or any other system with the specific structure $[\Psi^*(\mathbf{r}, t)\Psi(\mathbf{r}, t)]^{\text{specific}}$), would be more than a "state of matter" and has in addition a certain kind of mind (in analogy to that of human observers), it would not be able to deduce the physical law $i\hbar\partial/\partial t\Psi(\mathbf{r}, t) = -\hbar^2/(2m_0)\Delta\Psi(\mathbf{r}, t) + V(x, y, z, t)\Psi(\mathbf{r}, t)$ by means of the specific solution (model) $[\Psi^*(\mathbf{r}, t)\Psi(\mathbf{r}, t)]^{\text{hydrogen}}$. Due to its "states of mind" the hydrogen atom has perhaps the ability to look beyond space and time $[(\mathbf{r}, t)$-space$]$ because it contains more "structure" than that given by the information $\Psi(\mathbf{r}, t)^{\text{hydrogen}}$ in $(\mathbf{r}, t)$-space, which exclusively follows from $(\mathbf{p}, E)$-fluctuation in $(\mathbf{p}, E)$-space. This eventually opens the possibility for the hydrogen atom to recognize the general physical law $i\hbar\partial/\partial t\Psi(\mathbf{r}, t) = -\hbar^2/(2m_0)\Delta\Psi(\mathbf{r}, t) + V(x, y, z, t)\Psi(\mathbf{r}, t)$ on which $[\Psi^*(\mathbf{r}, t)\Psi(\mathbf{r}, t)]^{\text{hydrogen}}$ is based and which represents a "state of mind" as we have outlined above at the beginning of this section. However, a statement about this general physical law cannot be made on the basis of $[\Psi^*(\mathbf{r}, t)\Psi(\mathbf{r}, t)]^{\text{hydrogen}}$ alone.

In other words, within projection theory, we cannot make statements about the "states of mind" on the basis of the "states of matter". Clearly, due to our experience it is difficult to imagine how the mind of a hydrogen atom or another specific system could be realized and, therefore, such an assumption should rather be classified as nonsense.

#### 4.2.4.4. *Level of General Arguments*

All types of specific systems have to be characterized in this way. A human observer is able to recognize a general physical law but he cannot deduce such a law from its specific analytical and/or numerical solutions (models), and also not from the outcome of measurements. We know that $[\Psi^*(\mathbf{r},t)\Psi(\mathbf{r},t)]^{\text{specific}}$ and $[\Psi^*(\mathbf{p},E)\Psi(\mathbf{p},E)]_1^{\text{specific}}$, respectively, can be measured (Chapter 3), but this is not the case for a general physical law as, for example, the law given by Eq. (3.22). For the deduction of this physical law we have to use more *general arguments*, more general than the physical law itself. In conclusion, while the level of specific models is positioned *below* the level of this physical law, the level of these general arguments is positioned *above* the level of a general physical law. We will still discuss this point below.

#### 4.2.4.5. *Levels of Reality*

We would like to arrange the various levels vertically to each other in accordance with the degree of generality. Because the physical law is more general than the specific $(\mathbf{p},E)$-processes (the characteristics of all possible $\mathbf{p},E$-processes are equally dictated by the physical law), we may conclude that this physical law is positioned at a higher level than that on which the models for these specific $(\mathbf{p},E)$-processes are positioned. While the specific $(\mathbf{p},E)$-processes in $(\mathbf{p},E)$-space produce specific pictures in $(\mathbf{r},t)$-space with exactly the same information, the physical law is not depictable in $(\mathbf{r},t)$-space and, therefore, a general physical law [as, for example, $i\hbar\partial/\partial t\Psi(\mathbf{r},t) = -\hbar^2/(2m_0)\Delta\Psi(\mathbf{r},t) + V(x,y,z,t)\Psi(\mathbf{r},t)$] cannot be produced by $(\mathbf{p},E)$-processes. The general physical law is more than that and, therefore, the level of the general physical law must be above the level of the specific models that are responsible for the $(\mathbf{p},E)$-processes (Fig. 4.1).

Both, the laws of the specific models for the $(\mathbf{p},E)$-processes and the physical law, are descriptions of the same physical reality. Within projection theory, the "objects" of both levels are as real as the hard objects in front of us (trees, cars, etc.). Thus, we have to consider both levels in Fig. 4.1 as "levels of reality". In this connection, we pointed

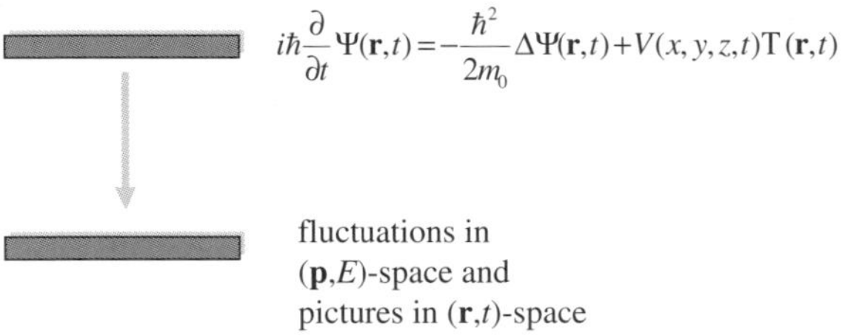


Fig. 4.1. When we vertically arrange the various levels in accordance with the degree of generality, then the level of the physical law $i\hbar\partial/\partial t\Psi(\mathbf{r},t) = -\hbar^2/(2m_0)\Delta\Psi(\mathbf{r},t)+V(x,y,z,t)\Psi(\mathbf{r},t)$ has to be positioned above the level of $(\mathbf{p},E)$-fluctuations describing the structure of the hard objects (stones, tree, animals, humans, etc.) in $(\mathbf{r},t)$-space. The physical law is more general than the laws for the $(\mathbf{p},E)$-fluctuations and cannot be represented in $(\mathbf{r},t)$-space.

out that within projection theory the hard objects in front of us and the physical laws are equally states of the observer's brain (Section 4.2.1). In the next section, we will still further study this point.

The objects of both levels have their origin in basic (objective) reality which is principally not accessible to an observer, i.e., only the objects on the various levels are recognizable to a human observer. The situation is summarized in Fig. 4.2 for the case of two levels.

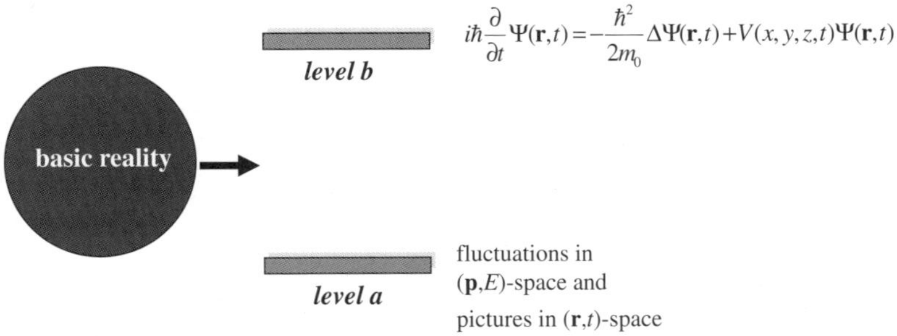


Fig. 4.2. Only the objects defined on the levels are accessible to an observer. They have their origin in basic (objective) reality which remains hidden to an observer. Like the hard objects (*level a*) also the physical laws [as, for example, $i\hbar\partial/\partial t\Psi(\mathbf{r},t) = -\hbar^2/(2m_0)\Delta\Psi(\mathbf{r},t)+V(x,y,z,t)\Psi(\mathbf{r},t)$, *level b*] have to be considered as *real objects* within projection theory (see in particular Section 4.2.1). In principle, we can have more than two levels. This point will be discussed in more detail below.

## 4.3. States of Mind

From our discussion in Sections. 4.1 and 4.2 directly follows that the "states of mind" cannot be reduced on physical processes [$(\mathbf{p}, E)$-fluctuations in $(\mathbf{p}, E)$-space], i.e., such specific processes are not qualified to describe the states of mind. Let us discuss this point in more detail.

### 4.3.1. *Brain Functions*

It has often been maintained that the states of mind can be reduced to physical real states. Such a conception cannot be correct from the viewpoint of projection theory. The deduction of a physical formula [as, for example, Eq. (3.22)] is obviously a production of mind and cannot be the result of specific physical processes [$(\mathbf{p}, E)$-fluctuations in $(\mathbf{p}, E)$-space] as we have outlined above. On the other hand, if there are no $(\mathbf{p}, E)$-fluctuations in $(\mathbf{p}, E)$-space we cannot have physical pictures in $(\mathbf{r}, t)$-space. This is a consequence of projection theory. In Section 4.2.3 we summarized the situation as follows. No $(\mathbf{p}, E)$-fluctuations in $(\mathbf{p}, E)$-space means that there cannot be a physical picture in $(\mathbf{r}, t)$-space. A physical picture is not defined without $(\mathbf{p}, E)$-fluctuations and vice versa! In fact, we are able to *think* about the physical law $i\hbar\partial/\partial t\Psi(\mathbf{r}, t) = -\hbar^2/(2m_0)\Delta\Psi(\mathbf{r}, t) + V(x, y, z, t)\Psi(\mathbf{r}, t)$ but the physical law itself is not depictable as a picture in $(\mathbf{r}, t)$-space.

Therefore, a physical law (formula) cannot be described by specific models (physical processes). The attempt to describe the products of mind by certain physical brain functions (neuronal activities) has within projection theory no basis at all. The reason for that is simple and has been worked out above. Any physical process, also those in connection with brain functions, is described by $(\mathbf{p}, E)$-fluctuations in $(\mathbf{p}, E)$-space and produces a concrete picture in $(\mathbf{r}, t)$-space. In the *reverse case*, we may argue as follows. Since the physical law $i\hbar\partial/\partial t\Psi(\mathbf{r}, t) = -\hbar^2/(2m_0)\Delta\Psi(\mathbf{r}, t) + V(x, y, z, t)\Psi(\mathbf{r}, t)$ (or other states of the mind) does not produce a concrete picture in $(\mathbf{r}, t)$-space, the physical law cannot be the result of physical brain functions (described by $(\mathbf{p}, E)$-fluctuations).

Since the origin of feelings and thinking — or quite general the products of mind — cannot be described as pictures in $(\mathbf{r}, t)$-space, their origin cannot be based on physical laws. The physical laws exclusively describe the hard objects (the objects in front of us but also atoms, molecules, etc.) in $(\mathbf{r}, t)$-space and exclusively come into play by the interactions [$(\mathbf{p}, E)$-fluctuations] in $(\mathbf{p}, E)$-space. However, the physical law itself is not depictable in $(\mathbf{r}, t)$-space as we have outlined in detail.

Remark

Clearly, we can have pictures in $(\mathbf{r}, t)$-space without $(\mathbf{p}, E)$-fluctuations, but this kind of pictures are not physical representations, i.e., the structures in such pictures do not come into play by physical-real processes, but these pictures are merely drawings and belong to the "products of imagination". Such drawings (for example, a work of art) may be arbitrary and must not satisfy physical laws.

### 4.3.2. *Conclusion: The "States of Mind" cannot be described by the "States of Matter"*

Within projection theory, the "products of mind" cannot be the outcome of physical neuronal activities but are more than that. The reason is obvious. A product of mind (as, for example, a physical formula) is positioned at a "level of reality" which is above that of the hard objects and the real physical processes, respectively (see also Fig. 4.1). As in the case of the hard objects in front of us, the physical laws also have to be considered as "objects" and also reflect certain features of basic (objective) reality. However, we know nothing about the structure of all these various kinds of objects (physical laws, hard objects) in basic reality since a picture-independent view is obviously not possible for a human observer (Chapter 2). The only thing we can say is that the object "physical law" is a more general representation of basic reality than that given by a hard object. A physical law is, as we have discussed above, a product of mind. In other words, we are able to think about a physical law [as, for example, $i\hbar\partial/\partial t\Psi(\mathbf{r}, t) = -\hbar^2/(2m_0)\Delta\Psi(\mathbf{r}, t) + V(x, y, z, t)\Psi(\mathbf{r}, t)$] but the physical law itself is not depictable as a picture in $(\mathbf{r}, t)$-space.

In summary, any effect which is not depictable in $(\mathbf{r}, t)$-space does not belong to the class of phenomena that we have defined as *physical* phenomena (states of matter). These are exclusively $(\mathbf{p}, E)$-fluctuations in $(\mathbf{p}, E)$-space within projection theory. Trees, cars, atoms, molecules, stars, etc. belong to these phenomena. However, we have discussed above that not all phenomena are depictable $(\mathbf{r}, t)$-space. Consequently, such phenomena cannot be described by physical processes (states of matter), that is, by $(\mathbf{p}, E)$-fluctuations in $(\mathbf{p}, E)$-space. In other words, when we try to explain the brain activities exclusively by physical processes [$(\mathbf{p}, E)$-fluctuations] we cannot seize phenomena that belong to the "states of mind", i.e., we are not able to recognize phenomena like physical laws, products of fantasy, feelings, etc. by means of the "states of matter". In other words, such appearances cannot be reduced on physical phenomena. The projection principle allows such kind of analysis and even allows detailed statements about possible brain functions.

Again, phenomena that are not depictable in $(\mathbf{r}, t)$-space also have to be considered as real, like those phenomena which produce a picture in $(\mathbf{r}, t)$-space (trees, cars, etc.), see in particular Section 4.1. Thus, we may state that not everything is matter or enveloped in matter. We will discuss this point in more detail in the next sections.

## 4.4. Outside World and Basic Reality

### 4.4.1. *Descriptions and Observations*

Basic reality exists independently of the observer. As we have already mentioned several times, basic reality is in principle not accessible in a direct way. Rather, it is observable or described by means of pictures on different levels, i.e., levels of reality; a picture-independent view is obviously not possible for a human observer (see also Chapter 2). Space and time do not belong to basic reality but have to be considered as auxiliary elements for the representation of the pictures and are elements of the brain within projection theory. These pictures on the different levels are in general not depictable in $(\mathbf{r}, t)$-space. We only obtain pictures in $(\mathbf{r}, t)$-space for the hard objects in front of us, atoms, molecules, etc. (states of matter) but not in connection with physical laws etc. (states of mind).

In Section 4.2, we have introduced certain levels of reality. We have positioned the physical laws on a level above the level for the specific models (Fig. 4.1), where the specific models come into existence through $(\mathbf{p}, E)$-fluctuations in $(\mathbf{p}, E)$-space corresponding to pictures in $(\mathbf{r}, t)$-space. These models should particularly be able to *describe* the hard objects in front of us, and these theoretical pictures for the hard objects are *descriptions* (not observations) which are located in the observer's head.

But what about the *observed* hard objects which appear spontaneously in front of us in connection within our *observations* (not descriptions) in everyday life? How can we classify these *observed* hard objects with respect to the levels of reality? Also these *observed* objects in front of us (trees, cars, etc.) are located in the brain of the observer (Section 4.1). We have only the impression that all these objects are in front of us (outside of us). That is, the hard objects appear as material elements embedded in an "outside world", and most people believe that this observed "outside world" is identical with basic (objective) reality. This is obviously a fallacy. That which we believe is the "outside world" does not exist in this form. We already discussed this point in Section 4.1. Let us continue to study this in the following.

### 4.4.2. *Everyday Life Experiences*

Everything that we see within our experience in everyday life is primarily in our head; it is not outside us. Persons, cars, trees, aeroplanes, sun, moon and stars are pictures in our brain. All these objects are pictures of reality produced by projections from basic reality and are located in our brain. However, we have the impression that all these things are located outside us, and we assume that these sensations produced by the brain are identical to reality itself, but this is not the case, as we have analyzed in Chapter 2. (In particular the statement by Brückner in Section 4.1 is instructive).

Again, everything we observe (persons, cars, trees, etc.) is primarily a picture located in our brain. It is only an impression (produced by the brain) that we are standing opposite these things. This interpretation is not correct but a fallacy.

Thus, not only the *described* pictures of the hard objects (trees, cars, etc.) are states of the observer's brain but also the *observed* objects in connection with our observations in everyday life are states of the observer's brain as well. The *described* pictures and the *observed* objects are equally located in the brain of the observer (Fig. 4.3).

There is no principal difference between them. Therefore, the *observed* objects can also be assigned to a certain level of reality as in the case of the described pictures of them. But where is this level positioned? It is the macroscopic level and should be below *level a* (Fig. 4.2). Before we discuss this point let us first quote some opinions in connection with the relationship which we believe is inside and that which is outside of us, simultaneously producing an internal image inside us with exactly the same structure.

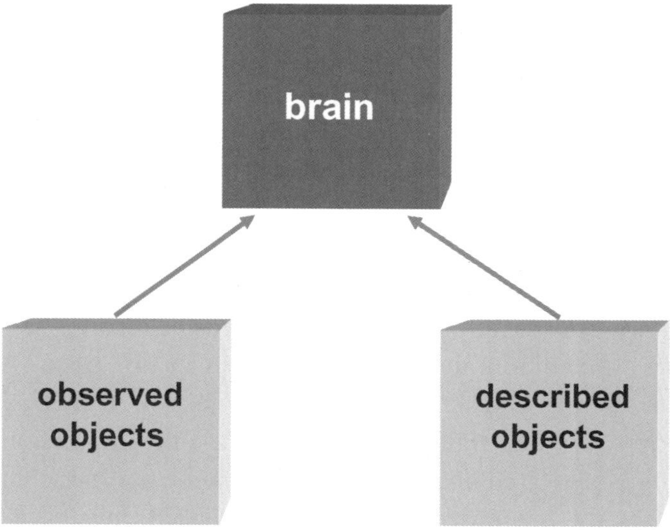


Fig. 4.3. Within projection theory, the *described* bodies (pictures) of the hard objects (trees, cars, etc.) and the *observed* objects are equally located in the brain.

It is usual to assume that all these things produced by our brain are identical with the things embedded in basic (objective) reality. That is, it is assumed that the phenomena in nature exist doubly. For example, the famous psychologist C. G. Jung was of the opinion that an event takes place twice, one in the head of the observer and the other identical event in the world outside. (See in particular Section 1.9. and Section 2.6).

In other words, Jung believes that the phenomena in nature exist doubly. However, it is naïve to assume that this is really the case. No, the assumption that the structure of the outside world is identical with the picture of reality, produced by the observer's brain, is merely a hypothesis and has to be considered as disproved by the principles of evolution. These important principles speak another language (see also Chapter 1) and we come to projection theory with its fictitious realities described in $(\mathbf{p}, E)$-space. Here the pictures of reality are obtained by transmitting the information about the things (trees, cars, etc.) from $(\mathbf{p}, E)$-space to $(\mathbf{r}, t)$-space. Again, the structures in $(\mathbf{p}, E)$-space are quite different from those in $(\mathbf{r}, t)$-space, and we can say nothing about the absolute character (structure) of all the things around us because basic (objective) reality remains hidden (Fig. 4.2). All these points have already been discussed above.

## Statement by Heinz Von Foerster

That the outside world cannot be of the composition that we have in front of us in connection with assumptionless observations, becomes particularly apparent when we realize that there is, for example, no light and no colour and also no sound and no music in the world outside. Since the world before our eyes is essentially made of light and colours, this observed world cannot be reality outside. What we call outside world exists only in our brain. In this connection the following comment is essential [24]:

> *The eminent cyberneticist Heinz Von Foerster points out that the human mind does not perceive what is, 'there', but what it believes should be there. We are able to see because our retina absorbs light from the outside of the world and conveys the signals to the brain.*

> *The same is true of all of our sensory receptors. However, our retinas do not see colour. They are 'blind', as Förster puts in, to the quality of their stimulation and are responsive only to their quantity. He states 'This should not come as a surprise, for indeed, 'out there' there is no light and no colour, there are only electromagnetic waves. 'Out there' there is sound and no music, there are only the periodic vibrations of air pressure. 'Out there' there is no heat and no cold, there are only moving molecules with more or less mean kinetic energy, and so on. Finally, for sure, 'out there' there is no pain. Since the physical nature of the stimulus — its quality — is not encoded into nervous activity, the fundamental question arises as to how does our brain conjures up the tremendous variety of colourful world as we experience it at any moment while awake, and sometimes in dreams while asleep'.*

In conclusion, light, as well as many other things, does not really exist in the outside reality. The world that we have in front of us in connection with assumptionless observations is mostly an invention of our brain. The statement by Heinz Von Foerster is close to what we have analyzed in connection with projection theory which is essentially based on the principles of evolution (Chapter 1). However, in contrast to Heinz Von Foerster within projection theory there are principally no statements about the world outside (basic reality) possible.

### 4.4.3. *Other Levels of Reality*

So far we have discussed two levels of reality: the level of general physical laws and the level of specific models. Both levels have been constructed by theoretical physics. Only the objects (features) defined on a certain level are accessible to an observer. They have their origin in basic (objective) reality and reflect certain features of it. The direct observation of basic reality is not possible for an observer. Therefore, such objects (features) are not inventions of the observer but have to be considered as discoveries, as observed objects so to speak. We will discuss this point below in connection with the notion "levels of observation".

#### 4.4.3.1. *Level of Macroscopic Objects*

We pointed out above that everything we observe (persons, cars, trees, etc.) is primarily a picture located in our brain. Thus, not only the *described* pictures of the hard objects (macroscopic objects like trees, cars, etc.) are states of the observer's brain but also the things that we *observe* in everyday life on the basis of the five senses are states of the observer's brain. The *described* pictures and the *observed* macroscopic objects are equally located in the brain of the observer (Fig. 4.3). The described pictures and the general physical laws represent conscious states. In contrast to that, the macroscopic objects that we observe spontaneously in everyday life come into existence in an unconscious way and appear as spontaneous pictures in front of us.

There is no principal difference between the described and the observed pictures. Therefore, such everyday life objects can also be assigned to a certain level of reality as in the case of the other two levels (Figs. 4.1 and 4.2). But where is this level positioned?

Within assumptionless observations (macroscopic level) the objects and phenomena are ordered corresponding to the direct, visual impression, i.e., as they appear spontaneously in front of us. Other information (as, for example, from interpretations and theoretical descriptions) has by definition no place on this level. Also the information from measurements does not belong to the level of macroscopic objects. Thus, the level of macroscopic objects (*level 1*) should be below *level b* and *level a* (Fig. 4.4).

#### 4.4.3.2. *Level of Basic Rules*

We can determine all the specific pictures $[\Psi^*(\mathbf{r},t)\Psi(\mathbf{r},t)]^{\text{specific}}$ [Eq. (4.2)] on the basis of the general physical law (3.22) but, on the other hand, we cannot deduce the scientific characteristics and rules, on which the physical law is based, on the basis of a specific model (picture) $[\Psi^*(\mathbf{r},t)\Psi(\mathbf{r},t)]^{\text{specific}}$. Again, the general physical law is more than one of its specific solutions. It contains potentially the characteristics of all possible physical systems.

In fact, the law $i\hbar\partial/\partial t\Psi(\mathbf{r},t) = -\hbar^2/(2m_0)\Delta\Psi(\mathbf{r},t)+V(x,y,z,t)$ $\Psi(\mathbf{r},t)$ [Eq. 3.22)] has been formulated on the basis of the projection

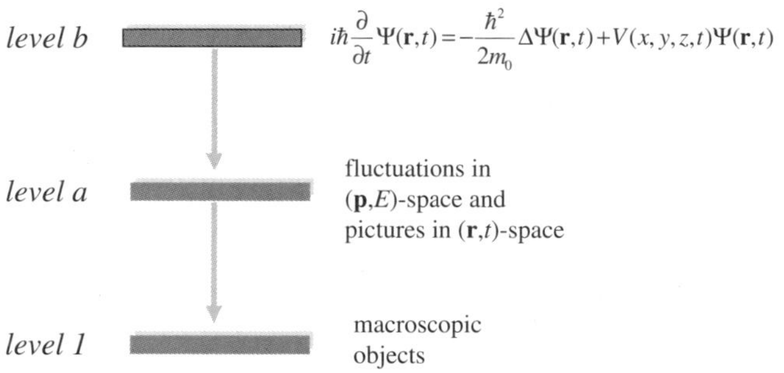


Fig. 4.4. The level of macroscopic objects (*level 1*) is positioned below *level a* and *level b*. Again, the various levels have been vertically arranged in accordance with the degree of generality.

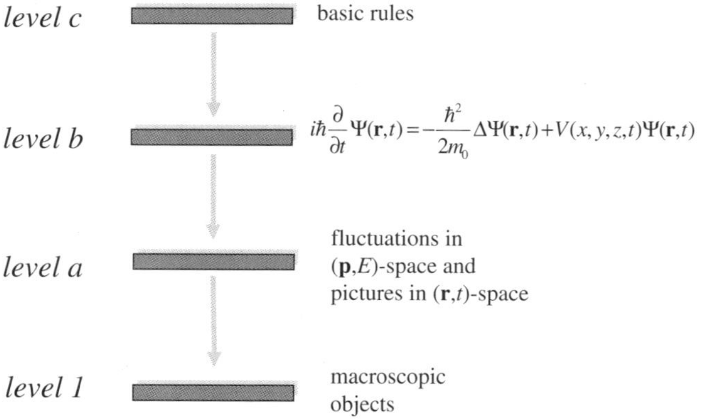


Fig. 4.5. Hierarchical level structure in accordance with the degree of generality.

principle and the operator rules deduced in [12]. These basic rules (projection principle, operator rules) are therefore positioned on a level (*level c* in Fig. 4.5) above *level b*. The basic rules are more general than the physical law $i\hbar\partial/\partial t\Psi(\mathbf{r},t) = -\hbar^2/(2m_0)\Delta\Psi(\mathbf{r},t) + V(x,y,z,t)\Psi(\mathbf{r},t)$. In principle, we could formulate other physical laws [different from $i\hbar\partial/\partial t\Psi(\mathbf{r},t) = -\hbar^2/(2m_0)\Delta\Psi(\mathbf{r},t) + V(x,y,z,t)\Psi(\mathbf{r},t)$] on the basis of these basic rules.

In conclusion, the *principle of projection* belongs to the "level of basic rules", but also the operators that have been deduced within the frame of the projection principle [12]. Various combinations for the

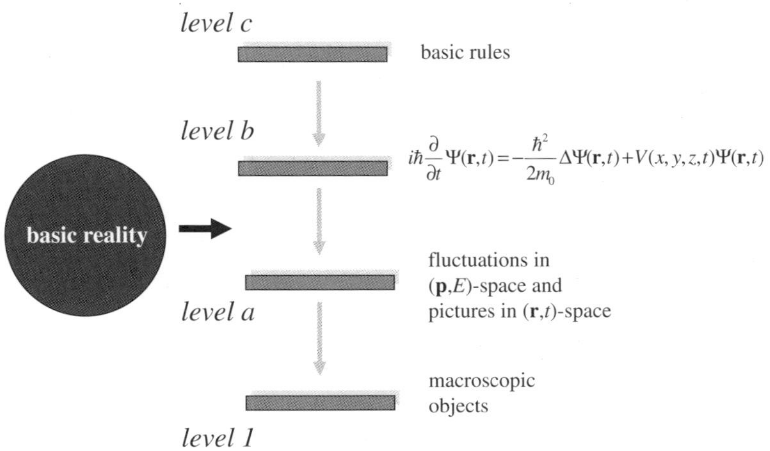


Fig. 4.6. All levels (basic rules, . . . , macroscopic objects) are connected to basic reality. The structure at a certain level describes a certain facet of basic reality. Basic reality itself is not observable.

variables and the corresponding operators lead to various forms of basic physical laws [as, for example, the law given by Eq. (3.22)] which are then positioned below the level of basic rules (*level c*) in Fig. 4.5.

All levels quoted in Fig. 4.5 are correlated with basic (objective) reality in the sense outlined above, and we come to Fig. 4.6.

### 4.4.3.3. *Level of Technological Developments*

We may define a *level 1'* on which only natural macroscopic objects like trees, stones, etc. are positioned, i.e., *level 1'* is part of *level 1* (level of macroscopic objects). Then we can define a level of technological developments. With the physical laws (*level a* and *level b*) and the natural material objects positioned on *level 1'*, we can construct technological systems (houses, cars, etc.) which also appear in front of us in an unconscious way. Therefore, such constructions must be positioned on a level between *level a* and *level 1'* (level of technological developments, it is *level a1'* in Fig. 4.7). The split of *level 1* leads to *level 1'* and *level a1'*: *level 1* → *level 1'* + *level a1'*.

If the relevant part of a technological system is realized in connection with the microscopic elements and the corresponding effects

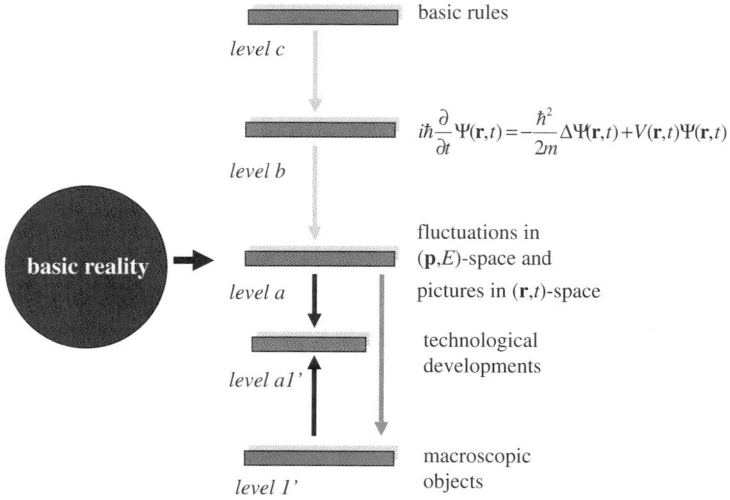


Fig. 4.7. Hierarchical level structure in accordance with the degree of generality. The level of technological developments (*level a1'*) is positioned between the level of natural material objects (*level 1'*) and *level a*.

typical for *level a* (as, for example, atoms which is not observable within the frame of assumptionless observations), this technological system is of course positioned at the same level, i.e., on level a and we have *level a1'* = *level a*. More details are given in below (Levels of Observation).

### 4.4.3.4. *The Various Levels Represent States of the Brain*

All levels are correlated with basic reality and we come from Fig. 4.5 to Figs. 4.6 and 4.7. As we have outlined above, the described pictures and the observed objects are equally located in the brain of the human observer (Fig. 4.3), and this is of course the case for all information on the various levels. All is represented by "states of the brain" and is experienced by the observer consciously or unconsciously depending on the level. In other words, within projection theory, the mind is the most direct thing in our experience. Arthur Eddington wrote in this connection: "*. . . Mind is the first and most direct thing in our experience*". (See also Section 2.6.) The reason for that is given within projection theory by the fact that basic reality is principally not accessible to the observer and, furthermore, that there is no space and no time outside and independent of the observer, but space and time exclusively appear in connection with brain functions (Section 3.17).

## 4.5. Objectivation Processes

We concluded that the "things" at the various levels represent states of the brain and reflect certain features of basic (objective) reality. Are these "things" discoveries or inventions? We do not know the details about the mechanisms within the observer's brain. However, we may state very generally that the observer picks up a certain information from basic reality (he discovers it) which is transformed by the observers brain into a picture of reality. It is a picture represented in space and time [$(\mathbf{r}, t)$-space] whereby space and time do not exist in basic (objective) reality (Chapter 2). That information, which is not despictable in $(\mathbf{r}, t)$-space, is given symbolically in the form of equations. Thus, the basic information is a discovery whereas the pictures and the equations should be considered as an invention of the observer. These discoveries are done by so-called objectivation processes, and we have to distinguish between conscious and unconscious objectivations. Let us briefly discuss these points.

### 4.5.1. *Objectivation by Thinking (Conscious Objectivation)*

Objectivation by thinking here means that a physical view of the world is tested via a dialogue with nature on a certain level of reality. In this way a world view can be improved and, if necessary, also be rejected. The theory should be reflected in many and, as far as possible, in all physical situations, and that has to be verified and this defines the process of objectivation. Such a procedure is meant when we talk of called "objectivation by thinking". In other words, within projection theory a certain world view can be objectified and discovered, respectively. A world view that is useful should be constantly reflected in many variations of thinking and different experimental configurations.

Similar criteria — perhaps not so severe — should be fulfilled in connection with the products of imagination (fantasy). A literary picture very often also sums up reality in a single "image" (here in the form of a metaphor). This, however, can be applied in many situations of life, and the same "image" (metaphor) is reflected in the experience of many people. We may therefore state that the author of a narration also found this "single image" by a process of objectivation, just as in the case of a physical theory. More details are given in [9].

### 4.5.2. *Constancy Phenomena (Unconscious Objectivation)*

Also in connection with the "hard objects", there are objectivation processes (see also [9] and the literature therein). A "hard object" observed by the individual is unambiguously recognizable even in the case of large variations in the environment. This is guaranteed by the so-called constancy phenomena whose objectivation performance is managed by a complex physiological apparatus. Constancy phenomena thus allow objectivation in an unconscious, non-intellectual way by the physiological apparatus.

Also the elements recorded by measuring instruments (belonging to *level a1*': technological developments; see Fig. 4.7) are observed unconsciously. At the level of measuring instruments the observer imposes restrictive conditions on nature. This is done by the construction of the specific measuring instruments, leading to construction conditions that are imposed by the observer in order to get a specific answer on a specific question. Due to the construction of the measuring instruments we register only a few specific signals from a multitude of possible events in nature. The selection or construction of a measuring instrument has, in a certain sense, the meaning of a constancy mechanism, which we discussed in connection with unconscious objectivation, where the objectifying performance comes into existence through a complex physiological apparatus [9]. The registration (observation) of signals in connection with measuring instruments also takes place automatically, i.e., they are recorded without our conscious help, even if the selection of a measuring instrument is a conscious act.

### 4.5.3. *Conclusion*

We have stated in Section 4.2 that objects belonging to the various levels of reality are equally *real*, and we observe all these objects at the various levels by a common principle: it is the principle of *objectivation*. This principle is valid at each level, i.e., it can be applied independently on which level an object is located. We distinguish between "objectivation by thinking" and "objectivation in an unconscious, non-intellectual way" (see in particular [9]). This feature, principle of objectivation, supports the view that there is no principal difference between "states of mind" and "states of matter"; there are only gradual differences.

## 4.6. Observations

### 4.6.1. *Objects on Various Levels*

For biological systems, different from human observers, the same principal scheme should be valid but, due to the evolution, the effects that emerge on the various levels should be different from those defined for human observers. We are caught in our own system and, therefore, it is difficult for a human observer to say something about other biological systems (animals). However, some interesting, important and surprising experiments exist which have been done within behavior research (see Section 2.9 and [9]).

The various levels of reality emerge properties that are qualitatively different from each other although the various levels reflect features of the same reality (basic reality). We may state that at each level of reality emerges a certain facet of basic reality.

The difference between the "objects" on the various levels is not only reflected in their theoretical description but also in their effect on us. In this connection, the kind of objectivation is of special importance (Section 4.5). In principle, the objectivation processes on the various levels are different from each other. Since the basic information on the various levels have to be considered as *discoveries* and not as inventions (Section 4.5) we may state that each "level of reality" is accompanied with a "level of observation"; because the features of the "objects" on the various levels are defined differently, the methods of observation must vary from level to level.

The process of objectivation emerges the "objects" on the various levels, and these produce certain level-specific feelings inside the observer. We feel the effect of "objects", which are objectified *unconsciously* (hard objects like trees, cars, etc.), differently from those that are objectified *consciously* (scientific laws, products of imagination, etc.). In Section 4.5 we have distinguished between "objectivation by thinking" and "objectivation in an unconscious, non-intellectual way". In principle, many objectivation methods should be possible depending on the number of levels and the features at the various levels.

### 4.6.2. *Levels of Observations*

The macroscopic objects, positioned on *level 1'*, are by definition exclusively observed by the observers five senses, and we do that within the frame of assumptionless observations (*level I'*, see Fig. 4.8). Again, within assumptionless observations the macroscopic objects and phenomena are ordered corresponding to the direct, visual impression, i.e., as they appear spontaneously in front of us.

In connection with technological developments (*level a1'*, level of artificial macroscopic objects) we need as observation methods (*level AI'*) not only the five senses but also certain instruments which

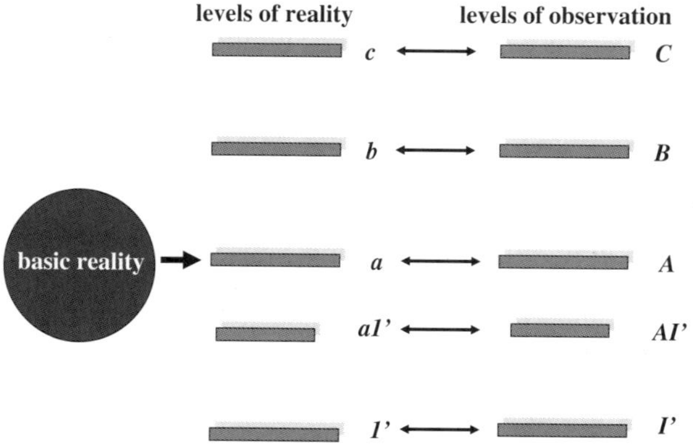


Fig. 4.8. Levels of reality (1'-c) in relation to the levels of observation (I'-C). The correspondence between the two types of levels is essential.

are necessary for the building up of the artificial macroscopic objects as, for example, houses.

Observations with respect to *level a* also need technological devices but their relevant parts must be tailor made with respect to the physical effects typical for this level, *level A* (see Fig. 4.8). The observation methods on *level A* are in general more sophisticated than those on *level AP* since we enter here the realm of microscopic objects as, for example, nanosystems. Clearly, at *level a* the objectivation by thinking comes into play because the models for the $(\mathbf{p}, E)$-fluctuations in $(\mathbf{p}, E)$-space and the pictures in $(\mathbf{r}, t)$-space, positioned on this level, have to be found and formulated, respectively.

The observation of the "objects" typical for *level b* [general physical laws as, for example, $i\hbar\partial/\partial t\Psi(\mathbf{r},t) = -\hbar^2/(2m_0)\Delta\Psi(\mathbf{r},t) + V(x,y,z,t)\Psi(\mathbf{r},t)$] and those typical for *level c* (basic rules) are observed by thinking (objectivation by thinking, Section 4.5), i.e., without material observation devices. In the objectivation of the basic rules (*level c*), we need other trains of thinking than in the objectivation of the general physical laws (*level b*). Clearly, the situation cannot be analyzed by means of models that are based on $(\mathbf{p}, E)$-fluctuations in $(\mathbf{p}, E)$-space and on pictures in $(\mathbf{r}, t)$-space. The laws positioned on *levels b* and on *level c* are not depictable in $(\mathbf{r}, t)$-space and represent "states of mind" which cannot be described by "states of matter", i.e., such laws cannot be produced by $(\mathbf{p}, E)$-processes in $(\mathbf{p}, E)$-space (see also the detailed discussion in Section 4.2).

### 4.6.3. *Other Biological Systems*

Concerning other biological systems let us repeat what we have said in Section 4.6.1: For biological systems, different from human observers, the same principal scheme should be valid but, due to the evolution, the effects that emerge on the various levels should be different from those defined for human observers. We are caught in our own system and, therefore, it is difficult for a human observer to say something about other biological systems (animals). However, some important investigations already exist (see in particular Section 2.9 and [9]).

### 4.6.4. *Concluding Statements*

In Section 4.5, we concluded that unconscious and conscious objectifying are analogous processes. They both select certain information from a diversity of "objects" and processes. Both objectifying procedures should be classified as observation processes.

We have said above that unconscious objectifying takes place on the level of the five senses (macroscopic level) and also in part on the level of measuring instruments, even when here the human consciousness is essentially involved. Conscious objectifying is objectifying through thinking and takes place at specific levels of reality. Here the "objects" are also selected from basic (objective) reality. This process, however, is exactly the same as the one which takes place on the level of the five senses and the level of measuring instruments. Konrad Lorenz has already pointed out the strict analogy between unconscious objectifying through the physiological apparatus and conscious, intellectual objectifying (see also [9] and the literature therein). Thinking is therefore a certain kind of observation procedure, and in fact a very versatile one. It is not new to state that one can think on different levels.

In this connection, we can consider the reading of a specialized book as an observation procedure. The various chapters of the book can be written on different levels, that is, one chapter might deal with factors which are less general than those treated in other chapters. The observing (by thinking) of the different chapters presupposes different ways of thinking. Thus, we come to the "way of thinking 1", "way of thinking 2", etc., which can be arranged according to the degree of generality of the levels of reality. Does the reading of a book not presuppose that the cognition system of the reader is an open system?

## 4.7. No Interactions within $(\mathbf{r},t)$-Space

In this section we would like to underline once more that we operate in conventional physics after principles that are different from those in projection theory. Within projection theory the world is analyzed on the basis of certain levels (levels of reality), and each level reflects a certain facet of basic reality. In contrast to that, within conventional physics the material world is embedded in space and time. Here we have one (absolute) view and no facets as in projection theory. In this section we would like to repeat the meaning of the conventional view from the point of view of projection theory.

### 4.7.1. *Classical Mechanics and Usual Quantum Theory become mere Calculation Schemes*

Classical mechanics and usual quantum theory are based on the idea that the interacting objects are embedded in space and time (represented by $\mathbf{r} = (x, y, z)$ and $\tau$). According to projection theory, a re-interpretation is necessary because here the objects are not embedded in space and time. From the point of view of projection theory, classical mechanics (and of course usual quantum theory) becomes a scheme which makes it possible to calculate the geometrical positions in the pictures "as if" the objects were interacting in space and time. It is not so but we "pretend" that it is so. Newton's mechanics (and usual quantum theory) is, under these circumstances, not more than a mere calculation scheme, directly derived from observations in everyday life. Let us briefly discuss an example.

### 4.7.2. *An Example*

For example, when a ball is thrown, the definite impression emerges that the geometrical position is moved by the strength of the thrower. Because of this concrete feeling of the observer the *position* of the ball is identified with a *material object* (ball). The geometrical position seems to exert a pressure on the palm of the throwers hand (for example, through gravity or muscle power). The whole scenario is consistent because we also have a picture of the thrower in front of us. That is, the palm also merely represents a geometric figure in $(\mathbf{r}, t)$-space, which is actually in basic reality together with the ball, and both elements interact in basic reality. We model this interaction process by $(\mathbf{p}, E)$-fluctuations within $(\mathbf{p}, E)$-space and the whole scenario is projected onto $(\mathbf{r}, t)$-space.

In summary, the process of throwing exclusively takes place in basic reality. The thrower and the material object (ball) interact within basic reality but we have only a picture of them in front of us. It is a picture within $(\mathbf{r}, t)$-space. There is a strong correlation between the relative positions of the palm and the ball and the real physical effect producing the impression that the material objects (palm and ball) are embedded in $(\mathbf{r}, t)$-space, but this is not the case. There are merely geometrical positions in $(\mathbf{r}, t)$-space.

The formulations "as if" and "pretend", used above in connection with the ball and the thrower, correspond to the situation as we perceive the world with our senses and behave towards it. The objects around us appear as material objects to be located in $(\mathbf{r}, t)$-space, although these objects are exclusively positioned as pictures in our brain. We discussed in Chapter 2 and also in Section 4.1 that it is characteristic for the process of seeing that our awareness does not register the picture inside the eye on the retina, but we have the impression that we are observing external objects (or rather, "as if" we were observing external objects). In other words, we have the impression that all these things, including the other observers, are outside in a common space–time [$(\mathbf{r}, t)$-space]. This is the situation at the "level of assumptionless observations" (Section 4.6). It is the physiological apparatus which creates this effect. So far, this "as if" is not artificial because it is realized in nature.

### 4.7.3. *Application of the Results Deduced in Projection Theory*

In accordance with projection theory one would say that there is a strong correlation between the geometrical position and the perceptions of the acting person, but the ball, the material object, is not actually embedded in $(\mathbf{r}, t)$-space. When the geometrical position of the ball is identical with the geometrical position of the observer's hand, a certain effect arises in the observer. There are definite relations between the actual event in basic (objective) reality and the geometrical positions in the picture.

This conception is nothing else than what we have discussed in Chapter 3 in connection with the wavefunction $\Psi(\mathbf{r}, t)$. Let us summarize the relevant facts.

$\Psi(\mathbf{r}, t)$ is determined at location $\mathbf{r}, t$ of space–time by all possible values $\mathbf{p}$ and $E$ $(-\infty < \mathbf{p}, E < \infty)$, which are given with the probability density $\Psi^*(\mathbf{p}, E)\Psi(\mathbf{p}, E)$. Therefore, $\Psi^*(\mathbf{r}, t)\Psi(\mathbf{r}, t)$ can only be interpreted in connection with the variables $\mathbf{p}$ and $E$. There is no other way: At time $\tau$, one of the possible values for $\mathbf{p}$ and for $E$ is present in the intervals $\mathbf{r}, \mathbf{r} + d\mathbf{r}$ and $t, t + dt$ with the probability density of $\Psi^*(\mathbf{r}, t)\Psi(\mathbf{r}, t)$. Since only the variables $\mathbf{p}$ and $E$ are accessible to measurements we can also state the following. The measurement of one of the possible values for $\mathbf{p}$ and for $E$ at time $\tau$ is done in the space–time intervals $\mathbf{r}, \mathbf{r} + d\mathbf{r}$ and $t, t + dt$ with the probability density of $\Psi^*(\mathbf{r}, t)\Psi(\mathbf{r}, t)$. A detailed analysis is given in Appendix A.

Let us mention once more that the variables $\mathbf{p}$ and $E$ do not belong to $(\mathbf{r}, t)$-space but to fictitious reality, which is embedded in $(\mathbf{p}, E)$-space and which approximates basic reality.

This is actually the situation in practical experiments. Signals are recorded with detectors in space and time, no more, no less. This has primarily nothing to do with a material particle localized in space.

The situation just described for $\Psi(\mathbf{r}, t)$ in connection with $\mathbf{p}$ and $E$ exactly works in the same way as in the case of the process of throwing where the hand and the ball are in the center. There is an interaction between the hand and the ball. When the geometrical position of the ball is identical with geometrical position of the observer's hand, a certain effect arises in the observer [due to the $(\mathbf{p}, E)$-fluctuations, i.e., the pair interaction between the hand and the ball].

## 4.7.4. *Conclusion*

Classical mechanics and usual quantum theory were very successful and are still very successful. But the idea that these theories describe facts that actually occur in this way in nature is certainly not correct when we judge the situation from the point of view of projection theory.

## 4.8. The General cannot be Deduced from the Particular

### 4.8.1. *Concept of Induction*

After the concept of induction, the general is concluded from the particular. If we base however our considerations on projection theory, the concept of induction cannot be valid. Within projection theory it is not possible to derive a physical law belonging to a certain level (say *level b*, Fig. 4.5) from the laws of another level (say *level a)* which is located in the hierarchy below *level b*. (In Section 4.4, we have arranged the levels of reality according to the degree of generality.) The structures (laws) of *level b* are by definition more general than those of *level a* and this means that *level a* cannot contain the "additions" which are necessary to pass from *level a* to *level b*. If it did, this would mean that the laws (structures) of *level b* were already contained on *level a*, which would entail that *level b* could not be located above *level a* in the hierarchy for the levels of reality. The situation in connection with *level a* and *level b* can be summarized as follows.

Any specific model is captured in $(\mathbf{r}, t)$-space $[(\mathbf{p}, E)$-space$]$, that is, within the framework of a specific model, one cannot leave $(\mathbf{r}, t)$-space $[(\mathbf{p}, E)$-space$]$ in order to describe a general physical law which is not depictable in $(\mathbf{r}, t)$-space. Therefore, both levels can principally not be joined. However, we may conclude a specific model from the general physical law, but we cannot conclude the general physical law from the specific model. The "concept of induction" is not valid. instead projection theory leads to the "principle of deduction".

### 4.8.2. *New Ideas are Needed*

Therefore, according to projection theory the laws cannot be derived by *induction* but only by *deduction*, i.e., projection theory only allows us to conclude the structures of a certain level from those of another level located above it. We discussed above that the laws of *level b* [as, for example, $i\hbar\partial/\partial t\Psi(\mathbf{r}, t) = -\hbar^2/(2m_0)\Delta\Psi(\mathbf{r}, t) + V(x, y, z, t)\Psi(\mathbf{r}, t)$] cannot be derived from specific models which are located on *level a*. This means that the creation of a new scientific law must be initiated by a principally new idea. From this idea, follows the deduction and the comparison with well-known structures which are below that level on which the new scientific law has been established.

The new idea is relevant, and a lot of examples demonstrate the validity of this principle. Newton's theory contains a completely new idea compared with Kepler's laws, i.e., the idea of force. It is impossible to deduce this new idea (either deductively or inductively) from Kepler's laws. From our point of view, the concept of force can therefore be considered as an "addition", i.e., it is an "addition" to Kepler's laws. The level of Newton's theory is therefore above that of Kepler's laws within the hierarchy; Kepler's laws can be deduced from Newton's theory. Newton himself believed mistakenly that he had only extended Kepler's laws by induction. However, Kepler's laws can be derived from Newton's mechanics by deduction. Newton's mechanics did not supplement Kepler's theory but completely superseded it. Newton's mechanics is more general than Kepler's theory.

### 4.8.3. *Where do Ideas come from?*

Important questions in this connection are: What is an idea? Where does it come from? Because of the strong analogy between "hard objects" and the products of mind (e.g., scientific laws), which we discussed above, we can establish ideas on certain levels of reality; they are not components of the individual cognition apparatus, as we already outlined in [9]. The cognition apparatus takes ideas from reality by objectivation. Thus, ideas are pictures of reality on certain levels of reality, just as "hard objects" are pictures of reality on a certain level.

### 4.8.4. *Positivism*

Positivism is a branch of philosophy which starts from the "positive", whereby the "positive" is understood as the certain and unquestionable. The positivist essentially limits his research to these points; metaphysical debates are from this point of view theoretically impossible and practically useless.

The basis for all knowledge for the positivist is therefore certainty and doubtlessness, whereby he believes that this can only be experienced by his senses. He therefore only learns something about the world by opening his eyes and ears. In his view, he is born without knowledge and all knowledge about the world is exclusively realized through the five senses. In other words, the positivist does not believe that there is more than his sensory impressions.

The positivist is basically a solipsist, i.e., a person that only recognizes the content of his own consciousness as existing. Karl Popper outlines in [25] that all people who are not solipsists, but are nevertheless positivists, have already made a compromise if they concede that other persons exist in addition to themselves.

Bertrand Russell went a step further with his "solipsism at the moment", for he does not even admit that there is a past and a future. Russell does not even trust the memory of past experience, nor the projection of the experience into the future, i.e., for him these elements are not sufficiently sure and reliable.

These ways of thinking, which form the basis of positivism, are certainly consistent in their logic, even including the "solipsism at the moment", as long as one recognizes that certainty and the unquestionable actually exist.

Karl Popper says very aptly [25] that one must understand positivism as a far-reaching generalization of the concept of induction, i.e., the conclusion of the general from the particular. In [25], we find the following comment:

> *"Positivism is actually the view that the idea of going from the particular to the general must be applied so strictly that we start from our direct observations, especially from our elementary feelings produced by the senses. From these experiences, we then gradually develop our knowledge of the world and our theories. That is positivism. Mach developed these conceptions very strongly, and this is particularly expressed in his famous book 'Die Analyse der Empfindungen'. Feelings are for him the elementary observational experiences, and in his early phase Einstein was also a follower of this view — although later he reacted very strongly against it, and then arrived at a completely different view of the world."*

Positivism is thus essentially based on the concept of induction, i.e., it is concluded from the particular to the general. However, experience tends to speak against the concept of induction, and in fact militantly so, that such a view can hardly be maintained. Too many breakthroughs in research have been achieved by deduction, that is, the particular has been inferred from the general. We already pointed out above that Kepler's laws can be derived from Newton's mechanics by deduction. Here are just a few more examples which have already been quoted in [8,9].

1. In connection with Albert Einstein, "mathematical intuition" was a relevant point and the starting point of his theoretical investigations. He was guided by the mathematical beauty of a theory. Using this theoretical approach he then deduced the various observable facts and compared them with experimental results.
2. The discovery of the DNS in 1953, certainly one of the greatest discoveries in the last century, was also made in a deductive way. Watson, Crick and Wilkins designed the macro-molecule step by step in advance and proved it experimentally afterwards.
3. In accordance with Kepler's first law, the planets move on trajectories, whose form is described by ellipses. The ellipse was a fundamentally new idea as a solution for the problem of orbital paths. It does not follow from the circle by induction. The orbit can be a circle if we assume that the elements of the world can only be experienced without exception through the senses, which is precisely the position of the positivists. Then, the intuitive basis for the movement of the planets is the movement of a wheel. Kepler was convinced of the harmony of the world. This conviction (and not the wheel) possibly helped him to objectify the ellipse as the path of orbit.

Experience tends to speak against the concept of induction. Such a view can hardly be maintained. However, for a lot of scientists the concept of induction is still convincing, and the reason for that is probably the belief that the facts can possibly be re-interpreted some day in favor of the concept of induction.

### 4.8.5. *Groping in all Directions*

In a discussion with Franz Kreuzer, Karl Popper underlines significantly that the concept of induction has to considered as wrong [25]:

> *"Groping in all directions. I do not favor the picture of science as gathering observations and distilling the laws from them in the same way, as Bacon said, as wine is pressed from grapes. In this conception the grapes are the observations and the wine obtained from them is the generalization, the theory. This conception is utterly wrong. It mechanizes the creative act of human thinking and inventing. To make that point clear is most important to me. Science proceeds in a different way, that is, it proceeds in such a way as to test ideas and world views. Science derives from myths. This can be seen very clearly in connection with the early scientists, i.e., the early Greek, pre-Socratic philosophers, who were still strongly influenced by myths."*

In other words, the concept of induction (the general is concluded from the particular) should not be correct but just the opposite, i.e., the particular is concluded from the general (concept of deduction).

"Groping in all direction" means testing a physical view of the world via a dialogue with nature on a certain level of reality. In this way, a world view can be improved and, if necessary, also be rejected. The theory should be reflected in many and, as far as possible, in all physical situations, and that has to be verified. Such a process can be called "objectivation by thinking".

A theory that is useful should be constantly reflected in many variations of thinking and different experimental configurations. This should also be valid for other fields. In Section 4.5.1, we remarked the following: "Similar criteria — perhaps not too severe — should be fulfilled in connection with the products of imagination (fantasy). In this connection, we outline the following. A literary picture very often also sums up reality in a single "image" (metaphor). This, however, can be applied in many situations of life, and the same "image" (metaphor) is reflected in the experience of many people. We may therefore state that also the author of a narration found this single "image" by a process of objectivation, just as in the case of a physical theory. This point is discussed in more detail in [8,9]".

### 4.8.6. *The Viewpoint of Usual Physics*

Within usual physics, it remains uncertain what an *idea* is at all. Generally speaking, in connection with the "world of ideas" there is an immense diversity of opinion and a huge variety of interpretations. There does not even exist a satisfactory interpretation of usual quantum theory up to the present day, although the essential features of the mathematical formalism of this theory was already constructed approximately 80 years ago.

What is the relationship between the world of the ideas (states of mind) and that what we call physical reality (states of matter)? Within usual physics, there is no clear concept and definition, respectively, to answer this and similar questions. This might probably also be the reason why the concept of induction is still used in literature.

### 4.8.7. *The Viewpoint of Projection Theory*

When the states of mind [as, for example, a product of fantasy or a physical law like $i\hbar\partial/\partial t\Psi(\mathbf{r},t) = -\hbar^2/(2m_0)\Delta\Psi(\mathbf{r},t) + V(x,y,z,t)$ $\Psi(\mathbf{r},t)$] are created by the physical brain activities, then it is not absurd at all to assume that the general physical law can be concluded from observed data, and the concept of induction would be valid. The general is concluded from the particular. However, as we have seen above, a lot of arguments and convincing examples speak against the validity of the concept of induction. Therefore, a physical law like $i\hbar\partial/\partial t\Psi(\mathbf{r},t) = -\hbar^2/(2m_0)\Delta\Psi(\mathbf{r},t) + V(x,y,z,t)\Psi(\mathbf{r},t)$ (state of mind) should not be created by physical brain activities, i.e., by its states of matter. There must be more than that. Such a far-reaching statement is however not acceptable for a lot of scientists. In their opinion there are not enough proofs for that. The standard conceptions are obviously too vague. Here projection theory helps further. We pointed out above that the states of mind cannot be produced by the physical brains functions (its states of matter).

Projection theory opens the possibility for a more detailed analysis. Here the "states of mind" and the "states of matter" describe equally basic reality in symbolic form on different levels, and we are able to distinguish clearly between these states, i.e., projection theory offers a solution in general form for the relevant questions in connection the "states of mind" and the "states of matter".

In particular, it turned out that the concept of induction cannot be valid within projection theory but exclusively the "principle of deduction". For example, the physical law $i\hbar\partial/\partial t\Psi(\mathbf{r},t) = -\hbar^2/(2m_0)\Delta\Psi(\mathbf{r},t) + V(x,y,z,t)\Psi(\mathbf{r},t)$ cannot be deduced from specific models but we need more general statements (basic rules). The situation is summarized in Fig. 4.9.

Therefore, also positivistic philosophies have no place here. Furthermore, we may state quite generally that strict materialistic

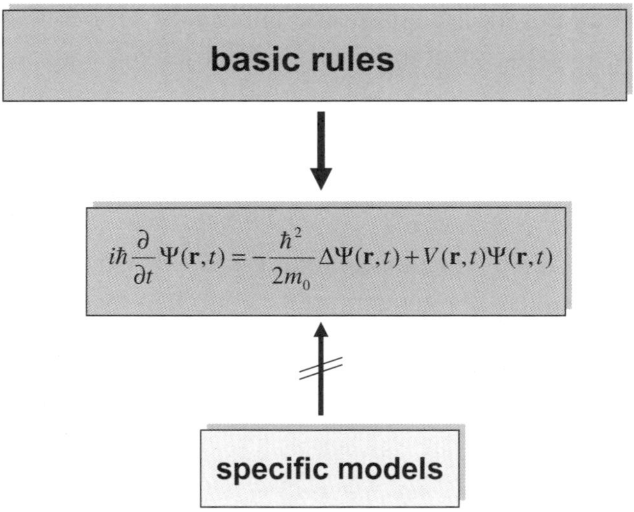


Fig. 4.9. A physical law [as, for example, $i\hbar\partial/\partial t\Psi(\mathbf{r},t) = -\hbar^2/(2m_0)\Delta\Psi(\mathbf{r},t) + V(x,y,z,t)\Psi(\mathbf{r},t)$] cannot be deduced from its specific solutions (models) but only from the information of a level that is positioned above the level on which the physical law is positioned. It is the level of the basic rules (projection principle, etc.). Consequence: Not the "principle of induction" is valid here but the "principle of deduction". Already the philosopher Karl Popper demanded to eliminate the "principle of induction"; after Popper only the "principle of deduction" should be applied.

philosophies (theories) seem to be overestimated from the point of view of projection theory. Projection theory obviously offers a more sophisticated view and goes distinctly beyond the conventional concepts without being in contradiction to experience. The main points of projection theory can be summarized as follows.

1. Basic reality exists independently of the observer and is in principle not accessible in a direct way.
2. Basic reality (world outside) is exclusively observable or describable by means of pictures on various levels, i.e., levels of reality; a picture-independent view is obviously not possible for a human observer.
3. Space and time do not belong to basic reality but have to be considered as auxiliary elements for the representation of the pictures and are, within projection theory, elements of the brain.
4. The hard objects (trees, stones, etc.) in front of us are states of the observers brain. We have only the impression that all these things are in front of us, i.e., outside of us. The hard objects, that we feel to be positioned in front of us, and the physical laws are equally states of the observer's brain. Nevertheless, they are different from each other and are therefore positioned on different levels.
5. All hard objects (material objects like trees, cars, atoms, molecules, etc.) are described by specific $(\mathbf{p}, E)$-processes in $(\mathbf{p}, E)$-space and always produce specific pictures (pattern) in $(\mathbf{r}, t)$-space; such systems have to be considered as "states of matter". On the other hand, a general law [as, for example, $i\hbar\partial/\partial t\Psi(\mathbf{r}, t) = -\hbar^2/(2m_0)\Delta\Psi(\mathbf{r}, t) + V(x, y, z, t)\Psi(\mathbf{r}, t)$] is not depictable in $(\mathbf{r}, t)$-space. Therefore, such a law cannot be produced by $(\mathbf{p}, E)$-processes in $(\mathbf{p}, E)$-space. We have defined these general physical laws as "states of mind".
6. Like the hard (material) objects, the general physical laws also have to considered as *real objects*. Physical laws are here *objects* in generalized form, more general than the hard objects. Both, the hard objects and the physical laws are within projection theory "states of the observer's brain".

7. The general physical law itself is not depicable in $(\mathbf{r}, t)$-space $[(\mathbf{p}, E)$-space)]; only specific solutions (in connection with modelling) $\Psi(\mathbf{r}, t)^{\text{specific}}$ $[\Psi(\mathbf{p}, E)^{\text{specific}}]$ can be pictured. Another kind of representation should be considered for the abstract general physical law $i\hbar\partial/\partial t\Psi(\mathbf{r}, t) = -\hbar^2/(2m_0)\Delta\Psi(\mathbf{r}, t) + V(x, y, z, t)\Psi(\mathbf{r}, t)$ which contains potentially the behavior of all possible physical systems, i.e., for all possible functions $V(\mathbf{r}, t)$, and this feature can be expressed symbolically by $\Psi = \Psi(V)$.
8. Any specific model is captured in $(\mathbf{r}, t)$-space $[(\mathbf{p}, E)$-space], that is, within the framework of a specific model one cannot leave $(\mathbf{r}, t)$-space $[(\mathbf{p}, E)$-space] in order to describe a general physical law, which is not despictable in $(\mathbf{r}, t)$-space. Therefore, both levels can principally not be joined. However, we may conclude the behavior of a specific model from the general physical law, but we cannot conclude the structure of a general physical law from a specific model. The "concept of induction" is not valid; projection theory leads to the "principle of deduction".
9. Space–time $[(\mathbf{r}, t)$-space] is tightly linked to $(\mathbf{p}, E)$-space and vice versa, and the statements, formulated above, particularly mean that the object "physical law" cannot be the outcome of $(\mathbf{p}, E)$-fluctuations as in the case of the hard objects in front of us. While the specific $(\mathbf{p}, E)$-processes in $(\mathbf{p}, E)$-space produce specific pictures in $(\mathbf{r}, t)$-space with exactly the same information, the physical law is not depictable in $(\mathbf{r}, t)$-space and, therefore, a general physical law cannot be produced by $(\mathbf{p}, E)$-processes, i.e., in connection with physical (material) brain functions.
10. Within projection theory, there is the following strict law. If there are no $(\mathbf{p}, E)$-fluctuations in $(\mathbf{p}, E)$-space there cannot be a physical picture in $(\mathbf{r}, t)$-space. A picture for a physically real process is not defined without $(\mathbf{p}, E)$-fluctuations and vice versa! Then, by definition, systems which are not depictable in $(\mathbf{r}, t)$-space, are not physical systems (are not material objects).
11. Clearly, we can have pictures in $(\mathbf{r}, t)$-space without $(\mathbf{p}, E)$-fluctuations, but these kinds of pictures are not physical representations but merely drawings and belong to the "products of

imagination". Such drawings (for example, an art work) may be arbitrary and must not satisfy physical laws.

12. Because a physical law is more general than the specific $(\mathbf{p}, E)$-processes (the behavior of all possible $(\mathbf{p}, E)$-processes are equally dictated by the physical law) we may conclude that this physical law is positioned at a higher level than that on which the models for these specific $(\mathbf{p}, E)$-processes are positioned.
13. Therefore, a physical law (formula) cannot be described by specific models (physical processes). The attempt to describe the products of mind by certain physical brain functions (neuronal activities) has no basis within projection theory. The reason for that is simple and has been worked out above: Any physical process, also those in connection with brain functions, is described by $(\mathbf{p}, E)$-fluctuations in $(\mathbf{p}, E)$-space and produces a concrete picture in $(\mathbf{r}, t)$-space. In the *reverse case*, we may argue as follows. Since the physical law $i\hbar\partial/\partial t\Psi(\mathbf{r}, t) = -\hbar^2/(2m_0)\Delta\Psi(\mathbf{r}, t) + V(x, y, z, t)\Psi(\mathbf{r}, t)$ (or other states of the mind) does not produce a picture in $(\mathbf{r}, t)$-space, the general physical law cannot be the result of physical brain functions [described by $(\mathbf{p}, E)$-fluctuations].
14. Since the origin of feelings and thinking or quite general the products of mind cannot be described as pictures in $(\mathbf{r}, t)$-space, their origin cannot be based on physical laws. The physical laws exclusively describe the hard objects (the objects in front of us but also atoms, molecules, etc.) in $(\mathbf{r}, t)$-space and exclusively come into play by the interactions [$(\mathbf{p}, E)$-fluctuations] in $(\mathbf{p}, E)$-space. However, the physical law itself is not depictable in $(\mathbf{r}, t)$-space, as we have outlined in detail. Clearly, we may express our fantasies in the form of imaginable pictures, but those pictures do not come into existence by $(\mathbf{p}, E)$-fluctuations.

## 4.9. Remarks on the Notion "World Equation"

A lot of scientists believe that we are able to develop a "world equation", and they are obviously firmly convinced that we will have this world equation soon (see in particular Chapter 1). Here world equation means the final mathematical code about the whole world (basic reality). Let us briefly discuss the term "world equation" in connection with projection theory, usual physics, and also within the framework of philosophy of science.

### 4.9.1. *Situation within Projection Theory*

In Section 4.4, we have discussed the features of five "levels of reality". The levels have been vertically arranged in accordance with the degree of generality (Fig. 4.7). The information on *level b* about basic reality is more general than that on *level a*. The various objects positioned at the various levels are equally real and are observed, i.e., they are primarily not invented.

Are these level-dependent "observations" qualified to make statements about the whole world, that is, are scientific statements possible in the form of a "world equation" containing the final code about the absolute, complete reality (basic realty)? This is definitely not possible within the framework of projection theory. Again, within this theory we know nothing about the structures in basic reality. Due to the principles of evolution we have to conclude that only species-relevant information are accessible to an observer. In other words, only a small part of the whole world is selected by the observer, and this restricted information is projected onto space–time leading to the species-relevant "pictures of reality". However, this species-relevant part of information is not qualified to make complete statements about the whole world (basic reality) in the form of a "world equation". The term "world equation" is nonsense here.

### 4.9.2. *World Equation within Usual Physics*

Within usual physics, we normally assume that the material world and everything else is embedded in space–time. Within the usual framework an observation is always a statement about absolute (objective) reality, and with each improvement of the experimental methods one believes to be a step closer to that what is often called "final solution" and "absolute truth", respectively. In other words, a lot of scientists are firmly convinced to have at some time the final code of the absolute, objective reality in their hands (Chapter 1). In fact, in this connection the notion "world equation" is in the centre of the discussions, and it is often believed to have this world equation soon. Instead of "world equation", the terms "theory of everything", "final explanation", "final theory", etc. are also used. However, the belief of certain scientists that there is the principle possibility for the construction of such a world equation should not be considered as realistic but rather reflects a certain kind of self-indulgence, and the belief to have such a final mathematical code soon is beyond that. All these tendencies should be declined, and this is particularly due the fact that the investigations within philosophy of science teaches just the opposite. Here philosophy of science (outlined in Chapter 1) agrees with the viewpoint of projection theory.

However, not only self-indulgence is probably the reason for this belief (that a final world equation is definable and, furthermore, to have it in the hands soon) but should also be due to the fundamental concept used in connection with usual physics. After that, the material world is embedded in space–time. This fundamental concept suggests that we have the basic (objective) reality in front of us, and it is often believed that this basic reality is directly accessible to the human observer. There are a lot of reasons to decline this view, in particular from the point of view of philosophy of science and of course from the viewpoint of projection theory. In fact, projection theory offers another, more realistic conception.

### 4.9.3. *World Equation within Theory of Science*

Within philosophy of science, the following question is often asked: Are electrons, quarks or other physical objects really existing entities? In other words, are they units of basic (objective) reality? Again, basic reality has to be understood as a single complex independent of the observer, that is, independent of the intellect, the language, our scheme of notations and theories. In literature instead of basic (objective) reality the term "reality in itself" is often used.

The "face" of an entity can be different on the various levels, and all levels are equally real within projection theory, and each "face" (picture of reality) reflects a certain property of basic reality at the level just under investigation. But none of these "pictures of reality" actually exist in this form in basic realty. The question which we have asked above "Are electrons, quarks or other physical objects really existing entities?" must therefore be answered negatively.

Within usual physics the projection principle does not exist. Nevertheless, general considerations within philosophy of science lead to the same result (clearly, without the use of the projection principle). Only on the basis of a scientific theory we can decide and define what entities exist and what entities do not exist, and the theory of course also dictates how these entities behave. However, within modern philosophy of science, we may not conclude that these theoretically defined entities actually exist in this form in basic reality. The scientific pictures have only a limited similarity with that what actually takes place in reality outside (basic reality). In this connection Nicholas Rescher wrote [4]:

> *"Scientific realism is the doctrine that science describes the real world, that the world actually is as science takes it to be, and that its furnishings are as science envisages them to be. If we want to know about the existence and the nature of heavy water or quarks, of man-eating molluscs or a luminiferous aether, we are referred to the natural sciences for the answer. On this realistic construction of scientific theorizing, the theoretical terms of natural science refer to real physical entities and describe their attributes and components. For example, the "electron spin" of atomic physics refers to a behavioral characteristic of a real, albeit unobservable, object — an electron. According to this currently fashionable theory, the declarations of science are — or will eventually become — factually true generalizations about the actual behaviour of objects that exist in the world. Is this "convergent realism" a tenable position? It is quite clear that it is not. There is clearly insufficient warrant for and little plausibility to the claim that the world indeed is as our science is correct science claims it to be — that we've got matters altogether right, so that our science is correct science and offers the definite "last word" on the*

*issues. We really cannot reasonably suppose that science as it now stands affords the real truth as regards its creatures-of-theory."*

This remark by Nicholas Rescher summarizes the results obtained within modern philosophy of science and is in close agreement to what have analyzed in connection with projection theory: we can nothing say about the absolute truth.

The statements in connection with philosophy of science are essentially based on the following important "general law" (Chapter 1): Our notions concerning basic reality are not supplemented in the course of time but they are re-formulated on the basis of completely new first principles. Since we can only decide on the basis of a scientific theory what kind of objects can exist and what objects cannot exist, the entities can come and go depending on the structure of the theory which can vary fundamentally in the course of time with respect to the *first principles.*

In other words, the progress in connection with physical laws is not dictated by an established absolute theoretical structure that is *supplemented* in the course of time. As within the framework of projection theory also philosophy of science teaches very clearly that there are no statements about the "absolute truth" possible but we can make only "estimations of the truth" but never final statements. Therefore, also within philosophy of science the notion "world equation" has no basis, just as in the case of projection theory. This of course supports projection theory. However, within projection theory we have "levels of reality" (see Fig. 4.7), and we have to say something about *observations* at the various levels. In Section 4.4, some principal remarks concerning level-specific observations are given.

## 4.10. On the Anthropic Principle

### 4.10.1. *Everything Seems to Revolve around Man*

The principle of usefulness is equivalent with another important principle — as little outside world as possible. In Chapter 1, we asked the following question: Has evolution developed man's sense organs so that they transmit as much information from reality outside as possible? The answer was "not at all". Just the opposite is correct. The strategy of nature (evolution) is to take up, by the sense organs, as little information from the outside world as possible. Reality outside (objective or absolute reality) is not assessed by "true" and "untrue" but by "favorable towards life" and "hostile towards life". Only that information from the outside world is transmitted that is absolutely necessary for the survival of the species (in this case man). In other words, "as little outside world as possible" is the rule. Thus, what we know about the world is man-specific: *man appears as the middle point within the observed world*. Everything seems to revolve around man. In the case of other biological systems, there are other observed worlds that are in general different from that of man, and this is because the conditions for survival are different from those of man. But all these specific realities have only a little to do with that what we call "objective reality"; objective reality outside is principally not accessible to man and also not to other biological systems. We already discussed all these points in Chapter 1.

All this is dictated by the principles of evolution and is valid not only for the primeval cell but also for all its descendants and therefore also for ourselves. Clearly, in the course of time we have extended our knowledge about the properties of the tangible environment more and more. However, in principle only those features of the objective reality outside are accessible to our perception apparatus which, in the meantime, we need in our stage of development. Also our brain has evolved on the basis of these principles, that is, not to understand the objective world but to survive. Hoimar von Ditfurth discussed this point convincingly [6].

In this connection, we have always to consider that there is an interlinkage of evolutionary processes, i.e., each further step in the evolution of a system is based on the steps before. So, the basic principle "as little information from the outside world as possible", is in particular valid at the beginning of the human evolution, but remains conserved even after the biological system has passed through various evolutionary steps, and this is because all further steps in the evolution are based on the first steps.

### 4.10.2. *Measurements*

What about measurements with physical devices? Are physical measurements suitable to make absolute statements in connection with basic reality possible? This is definitely not possible within projection theory: We have no "general" measuring instruments that would have access to absolute reality. The reason is simple. Also the physical devises are constructed by the observer within $(\mathbf{r}, t)$-space on the basis of a physical theory with $\mathbf{p}$ and $E$ as basis elements. The situation represented in Fig. 4.10. We know nothing about the device in basic reality but its geometrical structure appears as a projection from $(\mathbf{p}, E)$-space (fictitious reality). We have outlined in Chapter 3 that the geometrical picture of a material object in $(\mathbf{r}, t)$-space is always connected with specific $(\mathbf{p}, E)$-fluctuations in $(\mathbf{p}, E)$-space. In

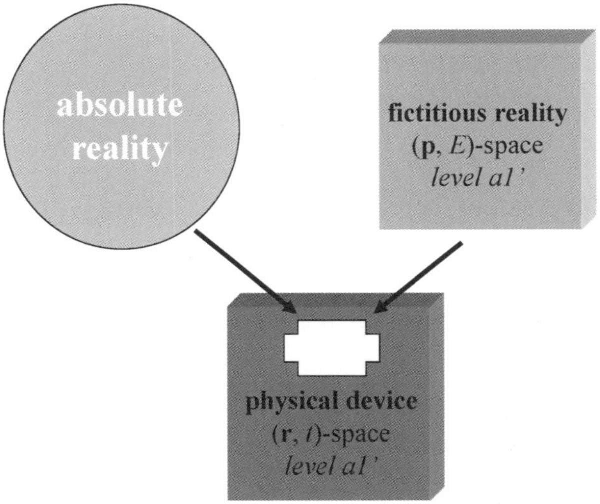


Fig. 4.10. On the role of physical devices. As in the case of all physical systems also the picture of a physical device is given in $(\mathbf{r}, t)$-space. We know nothing about its structure in basic reality but its geometrical structure has to be understood as a projection from $(\mathbf{p}, E)$-space. In the case of a measurement, the device interacts with a certain system (as, for example, an electron) and there is an exchange of momentum and energy. We have no access to a physical device in absolute reality.

the case of a measurement, the device interacts with the system under investigation (as, for example, an electron) and there is an exchange of momentum and energy:

$$\Delta \mathbf{p}_{\text{measurement}} \neq 0, \quad \Delta E_{\text{measurement}} \neq 0. \tag{4.5}$$

All that takes place on *level aI'* (see Fig. 4.7).

In conclusion, as in the case of the description of physical systems and also in the case of physical devices, we are strictly captured in, $(\mathbf{r}, t)$-space in connection with $(\mathbf{p}, E)$-fluctuations in $(\mathbf{p}, E)$-space. Clearly, also a physical device is a physical system and cannot be considered as an exceptional case. In other words, physical measurements are not suitable to make absolute statements with respect to basic reality. We have no "general" measuring instruments that would have access to absolute reality. Thus, with detectors (physical devices) we can only observe that what is in space and time [$(\mathbf{r}, t)$-space] and, as we have outlined above, there are exclusively human-specific information in $(\mathbf{r}, t)$-space. Let us still specify this point in the next section.

### 4.10.3. *Human-specific Information and Measurement*

In any case $(\mathbf{p}, E)$-fluctuations

Measurement within projection theory means the following (Chapter 3): The detection of an object is always connected with a certain structure $\Psi(\mathbf{r}, t)$ $[\Psi^*(\mathbf{r}, t)\Psi(\mathbf{r}, t)]$ in space and time [$(\mathbf{r}, t)$-space]. Due to the principles of evolution there are exclusively human-specific structures in $(\mathbf{r}, t)$-space and let us name them $\Psi(\mathbf{r}, t)_{\text{human-specific}}$ $\{[\Psi^*(\mathbf{r}, t)\Psi(\mathbf{r}, t)]_{\text{human-specific}}\}$ instead of $\Psi(\mathbf{r}, t)$ $[\Psi^*(\mathbf{r}, t)\Psi(\mathbf{r}, t)]$. In the case of $N$ structures in $(\mathbf{r}, t)$-space, we have $[\Psi(\mathbf{r}, t)_i]_{\text{human-specific}}$ $\{[\Psi^*(\mathbf{r}, t)\Psi(\mathbf{r}, t)_i]_{\text{human-specific}}\}$ with $i = 1, \ldots, N$. In particular, we have the following situation within projection theory (Chapter 3): The measurement at time $\tau$ of one of the possible values for $\mathbf{p}$ and for $E$ is done in the space–time intervals $\mathbf{r}, \mathbf{r} + d\mathbf{r}$ and $t, t + dt$ with the probability density

$$[\Psi^*(\mathbf{r}, t)\Psi(\mathbf{r}, t)_i]_{\text{human-specific}} \tag{4.6}$$

for the $i$th structure; the values $\mathbf{p}$ and $E$ $(-\infty < \mathbf{p}, E < \infty)$ are given with the probability density

$$[\Psi^*(\mathbf{p}, E)\Psi(\mathbf{p}, E)_i]_{\text{human-specific}}, \tag{4.7}$$

where $[\Psi(\mathbf{p}, E)_i]_{\text{human-specific}}$ and $[\Psi(\mathbf{r}, t)_i]_{\text{human-specific}}$ are connected by the Fourier transform

$$[\Psi(\mathbf{p}, E)_i]_{\text{human-specific}} = \frac{1}{(2\pi\hbar)^2}\int_{-\infty}^{\infty}[\Psi(\mathbf{r}, t)_i]_{\text{human-specific}} \times \exp\left\{-i\left[\frac{\mathbf{p}}{\hbar}\cdot\mathbf{r} - \frac{E}{\hbar}t\right]\right\} dtdxdydz. \tag{4.8}$$

As we have already pointed out above, the detection of the system, characterized by $[\Psi(\mathbf{r}, t)_i]_{\text{human-specific}}$, with a measuring instrument provides an interaction process and we have $\Delta\mathbf{p}_{\text{measurement}} \neq 0$, $\Delta E_{\text{measurement}} \neq 0$ [Eq. (4.5)]. Signals are recorded with detectors (physical devices) in space and time.

### Only a man-specific reality is observed

Then, also that what we measure with our measuring instruments can only reflect this man-specific observed reality and not the facts of the objective reality outside; the human observer is caught within his observed reality and can only construct his measuring instruments on the basis of this man-specific world. Thus, the measured parameters can only reflect the properties of this kind of reality. We can extend our knowledge about "man's world" by measurements but we never touch objective reality by these man-specific measurements.

Clearly, this man-specific reality is part of the objective world, but it is only a part of it and, furthermore, this tiny fraction of absolute reality must reflect a transformed reality because there is obviously no space–time in the objective (absolute) reality outside. As we have remarked several times, already Immanuel Kant (1724–1804) recognized this feature very clearly. Within projection theory this condition for space and time is applied the first time within the frame of a closed physical theory.

### 4.10.4. *Evolutionary Processes and Level Analysis*

Brain functions

Since also our brain has evolved on the basis of the principle "as little outside world as possible", also our theoretical pictures and world views, respectively, can only reflect this restricted man-specific view, which is quite in analogy to that what we measure. In other words, also our theoretical pictures and world views cannot describe the features of the objective world outside which remains completely hidden to all human beings. All these points follow from that, which we have outlined in Sections 4.1–4.9.

In summary, our visual impressions in everyday life as well as our statements in connection with measured facts and theoretical world views, must necessarily revolve around man. This is a result of projection theory in connection with the principles of evolution.

Man seems to be in the middle of the world but this is not true because we never observe objective reality. However, the principle "as little outside world as possible" makes clear that objective reality must be more complex than that what we have called "our observed reality" (see also Fig. 4.11).

In Section 4.4, we have pointed out that the observed reality can be analyzed in terms of levels. Since basic reality (absolute reality outside) is principally not accessible to a human observer, the only way to learn something about the outside world is to observe it on various levels of reality, which are vertically arranged in accordance with the degree of generality. The "objects" belonging to the various levels of reality are equally *real* and are exclusively states of the human brain, even the material objects that we observe in everyday life have to be considered as states of the brain. All elements (objects) on the various levels are observed by objectivation (Section 4.5). Although we know nothing about objective reality we may nevertheless state that no level can contain the complete information, that is, the information within objective reality must be principally larger than that on the specific levels and this feature is dictated by the principles of evolution. Our brain has evolved on the basis of the principle "as little outside

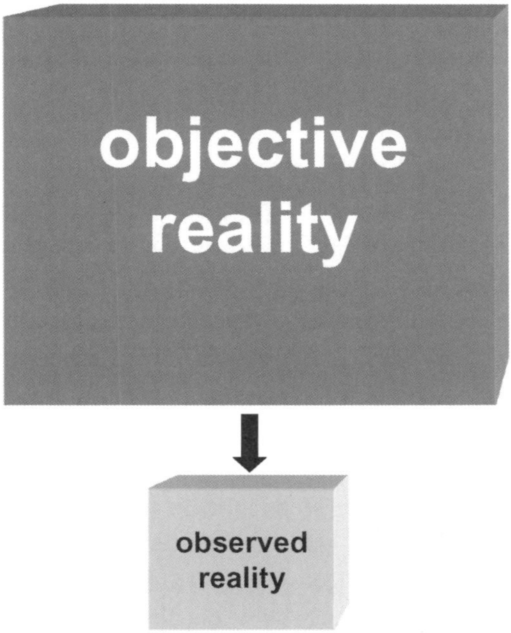


Fig. 4.11. We never observe "objective reality". The strategy of evolution is to take up as little information from the outside world (objective reality) as possible. Reality outside is not assessed by "true" and "untrue" but by "favorable towards life" and "hostile towards life". This necessarily means that objective reality must be more complex than that what we call "observed reality". The observed and theoretically described reality can only be a tiny and transformed fraction of absolute reality. From this point of view it is wrong to maintain that the objective laws of nature (objective reality) revolve around man's existence.

world as possible" and not in connection with that what is often called "absolute truth" and, as we have already discussed several times, the elements belonging to the various levels are just states of the brain.

In this connection, the following remark is for our understanding of the world relevant: A physical law [as, for example, $i\hbar\partial/\partial t\Psi(\mathbf{r},t) = -\hbar^2/(2m_0)\Delta\Psi(\mathbf{r},t) + V(x,y,z,t)\Psi(\mathbf{r},t)$, Eq. (3.22)] is, as we have discussed in Section 4.2, a product of mind and comes into play by brain functions. Thus, also this physical law can only be a reduced representation of what is actually the content of objective reality because our brain functions developed in the course of time on the basis of the principle "as little outside world as possible" and, therefore, a physical law cannot represent the "absolute truth".

Interlinkage of evolutionary processes

The law $i\hbar\partial/\partial t\Psi(\mathbf{r},t) = -\hbar^2/(2m_0)\Delta\Psi(\mathbf{r},t) + V(x,y,z,t)\Psi(\mathbf{r},t)$ (and of course other abstract relations) is more general than any specific solution (model) $\Psi(\mathbf{r},t) = \Psi(\mathbf{r},t)^{\text{specific}}$ because it contains the characteristics of all possible physical systems (see in particular the discussion in Section 4.2.3). A physical law is therefore located at a level (*level b*, Fig. 4.6) above the level of specific models (located at *level a*). On the other hand, the macroscopic level where everyday life experiences are located (*level 1*) is below *level a*.

In the first evolutionary steps, the lower levels in the hierarchy appeared, in particular that with the material objects represented in $(\mathbf{r},t)$-space (*level 1* in Fig. 4.6) which we observe in everyday life in an unconscious way. After that, the brain developed further and other, more sophisticated levels (*level a*, *level b*, etc, see Fig. 4.6) came into play in the course of evolution. In other words, after the evolution has developed the ability for the unconscious recognition of material objects, which are depictable in space and time [$(\mathbf{r},t)$-space], evolution developed those brain functions that are responsible for conscious recognition of abstract physical laws.

The general physical law $i\hbar\partial/\partial t\Psi(\mathbf{r},t) = -\hbar^2/(2m_0)\Delta\Psi(\mathbf{r},t) + V(x,y,z,t)\Psi(\mathbf{r},t)$ is not depictable in $(\mathbf{r},t)$-space [$(\mathbf{p},E)$-space)]. Only specific solutions (models) $\Psi(\mathbf{r},t)^{\text{specific}}[\Psi(\mathbf{p},E)^{\text{specific}}]$ can be pictured in $(\mathbf{r},t)$-space [$(\mathbf{p},E)$-space)] and of course all macroscopic objects.

Although a general physical law is not depictable in $(\mathbf{r},t)$-space, it is based on the space–time elements $\mathbf{r}$ and $t$ indicating the fact that there is an interlinkage of evolutionary processes, that is, the next step in this process is based on the steps before.

In particular, the analysis in terms of "reality levels" shows that objective reality must contain more information than what we have directly in front of us in the form of material objects. As we have extensively discussed above, these material objects (states) are observed in connection with everyday life experiences (assumptionless observations) but are also experienced with the help of measuring instruments. The situation is represented in Fig. 4.7.

### 4.10.5. *Conclusion*

Due to the principles of evolution, a human being is never able to observe facts within absolute reality. We can extend our knowledge about that what we call "world" by measurements and/or by thinking, but not with respect to the concrete structure of objective reality. This door remains principally closed.

Again, the human observer is caught within his observed reality. The world seems to be made for him but this impression is entirely due to the fact that the observed data are filtered out from absolute reality, and these observed data are species-specific. A human being discovers *his* world and not the objective reality outside. This is the reason why the observed universe appears just right for life.

The problems in connection with the anthropic principle appear when we assume that the world in front us or what we investigate by our theories reflects facts of the objective reality. That is however not the case. If we nevertheless do that we can erroneously come to the conclusion that this world is made for the purpose to guarantee man's existence. However, we have to be careful because all facts are selected from the objective world outside, and we have only a tiny and transformed piece of this objective reality in front of us.

All physical theories are necessarily based on those selected facts that are important for man and nothing else. In principle, other biological systems must experience another world, which is tailor-made for their purposes. When we assume that our observed facts reflect objective reality, we can of course come to the conclusion that man is in the center of the world, but this is obviously a fallacy as we have outlined above. Again, objective (absolute) reality is principally not accessible to a human being and also not to other biological systems because for all biological system, the principles of evolution are valid.

It is trivial to maintain that everything revolves around man when only those facts of the objective reality outside are selected which are relevant for a human being. Then, by definition, everything must revolve around man. That this universe appears to be fine-tuned for life is then a matter-of-course.

Then, the fine tuning of the physical laws and the fundamental constants means that we need for the description of life just such a complex system of fine-tuned physical laws and fundamental constants. Since the principles of evolution require "as little outside world as possible", we have to conclude that the objective realty outside is much more complex than what a human being is able to observe (Fig. 4.11). The cosmos contains the possibility for life but probably much more. It is not exclusively made for our life.

### 4.10.6. *Anthropic Principle: Solutions within usual Physics*

The anthropic principle has been intensively discussed in the past in connection with usual physics. Brandon Carter defined the (strong) anthropic principle as follows [26]: "*The universe (and hence the fundamental parameters on which it depends) must be such as to admit the creation of observers within it at some stage*". In other words, from the physical laws and fundamental constants of the universe must emerge life and intelligence.

Within usual physics, the anthropic principle is mainly discussed in connection with "multiple universes" and alternatively in terms of an "intelligent designer". Both explanations have often been criticized for being outside of accepted science. The *multiuniverse* defines an ensemble of worlds having all forms of physical laws and fundamental constants and we are just embedded within an universe that allows our existence. In the case of an *intelligent designer* (*creator*), there is only one world and the creator designed the world conciously so that life becomes possible. Otherwise the finetuning of the physical laws and fundamental constants becomes unprobable.

Clearly, within projection theory, there is no space for such pictures (multiuniverse, creator) in order to explain the fact that the world (physical laws and fundamental constants) seems to be made for a human being.

## 4.11. Summary

Basic reality, fictitious reality, pictures of reality

The world in front of us is represented as pictures in $(\mathbf{r}, t)$-space. The so-called "hard objects" (cars, trees, houses, etc) are described within projection theory by $(\mathbf{p}, E)$-fluctuations reflecting the interactions between the objects in $(\mathbf{p}, E)$-space. We get the pictures of the hard objects by transmitting the information from $(\mathbf{p}, E)$-space to $(\mathbf{r}, t)$-space.

We know nothing about basic reality, and we also know nothing about the process how this (principally unknown) information is transmitted and transformed from basic reality to the human brain where it appears as "picture of reality". These pictures with the hard objects are exclusively located in the brain of the observer, but we have the impression that all these objects are in front of us (outside of us). However, instead of basic reality, we can construct "fictitious realities" in $(\mathbf{p}, E)$-space, producing the information for the pictures of reality, in $(\mathbf{r}, t)$-space.

Space and time $[(\mathbf{r}, t)$-space$]$ do not belong to basic reality but have to be considered as auxiliary elements for the representation of the pictures. Thus, also space and time are elements of the brain.

In this connection, the following is relevant. A general physical law [as, for example, $i\hbar\partial/\partial t\Psi(\mathbf{r}, t) = -\hbar^2/(2m_0)\Delta\Psi(\mathbf{r}, t) + V(x, y, z, t)\Psi(\mathbf{r}, t)$, Eq. (3.22)] is not depictable in $(\mathbf{r}, t)$-space. Only specific solutions (models) for $\Psi(\mathbf{r}, t)$ of Eq. (3.22), that is $\Psi(\mathbf{r}, t) = \Psi(\mathbf{r}, t)^{\text{specific}}$, can be pictured in $(\mathbf{r}, t)$-space.

States of matter, states of mind

The following statement is relevant: $(\mathbf{r}, t)$-space is tightly linked to $(\mathbf{p}, E)$-space (and vice versa) and that means that the "object" physical law cannot be the outcome of $(\mathbf{p}, E)$-fluctuations as in the case of the hard objects in front of us.

Within projection theory, there is a strict law. If there are no $(\mathbf{p}, E)$-fluctuations in $(\mathbf{p}, E)$-space, there cannot be a physical picture in $(\mathbf{r}, t)$-space. A picture for a physically real process is not defined without

$(\mathbf{p}, E)$-fluctuations and vice versa! Then, by definition, systems that are not depictable in $(\mathbf{r}, t)$-space are not physical systems (are not material objects). This definition means that a general physical law itself [as, for example, Eq. (3.22)] is not a physical system because it is not depictable in $(\mathbf{r}, t)$-space [$(\mathbf{p}, E)$-space)]. It is not a physical system but it describes physical systems.

All objects that are based on specific $(\mathbf{p}, E)$-processes in $(\mathbf{p}, E)$-space always produce specific pictures in $(\mathbf{r}, t)$-space and define *material objects* (hard objects like trees, cars, atoms, molecules, etc.) and have to be considered as "states of matter". On the other hand, the general physical laws [as, for example, Eq. (3.22)] have to be considered as "states of mind".

It has often been supposed that the states of mind can be reduced to physical real states. Within projection theory, the deduction of a general physical formula is obviously a product of mind and cannot be the result of specific physical processes [$(\mathbf{p}, E)$-fluctuations in $(\mathbf{p}, E)$-space]. Within projection theory, we cannot have physical pictures in $(\mathbf{r}, t)$-space if there are no $(\mathbf{p}, E)$-fluctuations in $(\mathbf{p}, E)$-space. As we have remarked above, the following rule is valid. No $(\mathbf{p}, E)$-fluctuations in $(\mathbf{p}, E)$-space means that there cannot be a physical picture in $(\mathbf{r}, t)$-space. A physical picture is not defined without $(\mathbf{p}, E)$-fluctuations and vice versa! In fact, we are able to *think* about a physical law but the physical law itself is not depictable as a picture in $(\mathbf{r}, t)$-space and, therefore, a general physical law cannot be reduced to physically real states (states of matter).

In conclusion, any effect which is not depictable in $(\mathbf{r}, t)$-space does not belong to the class of phenomena that we have defined as *physical* phenomena (states of matter): these are exclusively described by $(\mathbf{p}, E)$-fluctuations in $(\mathbf{p}, E)$-space within projection theory. Trees, cars, atoms, molecules, stars, etc. belong to these phenomena. However, when we try to explain the brain activities exclusively by physical processes [$(\mathbf{p}, E)$-fluctuations] we cannot seize phenomena that belong to the "states of mind", i.e., we are not able to recognize phenomena like a physical law, products of fantasy, feelings, etc. by means of the "states of matter". In other words, such appearances cannot be reduced on physical phenomena within projection theory. The projection principle allows such kind of analysis and even allows detailed statements about possible brain functions.

Levels of reality

Basic reality (absolute reality outside) is principally not accessible to a human observer. But we observe the world on various levels of reality that are vertically arranged in accordance with the degree of generality.

The "objects" belonging to the various levels of reality are equally *real*, and we observe all these objects at the various levels by a common principle: it is the principle of *objectivation*. This method is valid at each level, i.e., it can be applied independently on which level an "object" is located. We distinguish between "objectivation by thinking" and "objectivation in an unconscious, non-intellectual way". This parallelism supports the view that there is no principal difference between "states of mind" and "states of matter". There are only gradual deviations.

The various levels of reality emerge properties that are qualitatively different from each other although the various levels reflect features of the same reality (facets of basic reality). The difference between the "objects" on the various levels is not only reflected in their theoretical description but also in their effect on us. In this connection, the kind of objectivation is of special importance In principle, the objectivation processes on the various levels are different from each other. Since the basic information on the various levels have to be considered as *discoveries* and not as inventions, we may state that each "level of reality" is accompanied with a "level of observation". Because the features of the "objects" on the various levels are defined differently, so the methods of observation must vary from level to level.

The process of objectivation emerges the "objects" on the various levels, and these produce certain level-specific effects and feelings inside the observer. We feel the effect of "objects", which are objectified *unconsciously* (hard objects like trees, cars, etc.), differently from those that are objectified *consciously* (scientific laws, products of imagination, etc.). We have distinguished between "objectivation by thinking" and "objectivation in an unconscious, non-intellectual way" but, in principle, other objectivation methods should be possible depending on the effect and the features of the "objects" on the levels, respectively.

Unconscious objectifying takes place on the level of the five senses (macroscopic level) and also in part on the level of measuring instruments, even when the human consciousness is essentially involved. Conscious objectifying is objectifying through thinking and takes place at specific levels of reality. Here the "objects" are also selected from basic (objective) reality. This process, however, is strictly analogous to that, which takes place on the level of the five senses and the level of measuring instruments. Konrad Lorenz already pointed out that there is a strict analogy between unconscious objectifying through the physiological apparatus and conscious, intellectual objectifying (see also [9] and the literature therein). Thinking is therefore a certain kind of observation procedure, and in fact a very versatile one. It is not new to say that one can think on different levels.

### Principle of deduction

It turned out that the "concept of induction" cannot be valid within projection theory but exclusively the "principle of deduction". For example, the general physical law $i\hbar\partial/\partial t\Psi(\mathbf{r},t) = -\hbar^2/(2m_0)\Delta\Psi(\mathbf{r},t) + V(x,y,z,t)\Psi(\mathbf{r},t)$ cannot be found from specific models but we need here more general statements (basic rules).

Therefore, positivistic philosophies have also no place here. Furthermore, we may state quite generally that strict materialistic philosophies (theories) seem to be overestimated from the point of view of projection theory. Projection theory obviously offers a more sophisticated view and goes distinctly beyond the conventional concepts, without being in contradiction to experience.

### World equation

A lot of scientists believe that we are able to develop a "world equation", and they are obviously firmly convinced that we will have this world equation soon. Here world equation means the final mathematical code about the whole world (basic reality). In this connection we discussed three viewpoints, the viewpoints of projection theory, of usual physics, and that of philosophy of science.

### Viewpoint of projection theory

Within projection theory the "observations" are level-dependent. Are these level-dependent "observations" qualified to make statements about the whole (complete) world, that is, to make scientific statements in the form of a "world equation" containing the final code about the absolute, complete reality (basic realty)? This is definitely not possible within the framework of projection theory. Again, within this theory we know nothing about basic reality. Due to the principles of evolution we have to conclude that only species-relevant information are accessible to an observer. In other words, only a small part of the whole world is selected by the observer, and this restricted information is projected onto space–time leading to the species-relevant "pictures of reality". However, this species-relevant part of information is not qualified to make complete statements about the whole world (basic reality) in the form of a "world equation". The term "world equation" is nonsense here.

### Viewpoint of usual physics

Within usual physics we normally assume that the material world and everything else is embedded in space-time. Within the usual framework, an observation is always a statement about absolute (objective) reality, and with each improvement of the experimental methods we believe to be a step closer to that what is often called "final solution" and "absolute truth", respectively. Thus, a lot of scientists are firmly convinced to have at some time the final code about the absolute (objective) reality in their hands. All these tendencies should be declined; they reflect a certain kind of self-indulgence, and the belief to have such a final mathematical code soon is beyond that.

### Viewpoint of philosophy of science

The viewpoint of projection theory is supported by the theory of science. As within the framework of projection theory, philosophy of science also teaches very clearly that there are no statements about the "absolute truth" possible because we can make only "estimations of the truth" but never final statements. Therefore, also within philosophy of science the notion "world equation" has no basis, just as in the case of projection theory. This of course supports projection theory.

## Anthropic principle

The human observer is caught within "his" observed specific reality. The world seems to be made for him but this impression is entirely due to the fact that the observed data are filtered out from absolute reality, and these observed data are species-specific. A human being discovers *his* world and not the objective reality outside. This is the reason why the observed universe appears just right for life.

The problems in connection with the anthropic principle appear when we assume that the world in front us or what we investigate by our theories reflects facts of the objective reality. If we however do that we can erroneously come to the conclusion that this world is made for the purpose to guarantee man's existence. However, we have to be careful because all facts are selected from the objective world outside, and we have only a tiny and transformed piece of this objective reality in front of us.

# 5

# Cosmological Constant and Physical Reality

So far we have discussed the following points: The Absolute Truth (Chapter 1), the Projection Principle (Chapter 2), Fictitious Realities (Chapter 3), Basic Reality and levels of Reality (Chapter 4). All the basic results which have been found in the four chapters can be used for the solution of the following big problem: It turned out that the quantum-theoretical treatment of the so-called cosmological constant leads to a disaster when we estimate the vacuum energy by means of quantum field theory. Some of the essential points given in this chapter have already been discussed in [27].

## 5.1. Introductory Remarks

Physical statements about the universe are presently formulated on the basis of General Theory of Relativity and also on quantum theory. While *local* systems as, for example, the solar system, can be treated very reliably on the basis of General Theory of Relativity, *global* cosmic structures are not so clear and easy to describe. Just the accelerated expansion of the whole cosmos has initiated an extensive and critical discussion about the structure of the field equations, and this discussion is obviously just at the beginning and the problems are by no means easy to solve; here the cosmological constant $\Lambda$ is in the center. In this connection, it is important to mention that quantum phenomena (in connection with vacuum energy) lead to serious contradictions and clearly show that we are still far away from a satisfactory picture for the whole cosmos.

The reason for this unsatisfactory situation is mainly due to the fact that the General Theory of Relativity and usual quantum theory, basis for cosmological descriptions, are mutually incompatible. Concerning this point, we find in [28] the following comment: "*Through years of research, physicists have experimentally confirmed to almost unimaginable accuracy virtually all predictions made by each of these theories* (*usual quantum theory and the General Theory of Relativity*). *But these same theoretical tools inexorably lead to another disturbing conclusion: As they are currently formulated, general relativity and quantum mechanics, cannot both be right. Both theories underlying the tremendous progress of physics during the last hundred years — progress that has explained the expansion of the heavens and the fundamental structure of matter — are mutually incompatible.*"

Both General Theory of Relativity and quantum theory lead to sets of laws that work fantastically. However, if we put them together we inevitably obtain irreconcilable differences. In [29], we find: "*It has been said that quantum field theory is the most accurate physical theory ever, being accurate to about one part in* $10^{11}$," writes Roger Penrose in "The Nature of Space and Time." "*However, I would like to point out that general relativity has, in a clear sense, now been tested to be correct to one part in* $10^{14}$ (*and this accuracy has apparently been limited merely by the accuracy of clocks on earth*)." In conclusion, both theories seem to work perfectly but, on the other hand, both theories are mutually exclusive.

This is clearly reflected in the quantum-theoretical treatment of the cosmological constant $\Lambda$, which leads to a disaster when we estimate the vacuum energy by means of quantum field theory. The problem in connection with the vacuum energy is not only a specific detail, and this disaster can obviously not be eliminated by invention of certain mechanisms within the frame of one of the two theories. In other words, the discrepancy in connection with vacuum energy and cosmological constant obviously reflects a basic fact which has probably its source in the very foundations of the two theories. We will discuss this point below in more detail.

It has often been argued that one of the theories must be wrong or even both. But "wrong" is not specific enough. It is probable that the reason for the discrepancies between General Theory of Relativity and quantum theory is due to the fact that both theories are based on foundations that are different from each other and not compatible. In order to avoid contradictions, both theories have probably to be based on one and the same principle, that is, we probably have to set "something" above the basics of quantum theory and theory of relativity. We will discuss this point in this chapter, and the projection principle will be of particular relevance.

## 5.2. The Cosmological Constant

Quantum theory and the Theory of Relativity have been developed independently of each other, and both theories differ considerably from each other in their conceptions and contents. But a realistic cosmology needs both, the description of relativistic (General Theory of Relativity) effects as well as the description of quantum phenomena. Because both theories, General Theory of Relativity and quantum theory, differ considerably from each other with respect to their basics, it is problematic to use the present forms of quantum theory and Theory of Relativity. As already mentioned, this is particularly recognizable in connection with the concept of vacuum leading to a quantum theoretical value for the cosmological constant which is not compatible at all with the basic structure of the General Theory of Relativity.

In modern cosmology the cosmological constant $\Lambda$ is of particular importance. How large is it? What energy effects are involved in its determination? Does it exist at all? Let us first discuss a basic problem in connection with $\Lambda$ that is often considered as the deepest mystery in physics.

### 5.2.1. *Quantum Field Theoretical Calculation of* $\Lambda$

It has often been argued that we cannot conclude with absolute certainty that the cosmological observations $\Lambda$ is zero: $\Lambda = 0$. An estimation leads to $|\Lambda| < 3 \times 10^{-52}\,\mathrm{m}^{-2}$. The cosmological constant can be interpreted as constant pressure which is repulsive for $\Lambda > 0$ and attractive for $\Lambda < 0$. $\Lambda$ is connected to a vacuum energy density by

$$\rho(\Lambda) = \frac{\Lambda c^2}{8\pi G}, \tag{5.1}$$

where $G$ is the gravitational constant and $c$ the velocity of light. Within the framework of quantum field theory we obtain in fact a non-vanishing energy density which is dependent on the used model but should take at least the value of

$$\rho = \frac{m_P}{l_P^3}, \tag{5.2}$$

where $m_P$ is the Planck mass ($m_P = 2.2 \times 10^{-8}$ kg) and $l_P$ is the Planck length ($l_P = 1.6 \times 10^{-35}$ m). From this we obtain for $\Lambda$ the value of $\Lambda = 9.8 \times 10^{70}\,\mathrm{m}^2$. Thus, on the basis of this specific estimation, we may conclude that the theoretical prediction deviates from the observed value by a factor of $10^{122}$. This is a lot and hardly imaginable. (In literature we often find the factor $10^{120}$ instead of $10^{122}$ or values around $10^{120}$ depending on the estimation.)

The factor $10^{122}$ is in fact a mind boggling number. Even the quotient

$$\frac{\text{mass of all atoms in the universe}}{\text{mass of one atom}}$$

does not approach $10^{122}$. In fact, if the vacuum would really contain all the energy that quantum field theory predicts, we would have cosmic features that have nothing to do with that what we really observe. In [29], we find the following instructive example: "*Indeed, if the vacuum contained all the energy physicists expect it to, it would be so repulsive that you wouldn't be able to see your hand in front of your nose. Even at the speed of light, the light from your hand wouldn't have time to reach your eyes before the expanding universe pulled it away. The fact that you can see anything at all, says Krauss, means that the energy of space cannot be so large.*" [29]

This discrepancy in connection with the cosmological constant is often considered as the deepest mystery in physics.

Are the zero point energies of quantum fields real? In connection with the quantum vacuum, the Casimir effect has often been used to demonstrate that the zero point energies of quantum fields are real. However, Jaffe showed [30] that Casimir forces can be determined without reference to such zero point energies.

In conclusion, if we combine the value of the vacuum energy resulting from usual quantum field theory with the General Theory of Relativity, the cosmological constant is obviously much too large and leads to cosmic features that we do not observe. However, this argument is based on the assumption that the General Theory of Relativity is correct.

### 5.2.2. *Emptying the Vacuum!?*

Because $\Lambda$ is too large, we may conclude that there is something wrong with the basic concept responsible for the quantum theoretical treatment of the vacuum. The vacuum obviously contains ballast that is not necessary.

From the point of view of modern physics, the vacuum has nothing to do with emptiness, just the opposite is the case: empty space (vacuum) is a "*hyperactive player, a prolific producer of jittering fields and virtual particles.*" [29] The vacuum is the most complex substance in the universe. The biggest challenge for theorists of all may simply be emptying the vacuum of all the trappings it has acquired over the past fifty years. "*They have filled the vacuum with so much garbage, there is not room for the cosmological constant,*" *said Leon Lederman,* "*Einstein freed us from the either. Now we need to get rid of (today's version of either) again. We need to sweep the vacuum clean.*" [29]

Since $\Lambda$ is equivalent to a certain vacuum energy density, we may conclude that in the case of an *empty* space (vacuum) the definition of a cosmological constant $\Lambda$ would make no sense and is excluded.

### 5.2.3. *Important Experimental Findings and Mechanisms*

Two basic experimental observations are relevant in connection with theoretical understanding: the geometry and the global dynamics (expansion) of the universe. Let us briefly summarize the main facts.

It turned out that the universe is flat, that is, the geometry of the cosmos is Euclidian. Within the General Theory of Relativity, the universe can only be flat if it has the appropriate mass and energy. However, ordinary matter and exotic dark matter only contribute approximately 35% (ordinary matter 5%, exotic dark matter 30%) and it was assumed that approximately 65% consists of dark energy. In contrast to ordinary and dark matter, the gravitation of dark energy is repulsive and shows the effect of antigravity and this effect is compatible with the second important observation in connection with the dynamics (expansion) of the universe. Let us also discuss this point.

The expansion of the universe is accelerated, that is, the universe expands faster and faster, rather than slowing down as normally expected, when we assume that the cosmos is governed by usual matter and gravitation. However, an accelerated expansion needs some sort of anti-gravity force first introduced by Einstein by means of the cosmological constant $\Lambda$. "*The discovery of this strange antigravity force has brought cosmology back to the confusion of the 1920s, when Einstein grappled with the idea that the universe was unstable. To deal with this calamity, Einstein altered his equations by putting in a fudge factor,* $\Lambda$, *that counteracted the effects of gravity. Just as the outward pressure of fusion energy counteracts the sun's gravitational collapse, the outward 'pressure' of* $\Lambda$ *counteracted the attractive forces between galaxies and galaxy clusters, keeping the universe in a stable equilibrium. There was no experimental support for* $\Lambda$, *no physical reason to believe that some sort of antigravity force existed, so when Hubble discovered the expansion of the universe, Einstein hastily withdraw the idea and later dubbed it the greatest blunder of his career. For seven decades,* $\Lambda$ *was left on the heap of discarded and discredited ideas. It simply did not fit into the way cosmologists thought the universe was put together. ... That changed suddenly and dramatically in 1998 when the supernova hunters discovered that the universe's expansion was accelerating rather than decelerating.*" [31]

In fact, a positive value for $\Lambda$ means that there is a certain repulsion and this could in principle explain the accelerated expansion of the universe extracted from supernova data in 1998.

In conclusion, the dark energy flattens out the cosmos and pushes matter apart. If the dark energy is identical with the vacuum energy, then the cosmological constant $\Lambda$ would do it. In this case the cosmological constant $\Lambda$ is expressed by Eq. (5.1). But, unfortunately, in this case we get an unacceptable large value for $\Lambda$ (see Section 5.2.1).

Therefore, many physicists believe that we have to find alternative explanations for the dark energy; one of them is the quintessence picture. However, also here we have to be careful with all theories that try to solve cosmological questions on the basis of quantum theory. From this point of view, all approaches for the explanation of the flatness of space and its accelerated expansion have to be considered with care since in all cases a cosmic problem is combined with quantum phenomena. This is the case for inflation theory but also for all the approaches that have been proposed to explain dark energy. An excellent review is given in [32] (*Dynamics of Dark Energy*). The following list of more or less exotic solutions in connection with the dark energy question is given in [32] Quintessence models; chameleon fields; a scalar field with a non-canonical kinetic term, called K-essence, which is based on K-inflation; "*modified gravity arising out of both string motivated or more generally, General Relativity modified actions which both have the effect of introducing large length scale corrections and modifying the late time evolution of the universe*" [32]; the effect of non-linearity in connection with the evolution equations; Chaplygin gases; tachyons arising in string theory; "*the same scalar field responsible for both inflation in the early universe and again today, known as quintessential inflation*" [32]; network of frustrated topological defects; phantom dark energy and ghost condensates; de Sitter vacua; "*the string landscape arising from the multiple numbers of vacua that exist when a string moduli are made stable as non-Abelian fluxes are turned on*" [32]; the cyclic universe; causal sets in the context of quantum gravity; direct anthropic arguments.

Nevertheless, the vacuum-energy-problem, due to quantum fluctuations, is not eliminated by the introduction of other mechanisms as, for example, the quintessence picture due to the large quantum vacuum energy density, the cosmological constant $\Lambda$ remains unacceptable large and no other mechanism can eliminate this problem. No doubt, we are confronted with a fundamental problem Again, this problem is often considered as the deepest mystery in physics.

### 5.2.4. *Proposals for the Elimination of the Vacuum Energy*

In other words, it makes not much sense to search for mechanisms to explain the flatness and the accelerated expansion of the universe without solving the quantum vacuum energy problem. So far, the problem could not be solved at all. In [29], the relevance of this point is expressed by the following text:

> "*As for theory, the cosmological constant poses perhaps the most interesting challenge of all. After all, the enormous energy of the vacuum is a product of that theory. If the theory is right, it has to be able to make the energy of the vacuum go away, gravitationally speaking. Something is wrong, as Witten puts it recently. It is not just a detail. It is hinting that there is a mechanism in nature we do not understand.*" [29]

We have to find a mechanism to make the energy of the vacuum go away or we have to find a new physical picture that eliminates the energy of the vacuum. In connection with this point, the following question is relevant: what is the level of description for these corrections? We have to select carefully but have to keep in mind that it is not just a detail we have to solve, as Witten and others stated. Therefore, the level of description for these corrections should be at the fundamental level. This is supported by the fact that other attempts to eliminate the vacuum energy did not lead to a positive result. Let us briefly discuss some of them.

#### 5.2.4.1. *Fine-Tuning*

No doubt, the energy of the vacuum is too large and must be wrong. "*There are ways to get the energy of the vacuum to disappear, but they are not worked out and tend to be fraught with problems. For example, as it turns out, different kinds of elementary particles add different kinds of energy to the vacuum. Some contribute positively, some negatively. According to some theories, they could conceivably cancel each other exactly (another kind of zero). If the galaxies really are accelerating, however, that only makes the problem worse. It would mean that the positive and negative contributions to the vacuum cancel out to 120 decimal places — but with a little negative pressure left over to push the stars around. It is much easier to get to exactly zero than to a tiny bit more or less than zero.*" [29]

What mechanism could function with such a precision? Such kind of fine tuning is not known in physics and can almost be classified as a transcendental process. Instead of that, Henning Genz demands a new physics. [33]

### 5.2.4.2. *Supersymmetry*

The vacuum energy shrinks to nothing in the case of supersymmetry because fermions and bosons cancel out, vacuum-wise. In other words, in the case of a perfectly symmetrical universe (supersymmetry), we just obtain what gravity says that it should vanish and we have no cosmological constant $\Lambda$. This would be the solution theorists are interested in. But unfortunately, we do not live in such a cosmos. We never did. We would not be able to exist in the case of a perfectly symmetrical universe.

### 5.2.4.3. *Strings*

String theory claims to be a theory for everything. Therefore, this theory should be able to solve the problems in connection with the cosmological constant $\Lambda$ immediately. But the string theory also could not offer a solution up to now. Strominger says rather typically, "*It is an embarrassment to all theoretical physics, and it is a particular embarrassment to string theorists because string theory is supposed to be a complete theory which describes everything. It's the problem I'd most like to solve. I've tried everything I can think of and I don't have anything to show for it.*" [29]

### 5.2.4.4. *Worm Holes*

It should be mentioned that the cosmological constant $\Lambda$ has also been discussed by Sidney Coleman in connection with quantum-sized worm holes that connect our universe with others. Here the local value for $\Lambda$ could disappear. However, this is probably not the way to solve the problem.

#### 5.2.4.5. *Projection Theory*

Projection theory solves the vacuum energy problem on the most fundamental level. Here we assume that matter, energy etc. are not embedded in space–time but the physical information about reality is projected on it. In this way we not only solve the cosmological constant problem but also other relevant points as, for example, the flatness of space. More details are outlined in Chapter 3 and also in Sections 5.4 and 5.8.

When we demand, as a condition, that the vacuum has to be free of everything, then the vacuum is really empty and we come to the demand of Leon Lederman (Section 5.2.2): "*We need to sweep the vacuum clean.*" [29] This demand must be the outcome of a new theory and needs a radical new physical approach because usual physics developed just in the opposite direction: to put more and more into the vacuum. Such a new physical ansatz is actually given by *projection theory* (see Chapter 3 and Section 5.4). In particular, we will recognize that the space can be flat without to need appropriate mass and energy densities as demanded within General Theory of Relativity (Section 5.2.3). This is a completely new situation and particularly means that the accelerated expansion of the cosmos must be due to specific processes and has nothing to do with vacuum effects as in the case of the cosmological constant $\Lambda$. $\Lambda$ would even not be definable in this case since we cannot have vacuum effects because no real masses etc. are embedded in space–time but only geometrical structures (Chapter 3).

### 5.2.5. *Another Peculiarity: Two Types of Energy*

"*... If something has matter or energy, gravity will respond by warping; gravity is the warping; if there's no warping, there's no matter or energy. So what does gravity "see" when it looks out into this fast writhing vacuum? Nothing. or almost nothing. For some unknown reason, all the activity of the vacuum is completely invisible to gravity. And yet, each jitter of the vacuum, each pair of particles that pops in and out, should be creating a small gravity well around it, just like a rock or a planet. The combined warping of all the gravity wells should be huge. You would think, said Strominger, that the whole universe would collapse because of the gravitational attraction of this sea of virtual particles.*" [29]

In connection with this strange situation (not to say impossible situation) the following is important to note: the vacuum energy (equivalent with the existence of the cosmological constant $\Lambda$) has definitely nothing to do with a certain kind of antigravity. Gravity means that there is a mutual attraction of matter and energy. On the other hand, antigravity means the same but merely the opposite: the mutual repulsion of matter and energy. However, the repulsive force in the universe is not due to matter and energy but comes entirely from empty space itself. It does not influence matter but stretches the space between the systems of matter in the universe. The effect is that it adds more and more nothing. Since this repulsive force comes from vacuum it grows with distance.

In conclusion, the energy that comes out of nothing behaves quite different from that coming out of something and this is strange. In [29], we find the following comment: "*I don't know a good heuristic argument about why it should be so crazy,*" *said Lawrence Krauss.* "*Why the energy comes out of nothing is so different from the energy that comes out of something.*"

All this seems to be artificial and is not convincing at all; this situation strongly suggests for us to eliminate this picture of space-filling quantum fields and quantum vacuum energy in order to get the vacuum clean and free of unacceptable peculiarities.

### 5.2.6. *Concluding Remarks Concerning* $\Lambda$

Within the framework of the present physical world view, the source for the accelerated expansion of the universe is the vacuum. The cosmological constant $\Lambda$ is equivalent to a vacuum energy density and, therefore, many physicists believe that $\Lambda$ is responsible for this acceleration. However, due to similar effects in the early universe $\Lambda$ should vary with time $\tau$ [$\Lambda = \Lambda(\tau)$] but this situation is not easy to treat within the framework of Einstein's field equations. Here $\tau$ is again the time of conventional physics, which is measured with our clocks.

The quintessence picture is another approach for the understanding of the accelerated expansion of the cosmos, and this picture is not based on vacuum effects as in the case of $\Lambda$ but describes the acceleration by an unknown energy source producing anti-gravitational effects. The quintessence picture is a new physical idea and defines a new force which is often called the "fifth force." The theoretical investigations in connection with the fifth force are just at the beginning and it is still an open question if this picture leads to a consistent framework.

The situation in connection with the cosmological constant $\Lambda$ demonstrates that there is something fundamentally wrong if we treat the problem within the frame of quantum field theory (Section 5.2.1). But what? This question is difficult to answer. The factor of $10^{122}$ in connection with $\Lambda$ (Section 5.2.1) is probably due to the fact that the conceptions and contents of General Theory of Relativity and quantum theory not only differ considerably in their basic structure, but are in particular not compatible. If we put a quantum result into the field equations of the General Theory of Relativity, the result is a catastrophe.

The formulas of General Theory of Relativity and quantum theory work fantastically in most cases but there are still open questions in connection with some basic points. For example, the problem of *time* is not satisfactorily solved at all within usual quantum theory. Here, time still behaves classically within all forms of usual quantum theory (see Chapter 3). Another example: *Mach's Principle* is not fulfilled within General Theory of Relativity and this is weighty. Both points, the nature of time and Mach's principle, are of fundamental importance for the understanding of the cosmos. We will recognize below that both points are tightly linked with the cosmological constant problem.

We already mentioned that General Theory of Relativity and quantum theory cannot be both right. What theory is correct and what not? But the situation is not so simple and, therefore, in general we have to assume that both theories require modifications although they are able to describe certain experimental material impressively. It would be desirable when both theories (General Theory of Relativity and quantum theory) could be based on the same basic principles producing a unification of gravitational and quantum effects. We will discuss this point in more detail below in connection with the introduction of such a unifying basic principle that is given by the projection principle introduced in Chapter 2. In particular, we have discussed in Chapters 2 and 3 the following basic question: How is the real world connected to space and time? Is the naive viewpoint really true that the real world is *embedded* in space–time? We have recognized in Chapters 2–4 that this viewpoint is quite questionable.

## 5.3. Critical Remarks on Basic Quantum Theory

It is usually assumed that the quantum theoretical value for the vacuum energy is wrong because the cosmological constant $\Lambda$ becomes unacceptably large in this case. Within the framework of these considerations, we work within quantum field theory. Here the laws of usual quantum theory are combined with the laws of Special Theory of Relativity. In this connection it is important to note that within quantum field theory, no mechanism could be found and no reasonable procedure could be introduced to reduce the value for the vacuum energy. The search for a solution of this big problem already lasts approximately two decades but, up to now, without any success. Therefore, we should conclude that the vacuum energy problem is tightly linked with the very foundations of physics. Due to the unsuccessful investigations on lower physical levels makes the step to the top level probably necessary. Then, the following question arises. Are our conceptions of basic physics really realistic enough for the solutions of problems of this calibre? Just in connection with usual quantum theory, a lot of questions have not been answered up to the present day. In Chapter 3, some of the problems have been listed. In addition, we would like to point out the relationship between energy $E$ and time $\tau$ within usual quantum theory.

### 5.3.1. *Basic Elements for Vacuum Description*

Our knowledge about the vacuum (space) properties is based on usual quantum theory and Special Theory of Relativity. Usual quantum theory delivers the uncertainty relations, in particular, the uncertainty relation for the energy $E$ and time $\tau$:

$$\Delta E \Delta \tau \geq \frac{\hbar}{2}. \tag{5.3}$$

From the Special Theory of Relativity, the formula $E = m_0 c^2$, the well-known relation between energy $E$ and mass $m_0$, is used. Both relations lead to the postulate of anti-matter. To every particle, there exists an antiparticle (it is, for example, a positron for an electron). Dirac actually considered the antiparticle as a very palpable hole in nothing. However, "*Dirac did not have all the details right. Now that physicists understand anti-matter better, it is no longer viewed as holes in nothing. A positron is a real particle in its own right. Still, as Wheeler points out, the modern antiparticle theory and Dirac's hole theory differ only in their imagery, not their mathematcs.*" [29]

Due to Eq. (5.3) the energy can fluctuate, and a particle–antiparticle pair may appear from the vacuum for a short time interval dictated by Eq. (5.3). The ground state energy of the vacuum can never be zero because Heisenberg's uncertainty relation (5.3) is assumed to be valid for all physical situations. Just these energy fluctuations, due to Eq. (5.3), are responsible for the large vacuum energy density and the impossible value for the cosmological constant $\Lambda$. What can we say about the validity of Eq. (5.3)? What are the limitations of this equation? Before we discuss this point in detail, we would like to give some remarks in connection with the Special Theory of Relativity.

### 5.3.2. *Remarks Concerning Special Theory of Relativity*

The Special Theory of Relativity is based on two postulates: (i) the principle of relativity and (ii) the principle of the constancy of the velocity of light. The laws derived from these two postulates are consistent, and this holds for the logical structure of Special Theory of Relativity. However, there is one weak point: empty space–time (vacuum) is *absolute* in character and this reflects an unphysical situation as already outlined by Ernst Mach. Within the Special Theory of Relativity, we have an absolute four-dimensional continuum and this situation is not better than within Newton's mechanics where we have an absolute three-dimensional continuum. However, we may assume that this absoluteness within the Special Theory of Relativity is not responsible for the large vacuum energy value that follows from quantum field theory.

It should be mentioned that the absoluteness of space–time can be eliminated without changing the postulates (i) and (ii) significantly. [34] Deviations of this non-absolute theory from Special Theory of Relativity are expected for sufficiently large masses, masses that are probably much larger than those of usual elementary particles. Therefore, we do not expect that also this non-absolute theory can solve the vacuum energy problem.

However, there could possibly be another problem. When we combine usual quantum theory and Special Theory of Relativity to quantum field theory, we could have a level inconsistency. In other words, the levels of descriptions might not be compatible, and this would be problematic. The mixing of physical laws that are positioned at different levels of description is in principle strictly forbidden. (We play golf on a golf course and tennis on a tennis court, but to play golf on a tennis court or tennis on a golf course is not possible.) Usual quantum theory and Special Theory of Relativity have been developed independently of each other, and both theories differ considerably from each other in their conceptions and contents. Therefore, we cannot expect that the levels of descriptions are the same with respect of usual quantum theory and the Special Theory of Relativity. However, it is difficult to construct a level hierarchy (see Chapter 4) in connection with usual quantum theory and Special Theory of Relativity.

We would like to assume here that the wrong vacuum energy value is not due to inconsistencies in connection with Special Theory of Relativity and its connection with usual quantum theory. In the next section, we will recognize that the laws of usual quantum theory are much more problematic, and they are possibly the reason for the wrong vacuum energy value.

### 5.3.3. *Remarks Concerning Usual Quantum Theory*

Special Theory of Relativity has been developed by Einstein within one step and was published in 1905. The essential points of its physical interpretation have been delivered simultaneously. The situation in usual quantum theory is different:

As we have already remarked in Chapter 3, the essential features of the mathematical formalism of (non-relativistic) quantum theory were constructed by Heisenberg and Schrödinger in 1925–1926. On the basis of this quantum-mechanical formalism an enormous number of effects in atomic physics, chemistry, solid state physics, etc. could be predicted and explained. However, more than 80 years after its formulation the interpretation of this formalism is "*by far the most controversial problem of current research in the foundations of physics and devides the community of physicists and philosophers of science into numerous opposing schools of thought.*" [15] In contrast to classical mechanics and Special Theory of Relativity, there is indeed an immense diversity of opinion and a huge variety of interpretations. The differences between the various interpretations are not only marginal but can be recognized by comparison of the *Copenhagen Interpretation* and the *Many-Worlds Theory*. It is well-known that both interpretations lead to pictures of reality which are completely different from each other.

## New aspects in connection with Bohr's principle of complementarity

The following new experimental finding makes the problem worse. There are new experimental results [35–39] which clearly indicate that Bohr's principle of complementarity cannot be considered as a general principle. These experiments show that Bohr's viewpoint in connection with the wave–particle duality is obviously not a general feature of nature, and if that is really the case, the situation is changed fundamentally.

The following question is important. Is matter (or light) a wave or a particle phenomenon? In answering this question, Bohr argued as follows. Only one of the two incompatible possibilities is realized, depending on the type of experiment performed on them, and we come to the notion of complementarity. Either we perform an experiment typical for waves or we choose an experimental arrangement that exclusively marks the particle aspect. Only in this way wave and corpuscular aspects of physical phenomena are not incompatible. In other words, not only wave and particle concepts are mutually exclusive but also the corresponding experimental arrangements. Just this point is obviously not always true as demonstrated in the case of light. There exist optical experimental arrangements [35–39] that can be used for the *simultaneous* detection of particle and wave aspects. If that is really the case, we have to change our quantum theoretical picture of reality fundamentally.

## 5.3.4. *Specific Problems*

There are obviously some specific serious problems in connection with usual quantum theory. These problems are often ignored, and this is mainly due to the fact that the empirical success of usual quantum theory is impressive. But due to the new experiments [37,39], which have been performed by two independent groups, we are confronted with a new situation, and we have to analyze critically the basics of conventional quantum theory (its mathematical apparatus as well as its interpretation). In Chapter 3 and [40], we already quoted some critical points which appear in connection with usual quantum theory. The main facts can be summarized as follows: 1. Schrödinger's equation could not be deduced. 2. The collapse of the wavefunction cannot be explained. 3. Time $\tau$ is still a classical parameter in usual quantum theory. 4. The use of a particle defined as a local existent (which is assumed within the frame of usual quantum theory) seem also to be problematic. More details concerning these four points are given in Chapter 3. All points need to be discussed in more detail. [40] Here we only want pick up point 1 and point 3 because these two topics are of particular relevance for the cosmological constant $\Lambda$ (vacuum energy).

### 5.3.4.1. *Schrödinger's Equation*

Schrödinger's equation

$$i\hbar\frac{\partial}{\partial\tau}\psi(\mathbf{r},\tau) = -\frac{\hbar^2}{2m_0}\Delta\psi(\mathbf{r},\tau) + U(x,y,z,\tau)\psi(\mathbf{r},\tau) \tag{5.4}$$

[see also Eq. (3.25)] could not be derived but has been assumed on the basis of "reasonable arguments." It can only be validated in connection with experiments, and here its success is without any doubt impressive but, unfortunately, we do not really know which physical reality is reflected by this equation. Thus, it is not or almost not possible to develop these equations systematically further.

In connection with Schrödinger's equation (5.4), the following question arises. What reality hides behind this equation? In order to come closer to an answer, let us introduce some postulates for the introduction of Eq. (5.4).

Schrödinger found his equation on the basis of the *classical* wave equation and de Broglie's wave–matter postulate. Another possibility for the introduction of Schrödinger's equation is to use the classical expression for the energy of an interacting particle

$$E = \frac{1}{2m_0}(p_x^2 + p_y^2 + p_z^2) + U(x, y, z, \tau) \tag{5.5}$$

and the following postulates. [41,42]

The time-dependent Schrödinger equation (5.4) can be obtained if the variables

$$\begin{array}{c} E \\ p_x, p_y, p_z \\ x, y, z, \tau \end{array} \tag{5.6}$$

of the classical equation (5.5) are replaced by

$$\begin{aligned} &\hat{E} = i\hbar\frac{\partial}{\partial\tau}, \\ &\hat{p}_x = -i\hbar\frac{\partial}{\partial x}, \quad \hat{p}_y = -i\hbar\frac{\partial}{\partial y}, \quad \hat{p}_z = -i\hbar\frac{\partial}{\partial z}, \\ &x, y, z, \tau. \end{aligned} \tag{5.7}$$

Hence, "keep the coordinates and time as they are and replace the energy and momenta by *operators*."

The source of this postulate is unknown. Within usual quantum theory, we cannot recognize what physical reality hides behind these laws. It is, for example, not at all clear why the coordinates $x, y, z$ are not treated on an even footing with the momenta $p_x, p_y, p_z$. We could also regard the coordinates as being operators equal to the coordinates, i.e.,

$$\hat{x} = x, \ \hat{y} = y, \ \hat{z} = z. \tag{5.8}$$

However, we would still have to calculate with these operators in just the same way as with ordinary numbers. [42]

Role of time within Schrödinger's theory

It is important to note once more that within Schrödinger's theory time $\tau$ is still a classical quantity, that is, time remains unchanged when we go from classical mechanics to quantum theory and behaves like an external parameter, as in Newton's physics. Schrödinger tried to modify his apparatus by the introduction of an operator for the time but without success. This is an important and principal point and we gave some remarks on this subject in [43]. All these points indicate that Schrödinger's theory is merely a phenomenological approach which works but we do not know its source.

### 5.3.4.2. *On the Symmetry between Space and Time*

In atomic physics, solid state physics, nanoscience etc. it is sufficient that quantum theory conforms to the Special Theory of Relativity (see, for example, [44]) because phenomena within the frame of these topics are hardly influenced by the laws of General Theory of Relativity, that is, by gravitation.

Since quantum theory and Special Theory of Relativity have been developed independently from each other, the following question is of relevance. Can quantum phenomena be treated fully relativistically in accordance with the basic laws of Special Theory of Relativity? Although the relativistic wave equations (e.g., Dirac's equation for the electron) are invariant under Lorentz transformation, the space coordinates $x$, $y$, $z$ and time $\tau$ are in its physical content definitely not symmetrical to each other and that is in contrast to the fundamental results of Special Theory of Relativity. This is due to the following facts.

(i) While the coordinates are *statistical* quantities, time does not behave statistically. We already mentioned above that time remains unchanged when we go from classical mechanics to quantum theory. This is clearly reflected in the fact that the coordinates can be *operators*, time is always a simple *parameter*. [12]

(ii) The determination of the eigenfunctions and eigenvalues is restricted on space and time is not involved in this process.

Louis de Broglie expressed this fact as follows: "The present quantum theory in all its versions takes time as the evolution parameter and therefore destroys the symmetry between space and time." [45]

Within the framework of Special Theory of Relativity, the world has to be considered as four-dimensional, and this is because time loses its independence which it still had within Newton's mechanics. The fourth of the equations of Lorentz transformations

$$\tau' = \frac{\tau - \upsilon x/c^2}{\sqrt{1 - \upsilon^2/c^2}} \tag{5.9}$$

shows that the time interval between two events in a moving frame of reference $S'$ (which moves relatively to a rest system, say $S$, with the constant velocity $\upsilon$) does not vanish in general, even when the time interval in $S$ becomes zero. The consequence of a pure "distance in space" in $S$ is an "interval of time" in $S'$, and this result can directly be read from Eq. (5.9). It is well known that Eq. (5.9) is obviously realistic, and this has been experimentally demonstrated by the slowing down of clocks.

In conclusion, in contrast to usual quantum theory within Special Theory of Relativity, space and time are tightly interrelated and symmetrical to each other and cannot be treated as independent quantities as is done in conventional quantum theory, even in the relativistic case.

That is in close connection with the fact that there is no uncertainty relation for the energy and time which would agree in its physical content with the uncertainty relation for the coordinates and momenta:

$$\delta p_x \delta x \geq \frac{\hbar}{2}, \quad \delta p_y \delta y \geq \frac{\hbar}{2}, \quad \delta p_z \delta z \geq \frac{\hbar}{2}. \tag{5.10}$$

An analogous relation for the time $\tau$ and a quantity which has the dimensions of energy is required from the point of view of Special

Theory of Relativity. The significance of the well-known relation [see Section 5.3.1, Eq. (5.3)]

$$\Delta E \Delta \tau \geq \frac{\hbar}{2}$$

is entirely different from that of (5.10). This difference is symbolically expressed by the use of $\Delta$ instead of $\delta$. In relations (5.10), the quantities $\delta p_x, \ldots$ and $\delta x, \ldots$ are the uncertainties in the values of the momenta and the coordinates at the same instant. As is well-known, this uncertainty means that the coordinates and momenta can never have entirely definite values simultaneously. The energy $E$, on the other hand, can be measured to any degree of accuracy at any instant. The quantity $\Delta E$ in (5.3) is the difference between two exactly measured values of the energy at two different instants and is not the uncertainty in the value of the energy at a given instant (see also the discussion in [44,45]).

This is a strange situation and is obviously due to the fact that within usual quantum theory time $\tau$ is still a classical quantity as we have already pointed out in connection with Schrödinger's equation (Section 5.3.4.1). This is an important point and might be the key for the understanding of the vacuum energy problem that is connected to the cosmological constant $\Lambda$ (Section 5.2). Before we start to discuss this point in more detail let us first cite some critical remarks with respect to $\Delta E \Delta \tau \geq \hbar/2$ [Eq. (5.3)].

### 5.3.4.3. *Bunge's Critique*

Since time within usual quantum theory is still a classical parameter, we should consider the energy-time relation $\Delta E \Delta \tau \geq \hbar/2$ [Eq. (5.3)] as a quasi-classical equation. The value of such an equation is however questionable.

In Mario Bunge's opinion, the energy-time relation (5.3) "*is a total stranger to quantum theory*" [46]. In particular, we find in [46]:

> "*This relation is made plausible by reference to some thought experiments, to radioactive decay, and to line breadths. But unlike the genuine indeterminacy relations,* (5.3) *has never been proved*

*from first principles. In other words,* (5.3) *does not belong to quantum theory, but is just a piece of doubtful heuristics.*

*The reason for this failure to incorporate* (5.3) *into quantum mechanics is the following. In this theory, as in every other known theory, time is a 'c number' and, more particularly a parameter, not a dynamical variable. Moreover, τ does not 'belong' (refer) to the system concerned. Even in relativistic theories the proper time, though relative to a frame of reference, is not a property of the system on the same footing as its mass or its momentum. In other words, τ does not belong to the family of operators in the Hilbert space associated to every microsystem. Therefore τ is not a random variable and its scatter vanishes identically ...*

*Consequently, no matter what the scatter in the energy may be, the inequality* (5.3) *does not hold. Also, it does not improve things to regard E, as is sometimes done, as the Hamiltonian of the system the standard deviation of the energy vanishes as well. In conclusion, the so-called fourth indeterminacy (or uncertainty) is a total stranger to the quantum theory although it can be found in works on this theory ....*

*In short, the fourth scatter relation is not deducible from the principles of the quantum theory, whether relativistic or not. But then why is it sometimes used, for example in the theory of line breadths? The reason is that it is not the same formula. Although it has the same typographical form, it has a different content. In particular, 'Δ' is not interpreted as a standard deviation (from what?) but as the half-life of the state ψ. But even thus reinterpreted, the formula is not deducible from the postulates of the general quantum theory. Mind, this has nothing to do with the question whether or not formula* (5.3) *is true under some suitable interpretation. Many other statements are true and yet they do not belong to quantum theory."*

#### 5.3.4.4. *Conclusion and Outlook*

In summary, the required symmetry between space and time (Special Theory of Relativity) is definitely not fulfilled within the framework of usual quantum theory (in its relativistic and its non-relativistic version). In particular, time $\tau$ remains a simple parameter in quantum mechanics. Thus, we may state that time is still a *classical* quantity within conventional quantum theory. From this point of view, usual quantum theory is not complete, and the relation $\Delta E \Delta \tau \geq \hbar/2$ is quite questionable. In Bunge's opinion, it is a total stranger in usual quantum theory. However, just this equation is responsible for the appearance of the vacuum energy (Section 5.3.1) leading to the problem in connection with the cosmological constant $\Lambda$ (Section 5.2).

It is demonstrated in [12,40] that the time is no longer a simple classical parameter within projection theory but becomes a real quantum variable and, furthermore, Schrödinger's equation could be derived within projection theory which is not possible in usual quantum theory (Section 5.3.4.1). In particular, within projection theory (Section 5.4) equation $\Delta E \Delta \tau \geq \hbar/2$ is not defined but we obtain here an uncertainty relation for the energy and time that is completely equivalent to the uncertainty relation for the coordinates and momenta [Eq. (5.10)].

We already remarked in Chapter 3 that the energy $E$ is strictly conserved within projection theory. A temporary violation of this conservation laws, as is possible within usual quantum theory, gives no sense within projection theory. Again, within usual quantum theory the conservation law for energy can be temporarily violated and this is due to equation $\Delta E \Delta \tau \geq \hbar/2$ which still contains time $\tau$ in classical form. This temporary violation of the conservation law for the energy is responsible for the unrealistic (wrong) vacuum energy leading to an unacceptable large value for the cosmological constant $\Lambda$.

In [12,40], we particularly pointed out that the "naive" point of view of conventional quantum theory cannot simply be extended but we have to include further basic facts, so far not directly considered in the basic description of physical systems. We recognized that we can learn something new about the relation between the "observer" and that what we call "reality" when we consider certain facts from biological evolution and behavior research.

## 5.4. Projection Theory and the Emptying of Space–Time

There are obviously some specific serious problems in connection with usual quantum theory. These problems are often ignored, and this is mainly due to the fact that the empirical success of usual quantum theory is impressive. But due to the new experiments [37,39], which have been performed by two independent groups, we are confronted with a new situation (see in particular Section 5.3.4), and we have to analyze critically the basics of conventional quantum theory (its mathematical apparatus as well as its interpretation). In Chapter 3, we already have quoted some critical points which appear in connection with usual quantum theory. We particularly emphasized that Eq. (5.3) $[\Delta E \Delta \tau \geq \hbar/2]$ is obviously a doubtful law.

In Chapter 3, we discussed some basic aspects of projection theory. In the next section, we will add some essential points that are relevant for the understanding of the cosmological constant. We recognized in Chapter 3 that the "naïve" point of view of conventional quantum theory, which is also reflected in the equation $\Delta E \Delta \tau \geq \hbar/2$, cannot be simply extended, but we have to include new basic facts.

### 5.4.1. *The Basic Principle*

In connection to the cosmological constant, the very basics of projection theory are relevant. Let us repeat the main facts, which we have already worked out in the preceding chapters. The most important features can be formulated as follows.

No doubt, there is an inside world and an outside world. The inside world is in the head of the observer representing the outside world within space and time. The picture inside the head contains geometrical positions. Within conventional physics, it is assumed that the structure of the outside world is exactly the same except for one point. The geometrical positions are replaced by masses. There is, in other words, a one-to-one-correspondence between the structures of the inside world and reality outside. However, this should not be realistic. Why? Here two points are of relevance.

1. The principles of evolution do not allow such a one-to-one-correspondence. The arguments are given in Chapters 1 and 2.
2. Due to the most basic properties of space and time, there should be no space–time in reality outside. The arguments are given in detail in Chapter 2.

Thus, we concluded that the real world cannot be embedded in space–time (container principle) but it can only be projected on it (projection principle), and we experience these projections as "pictures of reality."

In other words, space–time is not a container for the integration of physically real objects etc. Nothing can be inside space–time, no matter, no energy. In other words, within projection theory there is an emptying of space–time. Just this was required by Leon Lederman (see also Section 5.22). Space–time exclusively plays the role of an auxiliary element for the representation of the picture of reality. Space–time can only contain geometrical positions, trajectories etc. as in the case of a sheet of paper. As already said, all these points have already been explained and discussed in more detail in Chapters 2 and 3.

The following is important. We can say nothing about *basic reality* outside but we are able to introduce alternative realities (fictitious reality), which we have introduced in Chapter 3. All the details of basic reality have to be considered as cosmic secrets, which principally cannot be eliminated within projection theory.

### 5.4.2. *Projections from Alternative Realities onto Space–Time*

The projection principle has to be formulated by means of mathematical tools. We did that in [12]. Here we want only to give those facts that are relevant in connection with the cosmological constant. Let us already mention here that we are able to deduce within projection theory an uncertainty relation for the time and the energy which agrees in its physical content the uncertainty relation for the momentum $\mathbf{p}$ and position $\mathbf{r}$ [see Eq. (5.10)]. In Section 5.4.4, we will derive this new formula. In this connection (and also with respect to the cosmological constant and the vacuum energy), it is important to state that the energy is strictly conserved here, in contrast to those effects that are based on formula (5.3) [$\Delta E \Delta \tau \geq \hbar/2$] of conventional quantum theory. Eq. (5.3) does not exist in projection theory.

#### 5.4.2.1. *Other Spaces, Fourier-Space*

As we have pointed out in Section 3.2.1, instead of basic reality, we need alternative realities for the description of physical reality. Here alternative realities play the role of fictitious realities. Since the role of the variables, in particular for time, is important for our analysis in connection with the cosmological constant, let us repeat what we have stated in Section 3.2.1.

> "*It turned out* [12] *that the variables of fictitious reality must be given by the momentum* $\mathbf{p}$ *and the energy* $E$, *that is, fictitious reality is embedded in the so-called* $(\mathbf{p}, E)$*-space. The information that is embedded in* $(\mathbf{p}, E)$*-space is completely projected onto space and time. Since the variables* $\mathbf{p}$ *and* $E$ *reflect reality, where reality has be to identified with a certain system, the variables* $\mathbf{p}$ *and* $E$ *have to be considered as system-specific quantities. Thus, the projected information onto space and time must also be system-specific. This however inevitably means that the space–time frame, on which the information of* $(\mathbf{p}, E)$*-space is projected, may not contain the time* $\tau$. $\tau$ *is an external parameter and is in all cases measured by our clocks and, therefore, it cannot reflect system-specific properties. Thus, we have to introduce a system-specific time, and we would like to use the letter* $t$ *for it. In other words, the space–time frame on which the (fictitious) reality is projected is given by the variables* $\mathbf{r}$ *and* $t$. *In the following, we want to use the marking* $(\mathbf{r}, t)$*-space for the space–time frames, which are involved in the projections.*"

In summary, we have the outside world (basic reality) which is not accessible to the human observer. Therefore, the pictures in front of us are the most direct impression of the outside world. These impressions (pictures) have to be considered as the most essential information for a human observer. Thus, we may state the following. Within the memory of man all essential things are represented within the frame of pictures that appear spontaneously in front of us. This is the most basic statement. These pictures of reality are represented in $(\mathbf{r}, t)$-space and are described on the basis of processes which take place in fictitious reality. Each process is characterized by $(\mathbf{p}, E)$-fluctuations in $(\mathbf{p}, E)$-space (fictitious reality). The processes in $(\mathbf{p}, E)$-space are projected onto $(\mathbf{r}, t)$-space. As we have outlined in Chapter 2, the information transfer from basic reality to the picture of reality also has to be considered as a projection. Since basic reality (outside world) is not accessible to man, we are principally not able to describe the information transfer from basic reality to the picture of reality, and this is the reason why we have to construct fictitious realities, that is, alternative realities.

### 5.4.2.2. *Reality and its Picture*

In [12] and in Chapter 3, we characterized the information in $(\mathbf{r}, t)$-space by the wavefunction $\Psi(\mathbf{r}, t)$ and the corresponding information in $(\mathbf{p}, E)$-space by the wavefunction $\Psi(\mathbf{p}, E)$. The quantities $\Psi^*(\mathbf{r}, t)\Psi(\mathbf{r}, t)$ and $\Psi^*(\mathbf{p}, E)\Psi(\mathbf{p}, E)$ are probability densities (Chapter 3). As we already know from Section 3.3, $(\mathbf{r}, t)$-space and $(\mathbf{p}, E)$-space are equivalent as far as their information is concerned. Both spaces are connected by a Fourier transform and we get [12]

$$\Psi(\mathbf{r}, t) = \frac{1}{(2\pi\hbar)^2} \int_{-\infty}^{\infty} \Psi(\mathbf{p}, E) \exp\left\{i\left[\frac{\mathbf{p}}{\hbar} \cdot \mathbf{r} - \frac{E}{\hbar}t\right]\right\} dp_x p_y p_z dE. \tag{5.11}$$

The inverse transformation is

$$\Psi(\mathbf{p}, E) = \frac{1}{(2\pi\hbar)^2} \int_{-\infty}^{\infty} \Psi(\mathbf{r}, t) \exp\left\{-i\left[\frac{\mathbf{p}}{\hbar} \cdot \mathbf{r} - \frac{E}{\hbar}t\right]\right\} dtdxdydz \tag{5.12}$$

and we have

$$\int_{-\infty}^{\infty} \Psi^*(\mathbf{r}, t)\Psi(\mathbf{r}, t)dxdydzdt = \int_{-\infty}^{\infty} \Psi^*(\mathbf{p}, E)\Psi(\mathbf{p}, E)dp_x p_y p_z E. \tag{5.13}$$

If $\Psi(\mathbf{r}, t)$ or $\Psi(\mathbf{p}, E)$ is normalized to unity, Eq. (5.13) leads to 1.

We have the following situation. Both spaces [($\mathbf{r}$, $t$)-space, ($\mathbf{p}$, $E$)-space] are equivalent concerning their information about the physical system under investigation, [12] and this is due to the basic properties in connection with the Fourier transformation.

As outlined in Chapter 3, $\Psi(\mathbf{r}, t)$ is a certain "information" (wavefunction) in ($\mathbf{r}$, $t$)-space, and $\Psi(\mathbf{p}, E)$ is the equivalent "information" (wavefunction) in ($\mathbf{p}$, $E$)-space. As argued in [12] and in the references therein, the determination of $\Psi(\mathbf{r}, t)$ and $\Psi(\mathbf{p}, E)$, respectively, cannot be done on the basis of classical mechanics.

### 5.4.3. *Further Basic Properties*

Due to the equivalence of ($\mathbf{r}$, $t$)-space and ($\mathbf{p}$, $E$)-space, it gives only sense to express the properties of the world in one space of the both spaces, either we work ($\mathbf{r}$, $t$)-space in or in ($\mathbf{p}$, $E$)-space. Then, we have to know how the variables $\mathbf{p}$ and $E$ are expressed in ($\mathbf{r}$, $t$)-space and, on the other hand, if we would like to work in ($\mathbf{p}$, $E$)-space, we have to know how the variables $\mathbf{r}$ and $t$ are expressed in ($\mathbf{p}$, $E$)-space.

The structure of the Fourier transform means that it is not possible to give definite values for the coordinates $x$, $y$, $z$ and the time $t$ if $\mathbf{p}$ and $E$ take definite values. On the other hand, it is not possible to give definite values for $\mathbf{p}$ and $E$ if $\mathbf{r} = (x, y, z)$ and $t$ take definite values. For example, for the determination of $\Psi(\mathbf{r}, t)$ for definite values $\mathbf{r}$ and $t$, say $\mathbf{r}_1$ and $t_1$, we need all possible values for $\mathbf{p}$ and $E$, in principle from $-\infty$ to $\infty$. Thus, in the analysis of quantum phenomena given here the variables $\mathbf{p}$ and $E$, expressed in ($\mathbf{r}$, $t$)-space, cannot be simple numbers and, on the other hand, the variables $\mathbf{r}$ and $t$, expressed in ($\mathbf{p}$, $E$)-space, can also not be simple numbers. In particular, the projection principle directly leads to the following rules (see [12]).

(**r**, $t$)-space:

coordinates and time are *numbers* $(x, y, z, t)$,
momentum and energy are *operators* $(-i\hbar\partial/\partial x, -i\hbar\partial/\partial y, -i\hbar\partial/\partial z, i\hbar\partial/\partial t)$.

(**p**, $E$)-space:

coordinates and time are *operators* $(i\hbar\partial/\partial p_x, i\hbar\partial/\partial p_y, i\hbar\partial/\partial p_z, -i\hbar\partial/\partial E)$,
momentum and energy are *numbers* $(p_x, p_y, p_z, E)$.

Let us consider an example. Using (5.12) we can form the following identity:

$$-i\hbar\frac{\partial}{\partial E}\Psi(\mathbf{p}, E) = \frac{1}{(2\pi\hbar)^2}\int_{-\infty}^{\infty} t\Psi(\mathbf{r}, t) \times \exp\left\{-i\left[\frac{\mathbf{p}}{\hbar}\cdot\mathbf{r} - \frac{E}{\hbar}t\right]\right\} dxdydzdt. \tag{5.14}$$

Interpretation of (5.14): Any information given in (**r**, $t$)-space can be completely transformed into (**p**, $E$)-space, and vice versa. Both informations must be physically equivalent. We have two representations of the same thing. $\Psi(\mathbf{p}, E)$ in (5.14) is equivalent to $\Psi(\mathbf{r}, t)$, and vice versa (see [12]). Also $t\Psi(\mathbf{r}, t)$ and $-i\hbar\partial/\partial E\Psi(\mathbf{p}, E)$ in (5.14) must be equivalent. Thus, the operator

$$\hat{t} = -i\hbar\frac{\partial}{\partial E} \tag{5.15}$$

must be equivalent to the time $t$, i.e., time $t$ takes the form of an operator in (**p**, $E$)-space.

In the same way we can find, on the basis of (5.11) and (5.12), operators for $E$, **r** and **p**, and we come to the equations for the determination of $\Psi(\mathbf{p}, E)$ and $\Psi(\mathbf{r}, t)$ [Eqs. (3.23) and (3.22) (see in particular [12]).

(**p**, $E$)-space:

$$E\Psi(\mathbf{p}, E) = \frac{\mathbf{p}^2}{2m_0}\Psi(\mathbf{p}, E) + V\left(i\hbar\frac{\partial}{\partial p_x}, i\hbar\frac{\partial}{\partial p_y}, i\hbar\frac{\partial}{\partial p_z} - i\hbar\frac{\partial}{\partial E}\right)\Psi(\mathbf{p}, E).$$

$(\mathbf{r},\, t)$-space:

$$i\hbar\frac{\partial}{\partial t}\Psi(\mathbf{r},t) = -\frac{\hbar^2}{2m_0}\Delta\Psi(\mathbf{r},t) + V(x,y,z,t)\Psi(\mathbf{r},t).$$

Other equations follow directly from the projection principle, in particular, from the rules that we have formulated in this section. Such rules do not exist within usual quantum theory and, therefore, Schrödinger's equation could not be derived (Section 5.3.4.1). Projection theory delivers the key for the derivation of the basic equations (3.22) and (3.23), which are more general than Schrödinger's equations of conventional quantum theory. The projection principle allows us to derive certain rules (see [12] and the remarks above) for numbers and operators that lead to the above equations (3.22) and (3.23), and these rules are not known in conventional quantum theory. In Section 5.3.4.1 we came to Schrödinger's equation of usual quantum theory by using certain postulates, which are expressed by (5.7) but could not be derived.

We have pointed out above that $(\mathbf{r},\, t)$-space only contains geometrical positions etc. but never material objects. Thus, there can be no interactions in $(\mathbf{r},\, t)$-space. Therefore, the function $V(x,y,z,t)$ in Eq. (3.22) cannot be interpreted as potential energy, but $V(x,y,z,t)$ describes merely the correlations between the various geometrical positions. The interactions exclusively take place in $(\mathbf{p},\, E)$-space in form of $(\mathbf{p},\, E)$-fluctuations where we have to assume that $\mathbf{p}$ and $E$ are strictly conserved. A temporary violation of these conservation laws makes no sense within projection theory. Within usual quantum theory, the conservation law for energy can be temporarily violated and this is due to equation $\Delta E \Delta \tau \geq \hbar/2$ which contains time $\tau$ still in classical form (Section 5.3.4.1).

Again, the system described by $\Psi(\mathbf{p},E)$ $[\Psi(\mathbf{r},t)]$ interacts with its environment by an exchange of momentum and energy. It is therefore obvious that time $t$ cannot be an external parameter measured with our clocks but has to reflect a system-specific property. [12]

It must be emphasized once more that the fundamentally new nature of time within projection theory is particularly reflected in the fact that the time-coordinate $t$ is not always a simple number within projection theory, but must be an operator $-i\hbar\partial/\partial E$ within ($\mathbf{p}$, $E$)-space. This is of fundamental importance, and we have discussed the consequences in [12]. Within usual quantum theory time is always a simple *external* parameter, which we expressed by $\tau$, and $\tau$ has nothing to do with the process under investigation (see Sections 5.3.3 and 5.3.4).

Nevertheless, in order to be able to compare different time structures in the laboratory, we need a reference time structure which can be fixed arbitrarily but should be the same for all laboratories. Here we would like to define the reference time structure by our clocks that we use in everyday life, and let us denote this reference time with $\tau$.

## 5.4.4. *Uncertainty Relation for Time and Energy within Projection Theory*

### 5.4.4.1. *Mean Values*

Let us perform the analysis within ($\mathbf{r}$, $E$)-space. In this space, the energy is given by simple numbers and the time-coordinate plays the role of an operator (Section 5.4.3). The multiplication of the wavefunction $\Psi(\mathbf{r}, E)$ with its complex conjugate $\Psi^*(\mathbf{r}, E)$ determines the probability distribution of the variables $\mathbf{r}$ and $E$. $\Psi^*(\mathbf{r}, E)\Psi(\mathbf{r}, E)d\mathbf{r}dE$ is the probability that a measurement performed on the system leads to the values for the position vector $\mathbf{r}$ and the energy $E$ to be in the intervals $\mathbf{r}$, $\mathbf{r} + d\mathbf{r}$ and $E$, $E + dE$.

With (see [12])

$$\Psi(\mathbf{r}, E) = \frac{1}{(2\pi\hbar)^{1/2}} \int_{-\infty}^{\infty} \Psi(\mathbf{r}, t) \exp\left\{i\frac{E}{\hbar}t\right\} dt \tag{5.16}$$

we get

$$\int_{-\infty}^{\infty} \Psi^*(\mathbf{r}, t)\Psi(\mathbf{r}, t)dxdydzdt = \int_{-\infty}^{\infty} \Psi^*(\mathbf{r}, E)\Psi(\mathbf{r}, E)dxdydzE. \tag{5.17}$$

If $\Psi(\mathbf{r}, t)$ or $\Psi(\mathbf{r}, E)$ is normalized to unity, Eq. (5.17) leads to 1:

$$\int_{-\infty}^{\infty} \Psi^*(\mathbf{r}, E)\Psi(\mathbf{r}, E)dxdydzdE = 1.$$

In analogy to (3.22) and (3.23), the equation for the determination of $\Psi(\mathbf{r}, E)$ is given in $(\mathbf{r}, E)$-space by (see also [12]).

$(\mathbf{r}, E)$-space:

$$E\Psi(\mathbf{r}, E) = -\frac{\hbar^2}{2m_0}\Delta\Psi(\mathbf{r}, E) + V\left(x, y, z, -i\hbar\frac{\partial}{\partial E}\right)\Psi(\mathbf{r}, E). \quad (5.18)$$

The mean value of the energy is given by

$$\bar{E} = \int_{-\infty}^{\infty} E\Psi^*(\mathbf{r}, E)\Psi(\mathbf{r}, E)dxdydzdE = (\bar{E})_{\Psi(\mathbf{r},E)} \quad (5.19)$$

and the mean value of the time $t$ is, in analogy to (5.19), expressed by

$$\bar{t} = \int_{-\infty}^{\infty} t\psi\Psi^*(\mathbf{r}, t)\Psi(\mathbf{r}, t)dxdydzdt = (\bar{t})_{\Psi(\mathbf{r},t)}. \quad (5.20)$$

It is easy to show that $\bar{t}$ can also be expressed in terms of $\Psi(\mathbf{r}, E)$ and the operator $\hat{t} = -i\hbar\partial/\partial E$ (see Section 5.4.3). After some manipulations, we immediately find

$$\begin{aligned}\bar{t} &= \int_{-\infty}^{\infty} dxdydzdE\Psi^*(\mathbf{r}, E)\hat{t}\Psi(\mathbf{r}, E)\\ &= -i\hbar\int_{-\infty}^{\infty} dxdydzdE\Psi^*(\mathbf{r}, E)\frac{\partial}{\partial E}\Psi(\mathbf{r}, E) = (\bar{t})_{\Psi(\mathbf{r},E)}.\end{aligned} \quad (5.21)$$

In the case of the energy $E$, the mean value $\overline{E}$ can also be formulated by $\Psi(\mathbf{r}, t)$ and the operator $\widehat{E} = i\hbar\partial/\partial t$ (see Section 5.4.3) and we get

$$\begin{aligned}\overline{E} &= \int_{-\infty}^{\infty} dxdydzdt\Psi^*(\mathbf{r}, t)\hat{E}\Psi(\mathbf{r}, t)\\ &= i\hbar\int_{-\infty}^{\infty} dxdydzdt\Psi^*(\mathbf{r}, t)\frac{\partial}{\partial t}\Psi(\mathbf{r}, t) = (\overline{E})_{\Psi(\mathbf{r},t)}.\end{aligned} \quad (5.22)$$

When $t-\bar{t}$ and $E-\overline{E}$ are the deviations from the mean values $\bar{t}$ and $\overline{E}$, the scatter (uncertainty) $\delta t$ for the time $t$ is usually defined by

$$\delta t = \sqrt{\overline{(t-\bar{t})^2}} \tag{5.23}$$

and the scatter (uncertainty) $\delta E$ for the energy $E$ by

$$\delta E = \sqrt{\overline{(E-\overline{E})^2}}. \tag{5.24}$$

In the case of the wavefunctions $\Psi(\mathbf{r},t)$ and $\Psi(\mathbf{r},E)$ [see (5.19)–(5.22)] we have in general

$$\bar{t} = (\bar{t}\,)_{\Psi(\mathbf{r},E)} = (\bar{t}\,)_{\Psi(\mathbf{r},t)} \neq 0, \tag{5.25}$$

$$\overline{E} = (\overline{E}\,)_{\Psi(\mathbf{r},E)} = (\overline{E}\,)_{\Psi(\mathbf{r},t)} \neq 0. \tag{5.26}$$

### 5.4.4.2. *Shift of the Wavefunctions*

Let us shift the function $\Psi(\mathbf{r},E)$ by $\Delta_E$. Then we get

$$\begin{aligned}\psi(\mathbf{r},E) = \Psi(\mathbf{r},E+\Delta_E) &= \frac{1}{(2\pi\hbar)^{1/2}}\int_{-\infty}^{\infty}\Psi(\mathbf{r},t)\exp\left\{i\frac{E+\Delta_E}{\hbar}t\right\}dt\\ &= \frac{1}{(2\pi\hbar)^{1/2}}\int_{-\infty}^{\infty}\psi(\mathbf{r},t)\exp\left\{i\frac{E}{\hbar}t\right\}dt\end{aligned} \tag{5.27}$$

with

$$\psi(\mathbf{r},t) = \Psi(\mathbf{r},t)\exp\left\{it\frac{\Delta_E}{\hbar}\right\}. \tag{5.28}$$

On the other hand, we get from the inverse transformation formula

$$\psi(\mathbf{r},t) = \frac{1}{(2\pi\hbar)^{1/2}}\int_{-\infty}^{\infty}\psi(\mathbf{r},E)\exp\left\{-i\frac{E}{\hbar}t\right\}dE \tag{5.29}$$

and, if the function $\psi(\mathbf{r},t)$ is shifted by $\Delta_t$, we obtain

$$\phi(\mathbf{r},t)=\psi(\mathbf{r},t+\Delta_t) = \frac{1}{(2\pi\hbar)^{1/2}}\int_{-\infty}^{\infty}\phi(\mathbf{r},E)\exp\left\{-i\frac{E}{\hbar}t\right\}dE \tag{5.30}$$

with

$$\begin{aligned}\phi(\mathbf{r},E) &= \psi(\mathbf{r},E)\exp\left\{-i\frac{E}{\hbar}\Delta_t\right\}\\ &= \Psi(\mathbf{r},E+\Delta_E)\exp\left\{-i\frac{E}{\hbar}\Delta_t\right\}.\end{aligned} \tag{5.31}$$

With (5.28), we obtain

$$\begin{aligned}\phi(\mathbf{r},t) &= \psi(\mathbf{r},t+\Delta_t)\\ &= \Psi(\mathbf{r},t+\Delta_t)\exp\left\{i(t+\Delta_t)\frac{\Delta_E}{\hbar}\right\}\end{aligned} \tag{5.32}$$

and we get for (5.30) the following expression:

$$\Psi(\mathbf{r},t+\Delta_t)\ \exp\left\{i(t+\Delta_t)\frac{\Delta_E}{\hbar}\right\} = \frac{1}{(2\pi\hbar)^{1/2}}\int_{-\infty}^{\infty}\Psi(\mathbf{r},E+\Delta_E)$$
$$\times\ \exp\left\{-i\frac{E}{\hbar}\Delta_t\right\}\exp\left\{-i\frac{E}{\hbar}t\right\}dE. \tag{5.33}$$

Equation (5.33) can be used for the determination of the mean values for energy and time by applying the operators $\hat{E} = i\hbar\partial/\partial t$ and $\hat{t} = -i\hbar\partial/\partial E$. After some simple mathematical manipulations, we get

$$(\bar{t})_{\Psi(\mathbf{r},t)} - \Delta_t = (\bar{t})_{\Psi(\mathbf{r},t+\Delta_t)}, \tag{5.34}$$

where

$$\begin{aligned}(\bar{t})_{\Psi(\mathbf{r},E+\Delta_E)} &= (\bar{t})_{\Psi(\mathbf{r},t)}\\ &= (\bar{t})_{\Psi(\mathbf{r},E)},\end{aligned} \tag{5.35}$$

$$(\bar{t})_{\Psi(\mathbf{r},t+\Delta_t)} \neq (\bar{t})_{\Psi(\mathbf{r},t)} \tag{5.36}$$

with

$$(\bar{t})_{\Psi(\mathbf{r},t)} = \int_{-\infty}^{\infty} t\Psi^*(\mathbf{r},t)\Psi(\mathbf{r},t)dxdydzdt, \tag{5.37}$$

$$(\bar{t})_{\Psi(\mathbf{r},t+\Delta_t)} = \int_{-\infty}^{\infty} t\Psi^*(\mathbf{r},t+\Delta_t)\Psi(\mathbf{r},t+\Delta_t)dxdydzdt, \tag{5.38}$$

$$(\bar{t})_{\Psi(\mathbf{r},E)} = -i\hbar \int_{-\infty}^{\infty} dxdydzdE\Psi^*(\mathbf{r},E)\frac{\partial}{\partial E}\Psi(\mathbf{r},E), \tag{5.39}$$

$$(\bar{t})_{\psi(\mathbf{r},E+\Delta_E)} = -i\hbar \int_{-\infty}^{\infty} dxdydzdE\Psi^*(\mathbf{r},E+\Delta_E)\frac{\partial}{\partial E}\Psi(\mathbf{r},E+\Delta_E)). \tag{5.40}$$

On the other hand, we obtain [using the inverse transformation of (5.33)]

$$(\overline{E})_{\Psi(\mathbf{r},E)} - \Delta_E = (\overline{E})_{\Psi(\mathbf{r},E+\Delta_E)}, \tag{5.41}$$

where

$$\begin{aligned}(\overline{E})_{\Psi(\mathbf{r},t+\Delta_t)} &= (\overline{E})_{\Psi(\mathbf{r},E)} \\ &= (\overline{E})_{\Psi(\mathbf{r},t)},\end{aligned} \tag{5.42}$$

$$(\overline{E})_{\Psi(\mathbf{r},E)} \neq (\overline{E})_{\Psi(\mathbf{r},E+\Delta_E)} \tag{5.43}$$

with

$$(\overline{E})_{\Psi(\mathbf{r},E)} = \int_{-\infty}^{\infty} E\Psi^*(\mathbf{r},E)\Psi(\mathbf{r},E)dxdydzdE, \tag{5.44}$$

$$(\overline{E})_{\Psi(\mathbf{r},E+\Delta_E)} = \int_{-\infty}^{\infty} E\Psi^*(\mathbf{r},E+\Delta_E)\Psi(\mathbf{r},E+\Delta_E)dxdydzdE, \tag{5.45}$$

$$(\overline{E})_{\Psi(\mathbf{r},t)} = i\hbar \int_{-\infty}^{\infty} dxdydzdt\Psi^*(\mathbf{r},t)\frac{\partial}{\partial t}\Psi(\mathbf{r},t), \tag{5.46}$$

$$(\overline{E})_{\Psi(\mathbf{r},t+\Delta_t)} = i\hbar \int_{-\infty}^{\infty} dxdydzdt\Psi^*(\mathbf{r}, t+\Delta_t)\frac{\partial}{\partial t}\Psi(\mathbf{r}, t+\Delta_t). \quad (5.47)$$

In the case of [Eq. (5.41)]

$$\Delta_t = (\bar{t})_{\Psi(\mathbf{r},t)} \quad (5.48)$$

we have

$$(\bar{t})_{\Psi(\mathbf{r},t+\Delta_t)} = 0. \quad (5.49)$$

On the other, if we have [Eq. (53)]

$$\Delta_E = (\overline{E})_{\Psi(\mathbf{r},E)} \quad (5.50)$$

we get

$$(\overline{E})_{\Psi(\mathbf{r},E+\Delta_E)} = 0. \quad (5.51)$$

Using the definitions

$$\varphi(\mathbf{r}, t) = \Psi(\mathbf{r}, t+\Delta_t) \quad (5.52)$$

and

$$\varphi(\mathbf{r}, E) = \Psi(\mathbf{r}, E+\Delta_E) \quad (5.53)$$

we may continue our analysis for the case of $\bar{t} = \bar{t}_\varphi = 0$ and $\overline{E} = \overline{E}_\varphi = 0$ which is not a restriction because we used no approximation in the derivation of Eq. (5.33) (see also [12]). In conclusion, we may shift the functions $\Psi(\mathbf{r}, t)$ and $\Psi(\mathbf{r}, E)$ so that the mean values become zero without having to give up generality.

In conclusion, the analysis in this section showed that a shift of $\Psi(\mathbf{r}, E)$ [$\Psi(\mathbf{r}, t)$] has an effect on $\Psi(\mathbf{r}, t)$ [$\Psi(\mathbf{r}, E)$]. However, we have the following properties. The mean value $\overline{E}$ [$\bar{t}$] is changed but not $\bar{t}$ [$\overline{E}$]. The parameters $\Delta_E$ and $\Delta_t$ are independent of each other and can be chosen arbitrarily.

### 5.4.4.3. *Formulation of the Uncertainty Relation for Time and Energy*

With $\bar{t} = \bar{t}_{\varphi} = 0$, and $\bar{E} = \bar{E}_{\varphi} = 0$, we obtain instead of Eqs. (5.23) and (5.24)

$$\delta t = \sqrt{\overline{t^2}} \tag{5.54}$$

and

$$\delta E = \sqrt{\overline{E^2}}. \tag{5.55}$$

Let us investigate the following auxiliary integral:

$$A(b) = \int_{-\infty}^{\infty} \left| bE\varphi(\mathbf{r}, E) + \frac{\partial\varphi(\mathbf{r}, E)}{\partial E} \right|^2 dxdydzdE, \tag{5.56}$$

where $b$ is a real number and $\varphi(\mathbf{r}, E)$ is defined by Eq. (5.53). From (5.56), we obtain

$$\begin{aligned} A(b) &= b^2 \int_{-\infty}^{\infty} E^2\varphi^*(\mathbf{r}, E)\varphi(\mathbf{r}, E)dxdydzdE \\ &+ b \int_{-\infty}^{\infty} E \left\{ \frac{\partial\varphi^*(\mathbf{r}, E)}{\partial E}\varphi(\mathbf{r}, E) + \varphi^*(\mathbf{r}, E)\frac{\partial\varphi(\mathbf{r}, E)}{\partial E} \right\} dxdydzdE \\ &+ \int_{-\infty}^{\infty} \frac{\partial\varphi^*(\mathbf{r}, E)}{\partial E}\frac{\partial\varphi(\mathbf{r}, E)}{\partial E} dxdydzdE. \end{aligned} \tag{5.57}$$

Then, the function $A(b)$ takes the form

$$A(b) = b^2(\delta E)^2 + bB + \frac{(\delta t)^2}{\hbar^2} \geq 0, \tag{5.58}$$

where

$$(\delta E)^2 = \int_{-\nabla}^{\infty} E^2\varphi^*(\mathbf{r}, E)\varphi(\mathbf{r}, E)dxdydzdE, \tag{5.59}$$

$$
\begin{aligned}
(\delta t)^2 &= \int_{-\infty}^{\infty} \varphi^*(\mathbf{r}, E)\hat{t}^2 \varphi(\mathbf{r}, E)\, dxdydzdE \\
&= \int_{-\infty}^{\infty} \varphi^*(\mathbf{r}, E) \left(-i\hbar \frac{\partial}{\partial E}\right)^2 \varphi(\mathbf{r}, E)\, dxdydzdE \\
&= \int_{-\infty}^{\infty} t^2 \varphi^*(\mathbf{r}, t)\varphi(\mathbf{r}, t)\, dtdxdydz \\
&= \hbar^2 \int_{-\infty}^{\infty} \frac{\partial \varphi^*(\mathbf{r}, E)}{\partial E} \frac{\partial \varphi(\mathbf{r}, E)}{\partial E}\, dxdydzdE, \qquad (5.60)
\end{aligned}
$$

$$
B = \int_{-\infty}^{\infty} E \left\{ \frac{\partial \varphi^*(\mathbf{r}, E)}{\partial E} \varphi(\mathbf{r}, E) + \varphi^*(\mathbf{r}, E) \frac{\partial \varphi(\mathbf{r}, E)}{\partial E} \right\} dxdydzdE = -1. \qquad (5.61)
$$

Since $b$ is a real number and $A(b) \geq 0$, the solutions of

$$
A(b) = 0 \qquad (5.62)
$$

must be complex. This is the case if

$$
(\delta E)^2 (\delta t)^2 \geq \frac{\hbar^2}{4} \qquad (5.63)
$$

and we obtain the uncertainty relation for the energy $E$ and time $t$:

$$
\delta E \delta t \geq \frac{\hbar}{2}. \qquad (5.64)
$$

In contrast to Eq. (5.3), relation (5.64) agrees completely in its physical content with the uncertainty relation for the momentum and the coordinate. The required symmetry between space and time that is required by Special Theory of Relativity is definitely fulfilled within projection theory but not in usual quantum theory (in its relativistic and its non-relativistic version). We have discussed this point in Section 5.3.4.2.

Discussion concerning (5.64)

Let us consider the wavefunctions $\Psi(\mathbf{p},t)$ and $\Psi(\mathbf{p},E)$ at constant $\mathbf{p}=\mathbf{p}_0$. Then, we may write $\Psi(t) = \Psi(\mathbf{p}_0,t)$ and $\Psi(E) = \Psi(\mathbf{p}_0,E)$ and we have

$$
\begin{aligned}
&\tau_1 : \Psi^*(E_1)\Psi(E_1), \Psi^*(t_1)\Psi(t_1) \\
&\tau_2 : \Psi^*(E_2)\Psi(E_2), \Psi^*(t_2)\Psi(t_2) \\
&\tau_3 : \Psi^*(E_3)\Psi(E_3), \Psi^*(t_3)\psi(t_3) \\
&\quad\vdots
\end{aligned}
\tag{5.65}
$$

At time $\tau_i$, we measure with certain probabilities the energies $E_i$, $i = 1,2,3,\ldots$ and the system-specific times $t_i$, $i = 1,2,3,\ldots$ with $\tau_1 < \tau_2 < \tau_3 \ldots$. In contrast to $\tau$, the system-specific time $t$ is not directed. For example, if we measure $t_1$ at $\tau_1$, $t_1$ at $\tau_2$, $t_3$ at $\tau_3$ etc. we might have, for example, $t_1 > t_2$, $t_3 > t_2$ etc. (see also Section 3.2.3).

The wavefunction $\Psi(E)$ expresses the interaction of the system with its environment [described by $\Psi_{\text{env}}(E)$]. Interaction here again means that there is an exchange of energy. If the system has at time $\tau$ the energy $E_a$ and the environment the energy $E_b$ and if there is an energy exchange of $\Delta E$ we have, due to the conservation of energy, $E_a \pm \Delta E$ and $E_b \mp \Delta E$. The interaction exclusively takes place in $(\mathbf{p},E)$-space within projection theory. The information $\Psi(E)$ is projected on $(\mathbf{r},t)$-space and the resulting curves satisfy the uncertainty relation $\delta E \delta t \geq \hbar/2$ (see also Fig. 5.1) which is not known in usual quantum theory.

Or more generally, if the wavefunction $\Psi(\mathbf{p},E)$ expresses the interaction of the system with its environment [described by $\Psi_{\text{env}}(\mathbf{p},E)$]. Furthermore, if the system has at time $\tau$ the energy $E_a$ and the momentum of $\mathbf{p}_a$, the environment the energy $E_b$ and the momentum $\mathbf{p}_b$ and if there is an energy exchange of $\Delta^E$ and a momentum exchange of $\Delta^{\mathbf{p}}$ we have, due to the conservation of energy and momentum, $E_a \pm \Delta^E$, $\mathbf{p}_a \pm \Delta^{\mathbf{p}}$ and $E_b \mp \Delta^E$, $\mathbf{p} \mp \Delta^{\mathbf{p}}$. The information $\Psi(\mathbf{p},E)$ is projected on $(\mathbf{r},t)$-space and the resulting curves satisfy the uncertainty relations $\delta E \delta t \geq \hbar/2$ and $\delta p_x \delta x \geq \hbar/2$ etc. Again, within projection theory we have, instead of $\Delta E \Delta \tau \geq \hbar/2$, the law $\delta E \delta t \geq \hbar/2$ that is not known in usual quantum theory.

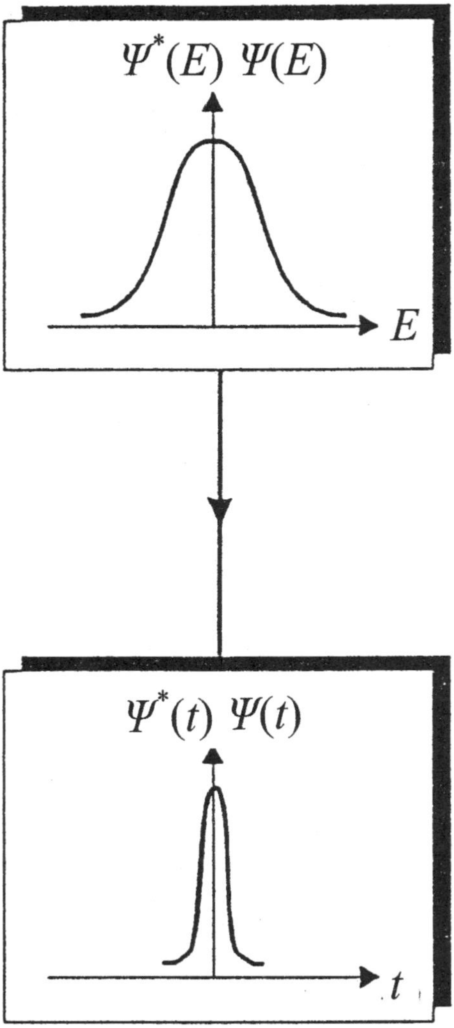


Fig. 5.1. $\Psi^*(E)\Psi(E)$ describes (fictitious) reality and $\Psi^*(t)\Psi(t)$ its picture. $\Psi(E)$ and $\Psi(t)$ are connected by a Fourier transform. Both, $E$ and $t$ fluctuate and the widths of the curves for $\Psi(E)$ and $\Psi(t)$ must satisfy the uncertainty relation $\delta E \delta t \geq \hbar/2$ which is not known in usual quantum theory. Due to the interaction of the system (described by $\Psi(E)$ in $(\mathbf{p}, E)$-space) with its environment, $E$ fluctuates where the energy is strictly conserved.

In conclusion, the interactions exclusively take place in $(\mathbf{p}, E)$-space in form of $(\mathbf{p}, E)$-fluctuations where we have to assume that $\mathbf{p}$ and $E$ are strictly conserved. A temporary violation of these conservation laws gives no sense within projection theory. Again, within usual quantum theory the conservation law for energy can be temporarily violated and this is due to equation $\Delta E \Delta \tau \geq \hbar/2$ which still contains time $\tau$ in classical form (Section 3.4.2).

In connection with time $t$, let us mention once more the following important feature. Despite the statistical fluctuations, at time $\tau$ the whole of time $t$ — *past, present and future* — is laid out frozen before us (Chapter 3):

$$\Psi^*(\mathbf{r}, t)\Psi(\mathbf{r}, t), \quad -\infty < \mathbf{r} < \infty, \quad -\infty < t < \infty.$$

This is valid for all times $\tau$. $\Psi^*(\mathbf{r}, t)\Psi(\mathbf{r}, t)$ will be systematically scanned by $\tau$ and only those values of $t$ can be observed which correspond with the reference time $\tau$ (see [12] and Chapter 3). Therefore, the sense of time $\tau$ within projection theory is to *select* a certain configuration of $\Psi^*(\mathbf{r}, t_k)\Psi(\mathbf{r}, t_k)$ with $t_k = \tau$ from the whole scenario. Since $\tau$ moves systematically from the past to the future, the system-specific time $t$ will also be observed as a directed quantity, but this is entirely due to the observer's selection process.

### 5.4.5. *Cosmological Constant within Projection Theory*

From specific estimations for the cosmological constant, we may conclude that the theoretical prediction deviates from the observed value by a factor of $10^{122}$. This is a lot and hardly imaginable. The basis for the quantum field theoretical calculation of $\Lambda$ is equation $\Delta E \Delta \tau \geq \hbar/2$, and this equation allows that the conservation law for the energy can be temporarily violated leading to the unacceptable (wrong) value for the cosmological constant $\Lambda$. Within projection theory we have, instead of $\Delta E \Delta \tau \geq \hbar/2$, the law $\delta E \delta t \geq \hbar/2$ and this changes the situation completely. Strictly speaking, a cosmological constant $\Lambda$ in the sense of General Theory of Relativity gives no sense within projection theory. The following points are relevant:

1. Within projection theory, space–time is not a container for the integration of physically real objects etc. Nothing can be inside space–time, no matter, no energy. In particular, there can be no vacuum energy. Space–time exclusively plays the role of an auxiliary element for the representation of the picture of reality. Space–time can only contain geometrical positions, trajectories etc. as in the case of a sheet of paper. All these points have been explained and discussed above and will still be studied further in the following.
2. The interactions exclusively take place in $(\mathbf{p}, E)$-space in form of $(\mathbf{p}, E)$-fluctuations where $\mathbf{p}$ and $E$ are strictly conserved. A temporary violation of these conservation laws gives no sense within projection theory. In contrast to $\Delta E \Delta \tau \geq \hbar/2$, equation $\delta E \delta t \geq \hbar/2$ does not lead to temporary violations of the conservation law for the energy.

For the basic understanding of quantum phenomena and the cosmological constant $\Lambda$, we used the features of space–time, which have been essentially discussed in Chapter 2. In the next sections, we will deepen the concept of space–time as it appears in connection with projection theory as well as in conventional physics.

## 5.5. Artificial Vacuum Effects!?

The large value for the quantum vacuum energy density, calculated on the basis of quantum field theory, leads to unrealistic results when we apply it to the General Theory of Relativity. It is usually assumed that the value of quantum field theory for the vacuum energy is wrong. However, within quantum field theory no mechanism could be found and no reasonable procedure could be introduced to reduce this value. The search for a solution for this big problem already lasts approximately two decades but, up to now, is still without any success. Therefore, we concluded in Section 5.3 that the vacuum energy problem is probably tightly-linked with the very foundations of physics. Due to the unsuccessful investigations on lower physical levels, this step to the top level is necessary. In Section 5.3, we analyzed the physical basics of quantum field theory.

We particularly discussed in Section 5.3 that relation $\Delta E \Delta \tau \geq \hbar/2$ [Eq. (5.3)] is quite questionable and this is because time $\tau$ is still a classical parameter in usual quantum theory. In Bunge's opinion relation $\Delta E \Delta \tau \geq \hbar/2$, a total stranger in usual quantum theory. But quantum field theory is mainly based on this equation and leads to the unacceptable large value for the vacuum energy and the cosmological constant $\Lambda$. Within projection theory we have, instead of $\Delta E \Delta \tau \geq \hbar/2$, the law $\delta E \delta t \geq \hbar/2$ that is not known in usual quantum theory. We have outlined in Section 5.4 that the $\Lambda$-problem can be considered to be solved within projection theory.

On the other hand, the General Theory of Relativity could also be the reason for the unrealistic outcome in connection with the large value for the vacuum energy. This possibility has not been seriously discussed in literature. Nevertheless, it cannot be assumed that General Theory of Relativity is a perfect, final approach. It makes however not much sense to ask what theory is the reason for this large discrepancy between a realistic value for the cosmological constant $\Lambda$ and what we obtain from the quantum vacuum energy density because we have to keep in mind that both theories, the General Theory of Relativity and quantum field theory, are inherently incompatible. We can hardly expect that modifications of only one theory (General Theory of Relativity or quantum field theory) could eliminate this bad situation. Therefore, we have to search for a principle that is valid for both phenomena, that is, for gravitational and quantum phenomena. But it is obvious that such a principle must hold for all phenomena in nature (including quantum and gravitational effects), taking place at the various levels of observation. This obviously means, on the other hand, that such a principle must be very basically anchored at the basic level of theoretical physics. In Section 5.4, we were successful with the following conception. The real world is not embedded in space–time (*container principle*) but is projected on it (*projection principle*). In this section, we would like to further investigate this conception.

## 5.5.1. *Mach's Principle*

The large value of quantum field theory for the vacuum energy points to a problem within quantum field theory. Quantum field theory has a vacuum problem. But the General Theory of Relativity also has a serious vacuum problem, and this is due to the fact that Mach's principle is obviously not fulfilled within the General Theory of Relativity. This principle has often been discussed in literature. In Chapter 3 (Section 3.13), we already introduced Mach's principle and we remarked the following. After Mach, space–time can never be the source for physically real effects, that is, space–time can never act on material objects giving them certain properties as, for example, inertia. According to Mach, a particle does not move in unaccelerated motion relative to space, but relative to the center of all the other masses in the universe. In fact, we can never observe such a space (space–time) because its elements (coordinates $x$, $y$, $z$ and time $\tau$) are in principle not accessible to empirical tests (see also the discussion in Chapter 2). This is also the case for the system-specific time $t$ (Section 5.4) which is not defined in General Theory of Relativity and Special Theory of Relativity but only in connection with projection theory. In Chapter 3 and Section 5.4, we have used $\tau$ as the reference time.

We can only say something about *distances in connection with masses*, and time intervals in connection with physical processes. In other words, an *empty* space–time as physical–theoretical conception should not exist. This is a fundamental point.

### 5.5.1.1. *Empty Space within General Theory of Relativity*

In conclusion, no realistic space–time theory should contain an empty space–time as a solution. Einstein thought that his field equations would fulfil this important and basic condition. However, in 1917, de Sitter gave a solution to Einstein' field equations which corresponds to an empty universe, i.e., within the framework of this solution, space–time could exist without matter, and this is in obvious contradiction to Mach's principle. This fact was annotated by Einstein's collaborator

Banesh Hoffmann as follows [47]:

> "*Barely had Einstein taken his pioneering step when in 1917 in neutral Holland de Sitter discovered a different solution to Einstein's cosmological equation. This was embarrassing. It showed that Einstein's equation did not lead to a unique model of the universe at all. Moreover, unlike Einstein's universe, de Sitter's was empty. It thus ran counter to Einstein's belief, an outgrowth of the idea of Mach, that matter and space–time are closely linked that neither should be able to exist without the other.*"

### 5.5.1.2. *The Effect of Inertia*

According to Mach, space–time can never be the source for physically real effects, as, for example, inertia. But also this condition is not fulfilled at all within the General Theory of Relativity. This has been demonstrated by de Sitter. In [48], we find the following text: "*In 1917, an eminent Dutch astronomer, Willem de Sitter, pointed out to Einstein that there was a finite valued solution of his field equations that gave the inertial mass of a particle even if it was the only one in the universe. In this case, the curved space–time of general relativity would be flat, that is the geodesic line passing through the particle would be straight. The lone particle would be guided along this geodesic line as if it was made of inertial matter. Einstein initially argued strongly against this solution. However, eventually he conceded that his interpretation of inertia could not therefore be due to other matter, as required by Mach's principle, because there was no other matter around in the de Sitter's example.*" [48]

### 5.5.1.3. *Where Does Inertia Come From?*

In Newton' physics, space and time are *absolute* quantities, they are independent from each other, and they may even exist when space is not filled with matter. Only with these space–time features Newton was able to construct a reasonable theory of motion.

However, the concept of an absolute space (and that of absolute time) has led to enormous intellectual difficulties which could not be eliminated by the theory of relativity. This point could be of basic relevance in connection with the definition of a cosmological constant.

Why does the concept of absolute space–time cause serious problems? Let us briefly quote the main points. Concerning the term *absolute*, note the following:

1. Absolute space was invented by Newton for the explanation of inertia. However, we do not know other phenomena for which absolute space would be responsible. So, the hypothesis of absolute space can only be proved by the phenomenon (inertia) for which it has been introduced. This is unsatisfactory and artificial.
2. The term "absolute" not only means that space is physically real but also "*...independent in its physical properties, having a physical effect, but not itself influenced by physical conditions.*" [49] This must also be considered as unsatisfactory and in particular unphysical.

Both points indicate that the concept of absolute space is actually an ill construction. Although Newton's mechanics was very successful (and it is still used in many calculations) a lot of physicists could not accept the concept of absolute space. This is demonstrated by the fact that scientists tried to solve this problem again and again up to the present day.

For our discussion in connection with the cosmological constant, it is important to note that General Theory of Relativity still contains the concept of absolute space. Within General Theory of Relativity, it is still possible to talk about the rotation of the entire matter of the universe relative to absolute space. Gödel's solution of Einstein's field equations leads to this result. Heckmann remarked [50]:

> "*This solution by Gödel describes a model of the world which is uniformly filled with matter. All points in it are equivalent, which is therefore homogeneous as in the cases mentioned up to now. It is infinitely large and rotates absolutely, but is not able to expand. At the beginning, in 1916, and still long time after that, Einstein himself believed that his theory would contain the relativity of all motions. Gödel's solution was the first solid evidence*

*that Einstein's belief was an error.... The absoluteness of space, which Newton claimed, and Einstein thought to have eliminated, is still contained in Einstein's theory, insofar as in Einstein's theory the concept of an absolute rotation is completely legitimate. Within Einstein' theory, it is still possible to talk about the rotation of the entire mass of the world relative to absolute space."* [50]

Concerning absolute space, Max Born wrote:

"*Indeed, the concept of absolute space is almost spiritualistic in character. If we ask, What is the cause for the centrifugal forces?, the answer is: "Absolute space." If, however, we ask what absolute space is and in what other way it expresses itself, no one can furnish an answer other than that absolute space is the cause of centrifugal forces but has no further properties. This consideration shows that space as the cause of physical occurrences must be eliminated from the world picture."* [51]

Furthermore, we find in this connection [51]:

"*Sound epistemological criticism refuses to accept such made-to-order hypothesis. They are too facile and are at odds with the aims of scientific research, which is to determine criteria for distinguishing its results from dreams of fancy. If the sheet of paper on which I have just written suddenly flies up from the table, I should be free to make the hypothesis that a ghost, say the spectre of Newton, had spirited it away. But common sense leads me instead to think of a draft coming from the open window because someone is entering by the door. Even if I do not feel the draft myself, this hypothesis is reasonable because it brings the phenomenon which is to be explained into a relationship with other observable events.*"

Although Mach's principle reflects a fundamental and important feature, it could not be realized within General Theory of Relativity up to the present day. Dehnen pointed out the following:

"*In those days Einstein had in mind that the structure of space and time is given completely by the particular distribution of matter in the world in accordance with his field equations of gravitation.*

*As a result of this, Mach's idea would be fulfilled simultaneously, after which the inertia of material bodies is determined by other masses in the world ... However, it should be emphasized that Einstein's vision that Mach's principle could be realized within the framework of the general theory of relativity failed, even by additional modifications of the original field equations ... The problem in connection with the absolute space–time within the framework of special theory of relativity — a relict from Newton's mechanics — is, in the General Theory of Relativity, still not solved ...*" [52]

Within General Theory of Relativity, there is no clear line between space–time and matter, and this is clearly against the facts. In the course of time, during the development of General Theory of Relativity, space–time took more and more the properties of matter which is in contradiction to what we can observe in connection with space and time. We can put a piece of matter on the table but not a piece of space–time. Kanitscheider remarked:

"*Although Einstein was led in the construction of his theory by Mach's idea of the ontological dominance of matter over space–time, he was even temporarily convinced that his theory had taken space and time the slightest trace of an objective reality. But the opposed tendency turned out to be the case. A tendency against Mach's principle became effective by a kind of inherent self-dynamics of the theory in the course of time of the development of the conception. Space–time became more and more an ontologically respectable entity, and it took over more and more the properties of material objects.*" [53]

In summary, General Theory of Relativity has a certain problem with space–time (vacuum) and this might be connected with the problem we have with respect to the cosmological constant $\Lambda$ (Section 5.2). Mach's principle is not fulfilled within General Theory of Relativity as it should be because this principle has to be considered as an important condition for any space–time theory, and Einstein was well aware of its importance. It is more than unsatisfactory that within General

Theory of Relativity, an empty space–time can exist, and even an absolute space–time is definable within this theory. In the course of the development of General Theory of Relativity, space–time took more and more the properties of material objects. Also this point reflects a critical situation and can basically not be accepted.

## 5.5.2. *Basic Features of Space and Time*

### 5.5.2.1. *Facts*

Within General Theory of Relativity, there is no clear line between space–time and matter as it should be. From certain formulations in literature, we may conclude that for some scientists there is no line at all. The "warp in space" and "matter" has often been considered as different aspects of the same thing. Already Arthur Eddington thought in this direction but it is, without further analysis, rather misleading and not appropriate. Instead we have to decipher the "code" that is anchored in the relationship between "space–time" and "matter," and this relationship is given by the following facts, which we have already listed in Chapter 2. We definitely cannot see, hear, smell, or taste space and time, that is, space and time (absolute or non-absolute) are not accessible to our senses. Also measuring instruments for the experimental determination of the space–time points $x$, $y$, $z$, and $\tau$ are not known and not even imaginable. We can only say something about *distances in connection with masses*, and time intervals in connection with physical processes.

Thus, space and time should never be the source for physically real effects as, for example, inertia. This is not fulfilled within General Theory of Relativity. Within Einstein' theory it is still possible to talk about the rotation of the entire mass of the world relative to absolute space (Gödel's universe, see Section 5.1.3 and [13]), and in this case the absolute space is the source of inertia.

If space and time cannot be the source for physically real effects, then also physically real effects due to a possible *cosmological constant* $\Lambda$ should be forbidden. This is an important point in connection with the discussion given in this chapter.

Since we can only say something about space distances in connection with masses, and time intervals in connection with physical processes, we have to conclude that "matter" and "space–time" are closely linked. Neither should be able to exist without the other. In other words, an *empty* space–time as physical-theoretical conception should not be existent (Chapter 2). But what kind of connection could exist between matter and space–time? As we have outlined in Section 2.3 (Chapter 2), this cannot be due to an interaction between the material bodies and space–time. No interaction between space–time and the sense organs or measuring instruments, which are made of material bodies, is possible.

In conclusion, space and time are not accessible by the human senses and no measuring instrument is able to make space-time-specific "clicks" but, on the other hand, we experience space–time in connection with material bodies and processes as a concrete phenomenon. Thus, we might assume that space–time has to be identified with a "substratum." However, this idea has never been considered within the framework of serious scientific argumentations. In the next section, we will briefly discuss Einstein's thoughts on a space–time-filling substratum, different from a substratum describing space–time.

#### 5.5.2.2. *Einstein's Substratum-Idea*

Einstein was always firmly convinced that space (space–time) can never be the source for physically real effects as, for example, inertia. In the case of only one particle in the universe Einstein field equations nevertheless allow inertial motion and the reason for that can only be the space. It is therefore not surprising that Einstein in 1920 discussed the possibility for the introduction of a space-filling substratum, an ether with specific properties (not in conflict with special theory of relativity, i.e., different from the electromagnetic ether he eliminated in his 1905 paper).

Why did Einstein propose this new sort of either? Probably to eliminate space (space–time) as an active cause for physically real effects (inertia, etc.). But Einstein's substratum-idea never found an echo in the scientific community. Einstein never mentioned this idea after 1920. However, Einstein's introduction of such a substratum (different from usual matter) possibly reflects his uncertainty about the phenomena space and time, even after he had developed theory of relativity.

The fact is, however, that space–time is often considered as a new *object* that can be seen as an independent object from the outside. That is legitimate and almost imperative within the General Theory of Relativity. We already discussed above that Einstein's field equations allow us to construct an empty space–time without matter (de Sitter universe) but this is in contradiction to all that we know about the characterization of space–time [see Sections 5.5.2.1 and 2.3 (Features 1 and 2)].

We have to find out why. We have to decipher this code. In our opinion, any theoretical picture of the cosmos must precisely fulfil this situation but concerning this point there is a certain ignorance and carelessness, respectively. Falk and Ruppel remarked:

> "... *But nevertheless we are accustomed to consider space as something special, namely as a substratum, in which all the things, as objects and fields, are embedded. Space is the house which takes up physical objects, so to speak. It forms the stage where the processes take place. It seems clear to everybody what one means when he speaks about empty space, i.e., space which is free of matter and fields. But is that really the case? It is astonishing again and again how easily we adopt conceptions and naturally and inevitably we hold them. Our conception of empty space certainly belongs to this category.*" [54]

In General Theory of Relativity, space–time and matter are tightly connected to each other and we have the following situation. The mass is responsible for the geometry of space–time and, on the other hand, the motion of mass is determined just by this geometry. Mass curves space–time, and space–time tells the mass how to move. In other words, there is a certain kind of symmetry between space–time and mass. Due to this features, it is difficult and impossible, respectively, to consider space–time within General Theory of Relativity as a nothing without any properties but, from this point of view, it is natural to consider space–time as a certain kind of substratum that is different from usual matter. We already mentioned that Einstein recognized this problem and that he still 1920 discussed the possibility for the introduction of such a space-filling substratum (we already mentioned that above). Concerning this point, Cole wrote:

> "*Einstein may have buried the ether, but ironically, he also came to praise it. In fact, in his later writings, he suggests that his four-dimensional space–time comprises a new sort of ether, refurbished and rehabilitated. According to the general theory of relativity, he wrote, space without ether is unthinkable —for in such a space there not only would be no propagation of light, but also — no basis for space–time intervals in the physical sense. But this ether may not be thought of as endowed with the qualities of ponderable media.*" [29]

If space–time cannot exist without such a space–time-filling substratum, we may also say that space–time has to be identified with such a substratum. Otherwise we have merely introduced a hidden substratum, different from matter, without answering the question what space–time really is.

However, the identification of space–time with such a substratum seems to be problematic, not to say impossible. The reason is simple: The elements of space and time are principally not observable and, therefore, the same must be true for the substratum. Its elements would also not be accessible to empirical tests (Feature 1, Section 2.3, Chapter 2). However, science is essentially based on observations. Without them we leave the realm of natural science. Any scientific approach may only contain *realistic* statements about space and time and these are given by the following facts, as we mentioned several times: We can only say something about distances in connection with material objects, and times in connection with physically real processes. Only these facts have to be considered within scientific theories where space and time are involved. If we identify space and time with a *substratum*, exactly the same arguments hold. Also this substratum has to be considered as a non-observable quantity. But a substratum with such a feature is a non-scientific term and this is not acceptable at all because the definition and consideration of non-scientific, metaphysical quantities within a scientific theory cannot be accepted and is therefore ill. For such a substratum, the same arguments are valid that we have used in connection with *absolute space* (see the discussion in Section 5.5.1.3 and, in particular, the critique by Max Born).

The fact is that we never observe the curvature of space directly but only what it does to matter. For example, Einstein's concept of curved space–time was first proved when light skimmed the sun during a solar eclipse. But this bending effect can also be described by Newton's theory, and it is not essential in this connection that Newton's theory can only give a qualitative description for this effect.

Matter curves space–time within General Theory of Relativity. How does it work? What is the mechanism for that? General Theory of Relativity does not answer this question. It is perhaps a wrong question. Within General Theory of Relativity, there is merely an "assignment" between matter and a curved space–time, and we have to ask what assignment means in this connection. Assignment simply means that there is a curvature of space–time *without* mechanism (process), that is, matter does not interact with the space–time-elements $x$, $y$, $z$ and $\tau$ in order to produce a curvature (Feature 1, Section 2.3, Chapter 2). We have outlined that General Theory of Relativity suggests the existence of a substratum, then the curvature of space–time should be due to an interaction process of usual matter interacting with this substratum. However, this would be a new kind of interaction not defined in General Theory of Relativity.

## Newton, Leibnitz, Kant

Clearly, Newton's space and time are also not directly visible but, nevertheless, he assumed that space and time are really out there — essential ingredients of the real world, as real as matter. However, Newton could not specify this point. Gottfried Wilhelm Leibnitz, a German philosopher and mathematician, declined Newton's view radically and argued that Newton's space and time cannot be more than an illusion and would have nothing to do with essential ingredients of the real world. Leibnitz clearly recognized that space and time were only a way of thinking about the things, a way of perceiving the relationships between the real things in the world. Here is a clear line, a clear qualitative difference between the real world and space–time, and this is obviously more realistic than Newton's view because space and time are in fact principally not observable, as we have outlined above. Newton's view does not clearly distinguish between reality and space–time as it should be due to the basic properties of space and time (the situation is obviously similar in General Theory of Relativity, see our discussion above).

Immanuel Kant thought exactly in the same direction as Leibnitz. However, both, Leibnitz and Kant, were not able to give mathematical formulations for their basic thoughts, that is, for the relationship between space–time and the real world. Newton's mechanics is a very well working mathematical theory and could describe and predict a lot of experimental material. This fact pushed Kant's and Leibnitz' principal ideas away into the background but, up to now, there is no serious solution within General Theory of Relativity for this basic and essential space–time problem.

### 5.5.3. *Conclusion*

Our discussions in this section indicate that not only quantum field theory has a vacuum problem (Sections 5.2 and 5.3) but also General Theory of Relativity. All these points must have consequences for the nature of space and time. In Section 2.4 (Chapter 2), we characterized the situation as follows:

> "*Due to Feature* 1, *space–time has to be considered as a 'nothing' in the physical sense but, on the other hand, we experience space and time as a real phenomenon. Clearly, this 'nothing' can hardly be occupied by physically real objects. It is problematic to embed the real world into such a metaphysical space–time. Within all modern physical theories (Newton's mechanics, Theory of Relativity, conventional quantum theory, etc.) real material bodies are the contents of an 'object' (space–time) that has to be considered as a nothing or as a metaphysical substratum different from matter.... This seems to be a contradiction. Two kinds of*

*objects are mixed which are obviously mutually incompatible. We can overcome this bad situation if we take the position of the philosopher Immanuel Kant: Space and time are exclusively features of our brain and the world outside is projected on it. Then, the material objects that occupy space and time can only be geometrical pictures* (*Fig.* 2.1). *In other words, space–time has to be considered as an auxiliary element for the representation of physically real processes.... Projection theory (Chapter 3) is based on this conception. If that is correct, the situation would be identical with that what we do in connection with our blackboards and notebooks: we draw pictures on them.*"

This position of projection theory is a realistic point of view for the deciphering of the "code" that is anchored in the relationship between "space–time" and "matter." We recognized above that space and time have exclusively to be considered as the auxiliary elements for the representation of real things and processes. The empirical facts obviously do not allow other possibilities. If space–time is really an auxiliary element it cannot give rise to physically real effects as, for example, inertia (General Theory of Relativity) and quantum fluctuations (quantum field theory). Such physically real effects seem to be artificial when we consider the basic features of space and time. In other words, from this point of view, both theories, quantum field theory and General Theory of Relativity, have a serious vacuum problem. Leon Lederman said in connection with the vacuum problem of quantum field theory (Section 5.2.3): "*Einstein freed us from the either. Now we need to get rid of* (*today's version of either*) *again. We need to sweep the vacuum clean.*" [29] We should take this statement seriously.

## 5.6. On the Observation of Physically Real Processes

Since space–time is obviously the important element in connection with observations, let us repeat and supplement in this section the main properties of space–time. We already stated above that within the memory of man all essential things are represented within the frame of pictures that appear spontaneously in front of us and are representations in space and time. This is the most basic statement, and it is independent of the projection principle.

Due to *Feature* 1 (Section 2.3, Chapter 2) we have to conclude that no physical real system is able to interact with space–time. We may therefore state that space–time has nothing to do with matter or another mysterious substratum. Due to *Feature* 1 space–time cannot be a real element of physical reality. On the other hand, space–time cannot be simply a nothing because we experience it as real phenomenon. This is not a contradiction at all but indicates the role of space–time in connection with the observation of physically real processes. On the one hand, space–time is entirely uninvolved in the process under investigation. On the other hand, we observe just this process on the basis of space–time. Then, only one conclusion is realistic: space–time is an auxiliary element for the representation of physically real processes. That inevitably means that the real world cannot be embedded in space–time (container principle) but we have to assume that the real world is projected onto space–time. Just this conception has been used within the framework of projection theory (Chapter 3), and we have called it *projection principle*. The projection principle has been worked out in Chapter 2. Let us give further arguments for this principle.

### 5.6.1. *Reality Outside and the Inner Picture*

We experience the world by our sense organs, that is, the observer interacts with reality outside. Information about reality outside flow via our senses into the body, and the brain forms a picture of it, and we obtain a "picture of reality." This is definitely a projection. In other words, we have a reality outside and an inner picture (picture of reality), and we are firmly convinced that the inner picture is identical with the structure in reality outside. In this connection we have emphasized in Section 1.9 (Chapter 1) the viewpoint of the well-known psychologist C. G. Jung. In the opinion of C. G. Jung, the structure of the inner picture is identical with that of reality outside. [7] Again, this statement by C. G. Jung suggests the following. Although there is a projection of reality onto space–time we may also state that the real world outside is embedded in space–time. But, due to *Feature 1*, this is strictly forbidden, as we have analyzed above. Strictly speaking, there can be no space–time in connection with reality outside because space–time can only be an auxiliary element for the representation of physically real processes that take place outside. It gives therefore no sense to assume that space–time is also an element of reality outside. The *unobserved* reality outside is free of space and time.

Thus, the structure of reality outside must be quite different from the structure of the inner picture that is always based on space and time. However, the inner picture is the most direct impression we have of reality outside. In Section 2.6 (Chapter 2), we cited Arthur Eddington. He wrote (Swarthmore Lecture (1929), *Science and the Unseen World*): "*In comparing the certainty of things, spiritual and things temporal, let us not forget this — Mind is the first and most direct thing in our experience; all else is remote inference* ..." This statement is close that what we have outlined above: Within the memory of man all essential things are represented within the frame of pictures that appear spontaneously in front of us and are representations in space and time. Again, the inner picture must be quite different from the structures in the outside world and this is because there can be no space–time in the outside world.

#### 5.6.1.1. *The Color Experiment*

Nevertheless, one might object that all the things (hard objects like houses, trees, cars, etc.) in front of us belong directly to the outside world. But this definitely not true! All what we have in front of us is designed by the brain. It is an invention of our brain and is merely a picture of reality but is not reality or basic reality. This can be demonstrated by a simple reproducible experiment (see also [8,9]).

If we look in good lighting at a black ground with, for example, three circular coloured areas on it (the figure is represented in [8]) approximately one minute and then look at a white surface (for example, a wallpaper), these figures will appear on this surface although they do not exist there. Furthermore, these figures do not have their original colours. Instead they appear as complementary to the original colours on the black field.

So much for the experiment. The figures on the white surface do not actually exist in the way observed by us anywhere, unless by chance. The forms of the figures (three full circles) on the white and the black surfaces are the same, but only in the form, not in the color.

It is relatively easy to explain this effect. In [6], we find: "*The phenomenon is due to a sign of fatigue in specific color receptors on our retina. The impression 'white' arises by the uniform excitation of all color receptors. If we stare at the object for an extended time, some of these color receptors will tire leaving the activity of those receptors which have not been strained to predominate and therefore allowing their color value to dominate temporarily the foreground we are experiencing.*"

The color experiment is reproducible. Under normal circumstances anybody is able to experience this phenomenon so that to this extent the guarantee of objectivity is given.

#### 5.6.1.2. *What is Outside, What is Inside?*

We see the colored figure *outside* on the white surface. The explanation of the effect is however given in terms of color receptors which are inside the head. This means that the figures cannot be on the white surface *outside* the observer and, in particular, this also means that everything (white surface, figures, and everything else that appears in front of us) is *inside* the head. We only have the *impression* that everything is situated in a space *outside* us.

In conclusion, everything that we see is primarily in our head. It is not outside us. (We already discussed this effect in Chapter 2.) Not only the colored figures but also persons, cars, aeroplanes, the sun, moon and stars are pictures of reality in our brain. We have only the impression that all these things are located outside us. We discussed in Chapter 2 and also in Section 4.4 that it is characteristic for the process of seeing that our awareness does not register the picture inside the eye on the retina, but we have the impression that we are observing external objects (or rather, "as if" we were observing external objects).

We normally assume that these sensations produced by the brain are identical with reality itself, but this should not be the case as we have argued above since space–time cannot be outside the brain because space–time has to be considered as an auxiliary element for the representation of physically real processes. In other words, the outside world, the material bodies, cannot be embedded in space–time. That in particular means that not only the things in front us (cars, houses, trees, etc.) are in our head but also space–time, where all these things are positioned. We have only the impression that all these "hard objects," together with space–time, are located outside us.

Material reality is the cause that we have a picture in front of us. All objects in such a picture are immaterial. The picture itself, i.e., the world before our eyes, is not material. The picture only contains geometrical positions, shapes and colors.

Space and time are obviously elements of the brain. They come into existence due to specific brain functions. This is the reason why we cannot put space and time on the table as we can in the case of matter (Chapter 2). Concerning time, the following remarks are instructive [55]: "*As mentioned, the concept of time cannot actually be understood. We are accustomed to think that time is something which can be found or which one has. But it has actually no existence. The physicist says that time is something that can be measured in one or another way by a clock. But what does the clock measure? Just the time! Regarding this problem, Harald Bo says in his interesting writings: Does a day exist? In any case not the whole day because the beginning already disappeared before the end comes. The same is true for an hour and for a second. The only time-interval which can exist is where the beginning and the end are so close together that they are identical. That means a quantum zero-time ... We cannot put the last day on the table or somewhere else. We have no time in the physical sense. What we have is a psychological time.*" [55]

## 5.6.2. *Effects of Evolution and Basic Reality*

From our analysis of space–time, it follows that the structures in the picture must be different from those in outside reality. This is compatible with the principles of evolution. In Section 1.9.1 (Chapter 1), we justified the position that there can be no one-to-one correspondence between the inner picture and the world outside.

All these facts mean that we cannot know how reality is actually constructed, i.e., what is its structure, independent of the observer's perception apparatus. It is probable that we may never know because we are always caught in our own system. Basic reality, that is, reality which exists external to the biological system, seems in principle to be remain hidden from our view.

In the case of a projected reality (onto space-time) nature is organized in another way than in the case where reality is embedded in space and time. Therefore, we have to consider the projection effect in the analytical treatment. This can only be done by fictitious realities (Chapter 3) on the basis of what we can know, and this is exclusively the picture of reality given in space and time, which we have called $(\mathbf{r}, t)$-space. Then, fictitious reality is given in terms of momentum $\mathbf{p}$ and energy $E$ and we have called the corresponding space $(\mathbf{p}, E)$-space. All the other details are given in Chapter 3.

The picture of reality must be species-dependent. In other words, we have to conclude that the actions of other biological systems are in general based on a picture of reality that is different from that of the human observer. Wolfgang Schleidt performed some interesting experiments and actually showed that the perception apparatus of a turkey is quite different from that of humans, and this obviously leads to the effect that the pictures of reality of both types of biological systems must be different from each other. More details are given in Chapter 2.

### 5.6.3. *Kant's Philosophy*

Already the philosopher Immanuel Kant (1724–1804) investigated the relation between true and perceived reality. Kant's ideas are close to that what we have developed in the preceding chapters. However, there are nevertheless considerable differences concerning statements about reality outside. Let us briefly discuss Kant's ideas.

Kant argued that we cannot make statements about the true reality outside. According to him all things we observe are located within space–time and these elements, space and time, are located inside the observer. In Kant's opinion, a human observer can nothing say about the structure of the outside world. In particular, in his opinion a human observer is *not* able to give answers on the following questions. Is there a one-to-one correspondence between the structures outside and those in the brain of the observer? Does the information in the pictures located in the brain reflect the complete information about reality outside? This is because Kant knew nothing about the principles of evolution but this is an essential point in answering such questions.

According to Immanuel Kant, space and time are not empirical concepts, which are determined by abstraction from experience. Experiences become possible at all only through the concepts of space and time. According to Kant, space and time are not objects, but have to be considered as preconditions for the possibility of all experience. Although in Kant's opinion space and time are not empirical concepts, they nevertheless have empirical reality. This is because all things which we observe are located in space and time. The structure of space and time is therefore reflected in the empirical objects. Kant denied the existence of a space and a time independent of brain functions (observations in everyday life, thinking).

According to Kant, space and time are located inside the observer. Whether space and time are also elements of actual (fundamental) reality outside remains principally an open question within Kant's point of view.

Kant's perspective is without any doubt important, not only in connection with philosophical questions. But what are the consequences for physics? If we take Kant's view seriously, then the physical laws as, for example, Newton's gravitational law, are merely pictures in the head of the observer and there is principally no way to express it for the reality outside. Nothing can be said about the processes in the outside world. (This is in contrast to "projection theory" outlined above (Chapter 3) where we can construct fictitious realities.) If the gravitational law (and all the other physical laws) is merely a picture in the head we get a problem because there can be no gravitational forces in the head of the observer. In conclusion, Kant's thoughts can lead to considerable problems when we apply them to physics. Barrow remarked [56]: "*We can see that Kant's perspective is worrying for the scientific view of the world.*" However, Kant's perspective has not taken so seriously in physics and there are other positions. The situation is well-analyzed by Barrow [56]: "*There are two poles about the relationship between true reality and perceived reality. At one extreme, we find 'realists', who regard the filtering of information about the world by mental categories to be a harmless complication that has no significant effect upon the character of the true reality 'out there.' Even when it makes a big difference, we can often understand enough about the cognitive processes involved to recognize when they are being biased, and make some appropriate correction. At the other extreme, we find 'anti-realists,' who would deny us any knowledge of that elusive true reality at all. In between these two extremes, you will find a spectrum of compromise positions extensive enough to fill any philosopher's library: each apportions a different weight to the distortion of true reality by our senses.*" [56]

In other words, there are no criteria to decide about the true nature of absolute reality. The realists cannot disprove the anti-realists and vice versa. The realists more or less assume that there is a one-to-one-correspondence between true reality and perceived reality (picture). The anti-realists maintain that we can say nothing about true reality. However, when we consider the basic facts of biological evolution, both viewpoints do not seem to be realistic. Evolution teaches us that in connection with humans and animals, cognition does not play the important role in nature but the differentiation between "favorable towards survival" and "hostile towards survival" does, at least at the early phase of evolution (see in particular Chapters 1 and 2). Each picture of reality (perceived reality) is tailor made to these characteristics. Since the conditions for survival are different for different biological systems, the perceived realities are also different for different biological systems. Wolfgang Schleidt's experiments (Section 2.9) with a turkey showed that very impressively.

The picture of reality designed by the individual (unconsciously) has to be correct but it may only contain, for economic reasons, information which is absolutely necessary for survival; everything else is unnecessary. The picture of reality does not have to be complete and true (in the sense of a precise reproduction) but restricted and reliable.

Furthermore, we learned from Schleid's experiments (Chapter 2) that the conception of the world of man and that of turkey are on the one hand different from each other, they are on the other hand correct in each case. This means that neither of these two conceptions of the world can be true in the sense that they are faithful reproductions of nature. There can be no one-to-one correspondence between the structures in the picture and those in true reality. Objective reality must be different from the images from which biological systems are constructed.

The statement that there can be no one-to-one correspondence is, on the one hand, against the realists and, on the other hand, it is simultaneously against the position of anti-realists because it is a statement about true reality.

The main features of Kant's philosophy are summarized in the first scheme and those of projection theory in the second scheme.

## KANT'S PHILOSOPHY

↓

**PICTURE OF REALITY**
**in space-time**
**(brain functions),**

no further statements

**BASIC REALITY**

no statements possible

**FICTITIOUS REALITY**

not defined

**PROJECTION THEORY**

↓

**PICTURE OF REALITY**
**in space-time**
**(brain functions),**

further statements possible:
no one-to-one-correspondence,
selected information,
pictures in symbolic form

**BASIC REALITY**

further statements possible:
no space-time,
no one-to-one-correspondence,
no specific statements about its structure possible

**FICTITIOUS REALITY**

defined by ($\mathbf{p}$,$E$)-space

#### 5.6.3.1. *Remarks on Newton's View*

Newton discovered his mechanics approximately two hundred years before Kant developed his perspective. Newton did not distinguish between "true reality" and "perceived reality (picture of reality)" and he based his theory directly on that what was directly in front of him: space and objects, which were more or less separated in space. The big success of Newton's theory pushed Kant's perspective into the background, and all the further developments in physics were based on Newton's view without to consider Kant's basic thoughts which merely remained the subject of philosophy up to the present day.

### 5.6.3.2. *Consequences*

Statements about the world are always species-specific, i.e., any world view must always be superimposed by certain properties of the biological system (man, turkey, etc.) that makes the statements, consciously or unconsciously. Therefore, it is principally not possible to give statements about the absolute reality since absolute reality is by definition independent of the observer's biological structure, and there is obviously no way to eliminate these species-specific factors. Reality is projected on the observer's cognition system leading to species-specific world views ("pictures of reality"). The observer constructs (consciously or unconsciously) such a picture on the basis of a "frame" which is in the case of a human observer given by our well-experienced space–time, but, in principle, this frame can be different for different biological systems.

There is the *realist* who believes that the structures in the picture and those in the world outside are identical or almost identical. On the other hand, the *idealist* states that we can say nothing about that what we call reality (Kant's point of view). However, we concluded from Schleidt's experiments with turkeys that there can be no "one-to-one-correspondence" between the picture and the reality outside. Already the statement "no one-to-one correspondence" contradicts Kant. There are obviously more statements about the outside world possible than Kant thought, and this should be due to the fact that Kant knew nothing about evolution.

But what does "no one-to-one correspondence" mean? If both, the things of the outside world and the structures in the picture, are equally embedded in space and time, we could assume that a complicated situation in the outside world is simplified by the elimination of those things which are unnecessary for a certain species. Then, the remaining species-relevant facts are merely arranged in the picture in another way than those in the outside world. However, we have learned above in connection with our space–time analysis (see in particular Chapter 2) that this should not be the case. The reality outside is obviously not embedded in space and time.

In the classical description (often called Galilean view) such questions are eliminated. The laws are directly based on the impressions of everyday life, but since these laws are to be considered as independent of the observer, it is assumed that the structures formulated by the theory are identical with those in reality outside. In quantum theory, it was the first time in physics that the observer plays a more active role and is explicitly needed for the understanding of the physical laws (Copenhagen Interpretation), but in another way than the Schleidt's experiments suggest.

## 5.7. Curved Spaces

We have stated above that we cannot make specific statements about *basic reality*; only very general features could be deduced above. In particular, we are principally not able to describe basic reality mathematically. However, we may construct *fictitious realities* on the basis of the pictures variables (Chapter 3). It is the $(\mathbf{p}, E)$-space from the point of view of $(\mathbf{r}, t)$-space. Both spaces, $(\mathbf{r}, t)$-space and $(\mathbf{p}, E)$-space, are different from each other but are equivalent concerning their physical information (see also Chapter 3).

This particularly means that not only the variables $\mathbf{r}$ and $t$ play the role of auxiliary elements but also the variables $\mathbf{p}$ and $E$. In other words, the variables $\mathbf{p}$ and $E$ cannot be real quantities in the reality outside but are exclusively elements of the brain as it is the case in connection with $\mathbf{r}$ and $t$. From this point of view energy, $E$ does not really exist. This is a fundamental conclusion, and this result also follows from general consideration *without* projection theory. Details are given in Section 3.6.1 (Chapter 3). In particular, the statements by Christian von Bayer and Edgar Lüscher are relevant.

Dark energy etc. is assumed to actually exist in reality outside but if we take the statements by Christian von Bayer and Edgar Lüscher seriously, this cannot be the case. How can reality be embedded in space–time when the energy is an auxiliary element? This is hardly possible. The way out of this dilemma is to use projection theory (or a similar theory which is however not known up to now) where the relationship between basic reality, the picture of reality and fictitious reality are defined without such contradictions.

### 5.7.1. *Re-Interpretation*

Let us consider a particle of mass $m_0$ in an external field, and let us treat this problem only in one dimension using the coordinate $x$. Then, within conventional classical mechanics (reality is embedded in space and time) the total energy $E_{\text{total}}$ is given by the kinetic energy $E_k$ and the potential $U(x)$:

$$E_{\text{total}} = \frac{p_x^2}{2m_0} + U(x), \tag{5.66}$$

where $E_{\text{total}}$ remains constant, that is, when $U(x)$ increases the kinetic energy $E_k$ must decrease and vice versa. However, within projection theory there can be no potential energy $U(x)$ because any energy can only be an element of $(\mathbf{p}, E)$-space. Thus, we need another argumentation.

Is it possible to re-interpret Eq. (5.66)? Can we simply transform Eq. (5.66) so that reality is no longer embedded in space–time but is projected on it? We have investigated this point generally (see Appendix B) and in the case of Eq. (5.66) such a re-interpretation should be possible. However, the term potential energy, that is, the function $U(x)$, may no longer appear and we have to shift the interaction problem on $(\mathbf{p}, E)$-space as it is necessary within projection theory. In $(\mathbf{p}, E)$-space, the system with mass $m_0$ exchanges momentum and energy with another system. In other words, the kinetic energy $E_k = E$ is the energy of the system $[E = E_{\text{total}} - U(x)]$ and we have $p_x$-fluctuations $\Delta p_x$ and $E$-fluctuations $\Delta E$ [instead of the potential $U(x)$] changing the energy $E$ of the system in the course of time $\tau$. (Note, that the system-specific time $t$ is not defined. We only want to re-interpret conventional physics.) Therefore, the energy $E$ and the momentum $p_x$ vary with time $\tau$ and the following relation is valid:

$$\tau : E = \frac{p_x^2}{2m_0}. \tag{5.67}$$

This is the situation in $(\mathbf{p}, E)$-space. But what about the situation in $(\mathbf{r}, t)$-space? It is shown in Appendix B that the system-specific time $t$ must be identified with the reference $\tau$ in the classical case: $t = \tau$.

Thus, instead of $(\mathbf{r}, t)$-space we will use in this section the expression $(\mathbf{r}, \tau)$-space.

There is no momentum, no energy, no mass $m_0$ in $(\mathbf{r}, \tau)$-space, but we may "only" have geometrical positions in $(\mathbf{r}, \tau)$-space. From Eq. (5.67), we immediately find

$$\frac{dE}{dp_x} = \upsilon = \frac{dx}{d\tau} \tag{5.68}$$

with $p_x = m_0\upsilon$, where $\upsilon$ is the velocity of the system. The term $dE/dp_x$ in Eq. (5.68) belongs to $(\mathbf{p}, E)$-space, and the term $dx/d\tau$ in Eq. (5.68) describes the system in $(\mathbf{r}, \tau)$-space. Then, the term $dE/dp_x$ corresponds to "reality" and $dx/d\tau$ is an element of the "picture of realty":

$$\frac{dE}{dp_x}(\text{reality}) \rightarrow \frac{dx}{d\tau}(\text{picture of reality, geomertical positions}). \tag{5.69}$$

From Eq. (5.68) directly follows that both spaces, $(\mathbf{p}, E)$-space and $(\mathbf{r}, \tau)$-space, contain exactly the same information, that is, both spaces are equivalent as in the quantum-theoretical case (Chapter 3). However, here we have a trivial transformation law that transforms the information from $(\mathbf{p}, E)$-space to $(\mathbf{r}, \tau)$-space (see Appendix B). Both spaces are connected by the function $U(x)$: $U$ has the dimensions of an energy and belongs to $(\mathbf{p}, E)$-space and $x$ is a geometrical position and belongs to $(\mathbf{r}, t)$-space. More details are given in Appendix B.

If we assume that the world is embedded in $(\mathbf{r}, \tau)$-space (Newton's mechanics, Special Theory of Relativity, General Theory of Relativity, usual quantum theory) the function $U(x)$ means that we have at point $x$ the potential $U$, that is, the potential is really there, independent on any observation procedure. In the other case (projection theory), where the world is projected onto space–time $[(\mathbf{r}, \tau)$-space$]$, there is no potential $U$ at point $x$ because $U$ belongs to $(\mathbf{p}, E)$-space. Or more precisely, instead of $U$, we have a function that describes the interaction (here $E$-fluctuations) in $(\mathbf{p}, E)$-space. Nevertheless, a human feels it at point $x$ and also measures it at point $x$. The reason is simple: humans and measuring instruments are also pictured in $(\mathbf{r}, t)$-space $[(\mathbf{r}, \tau)$-space in the classical limit$]$ but the interaction due to $U$ takes place in $(\mathbf{p}, E)$-space. This is close to what we stated in connection with projection theory (Chapter 3). "*The measurement of one of the possible values for* $\mathbf{p}$ *and for E is done in the space–time intervals* $\mathbf{r}, \mathbf{r} + d\mathbf{r}$ *and* $t, t + dt$ *with the probability density of* $\Psi^*(\mathbf{r}, t)\Psi(\mathbf{r}, t)$."

This indicates in a certain sense that Newton's theory can be reinterpreted, i.e., we are able to change the view. Instead of assuming that the world is embedded in space–time (container principle) we may also apply the projection principle. However, from the point of view of projection theory, the introduction of $U(x)$ seems to be artificial and gives, physically, not much sense. This construction, concerning $U(x)$, does not meet the features of projection theory. Here we only have interactions in the form of $(\mathbf{p}, E)$-fluctuations in $(\mathbf{p}, E)$-space while $(\mathbf{r}, t)$-space only contains geometrical positions and trajectories but never real physical objects.

In the case of General Theory of Relativity, we have a similar situation and this can be easily demonstrated by Einstein's famous field equation which has the following form:

$$R_{\mu\nu} - \frac{1}{2} g_{\mu\nu} R = -\kappa T_{\mu\nu} \tag{5.70}$$

with $\kappa = 8\pi G$, where $G$ is Newton's gravitational constant. The elements $g_{\mu\nu}$ form Riemann's metric tensor and are responsible for the geometry. $T_{\mu\nu}$ is the momentum–energy tensor, and $R_{\mu\nu}$ is the Ricci tensor. $R$ is a scale factor which determines the size of the universe. Equation (5.70) allows the determination of the elements $g_{\mu\nu}$ as a function of the momentum–energy distribution. In other words, reality $[(\mathbf{p}, E)$-distribution$]$ determines the properties of $(\mathbf{r}, t)$-space, i.e., its geometry which is, as we know, in general curved (non-Euclidian) within General Theory of Relativity.

It is surprising that this scheme is exactly the same as in the case of projection theory. Within projection theory, quantum phenomena are formulated within $(\mathbf{p}, E)$-space ("reality") and the projection of this information leads to the "picture of realty," however within a Euclidean geometry.

Within General Theory of Relativity, we have $(\mathbf{p}, E)$-distributions leading to a curvature of space–time, that is, each specific curvature reflects a specific $(\mathbf{p}, E)$-distribution. Thus, "reality" is given by a $(\mathbf{p}, E)$-distribution and this produces a curved space–time which can therefore be considered as the "picture of reality." It reflects the $(\mathbf{p}, E)$-distribution by a specific curved space. Then, we have

$$T_{\mu\nu}(\text{reality}) \rightarrow g_{\mu\nu}(\text{picture of reality}). \tag{5.71}$$

That is, instead of assuming that reality is embedded within space–time, we may also here re-interpret the facts.

It is important to note that the gravitational energy does not appear in the $(\mathbf{p}, E)$-distribution for the determination of the geometry because the gravitational energy is included in the metric expressed by the elements $g_{\mu\nu}$.

If we assume that the world is embedded in $(\mathbf{r}, t)$-space (Newton's mechanics, Special Theory of Relativity, General Theory of Relativity, usual quantum theory), gravitational energy means that we have at a certain point, say $x$, a certain energy, i.e., this (gravitational) energy is really there, independent on any observation procedure. In the other case (projection theory) where the world is projected onto space–time $[(\mathbf{r}, t)$-space$]$, there is no gravitational energy at point $x$ [because any kind of energy belongs to $(\mathbf{p}, E)$-space] but a human feels it there and also measures it at point $x$. Again, the reason is simple. Humans and measuring instruments are also only pictured in $(\mathbf{r}, t)$-space but the interaction with the gravitational field takes place in $(\mathbf{p}, E)$-space. Also this is close to what we stated in connection with projection theory (Chapter 3). "*A measurement for* **p** *and for E is done with a measuring apparatus positioned in space–time.*"

It is interesting and important to note that this curved space–time within General Theory of Relativity is only dependent on the *energy* and *momentum* and no other properties are involved. Exactly the same is the case in the formulation of quantum phenomena within projection theory (Chapter 3).

Einstein introduced the cosmological constant $\Lambda$ by a simple extension of his field equation. Instead of Eq. (5.70) we get

$$R_{\mu\nu} - \frac{1}{2}g_{\mu\nu}R - \Lambda g_{\mu\nu} = -\kappa T_{\mu\nu}. \tag{5.72}$$

The introduction of the cosmological constant $\Lambda$ does not change the view in connection with a re-interpretation $\Lambda$ corresponds to an energy density [see Eq. (5.1)] and this has an influence on the behavior of the geometry. From the point of view of projection theory, here the $\Lambda$-energies have to be considered as elements of $(\mathbf{p}, E)$-space producing a certain kind of geometry in $(\mathbf{r}, t)$-space.

This indicates also here that General Theory of Relativity can be re-interpreted in a certain sense, i.e., we are able to change the view. Instead of assuming that the world is embedded in space–time, we may also apply the projection viewpoint. However, we must say also here that from the point of view of projection theory the introduction of gravitational energy by a curved space–time seems to be artificial and makes, physically, not much sense. This construction, concerning gravitational energy (curved space–time), does not meet the features of projection theory. Here we only have energies and interactions in the form of $(\mathbf{p}, E)$-fluctuations in $(\mathbf{p}, E)$-space while $(\mathbf{r}, t)$-space only contains geometrical positions and trajectories but never real physical objects.

Equations (5.70) and (5.72) describe the connections between the $(\mathbf{p}, E)$-distribution and the geometry (metric). A certain $(\mathbf{p}, E)$-distribution produces a curved space–time. However, both equations do not tell us how the $(\mathbf{p}, E)$-distribution does it. What is the process in nature to come from the momentum–energy tensor $T_{\mu\nu}$ to the elements $g_{\mu\nu}$? One would expect that there is an interaction between the real quantities and space–time resulting in a curvature. Such an interaction between the real quantities and space–time is however not imaginable at all because this would imply that space–time consists of a certain kind of mysterious substratum (Section 5.5.2.2). We have already pointed in Section 5.5.2.2 that the connection between the $(\mathbf{p}, E)$-distribution and the geometry (metric) is not explicitly described within General Theory of Relativity but is merely an "assignment."

However, such a feature, the assignment, is undoubtedly more characteristic for projections like those developed within projection theory where space–time plays the role of an auxiliary element as it should be due to the facts.

### 5.7.2. *Metric within Projection Theory*

Space–time is in the head of the observer, its existence is due to brain functions, and reality outside is projected on this frame leading to different pictures of reality. From this viewpoint, it is hardly possible that the space–time frame itself, its metric, is different for different processes taking place outside. Only the projected structures on this frame with fixed metric are different for different processes in the outside reality. It gives no sense to vary the space–time properties (metric) simultaneously. In other, words, the geometry of space–time, its metric, should remain constant, i.e., it should be either Euclidean or constantly curved.

The "picture of reality" can be compared with a blank piece of paper (bumpy or even) on which geometrical positions and/or trajectories are drawn with a pencil. It is not possible that such a perfectly featureless piece of paper can have effects on what we draw on it. No doubt, this conception is quite different from that of General Theory of Relativity.

### 5.7.3. *Poincaré*

This view is compatible with Poincaré's analysis about space and time, in particular about flat (Euclidean) and curved spaces. Let us briefly summarize Poincaré's very basic ideas that have to be considered as facts and can hardly be denied by other argumentations. Poincaré (1854–1912) based his assertion on that what we have already stated above (Section 2.3, Feature 2): "*We can only say something about distances in connection with masses, and time intervals in connection with physical processes.*"

Here we should start with Gauss (1777–1855) who tried to find out the geometry of the space of our experience. Is it Euclidean or not?

Max Jammer wrote [57]:

> "*He* (*Gauss*) *tried to measure directly by an ordinary triangulation with surveying equipment whether the sum of the angles of a large triangle amounts to two right angles or not. Accordingly, he surveyed a triangle formed by three mountains, the Brocken, The Higher Hagen, and the Inselberg with sides measuring* 69, 85, *and* 107 km. *Needless to say that he did not detect any deviation from 180° within the margin of error and thus concluded that the structure of actual space is Euclidean as far as experience can show.*" [57]

However, Poincaré pointed out that such kind of experiments and any other for the determination of geometries have no meaning. Max Jammer remarked [57]:

> "*It was only toward the turn of the century that Poincaré demonstrated once for all the futility of this controversy and the fallacy of any attempt to discover by experiment which of the mutually exclusive geometries applies to real space. Measurement, he insists, is never a space itself, but always of empirically given physical objects in space, whether rigid rods or light rays. Regarding the structure of space as such, experiment can tell us nothing; it can tell us only of the relations that hold among material objects. Suppose, Poincaré says, a deviation from two right angles had occurred in the triangulation carried out by Gauss, would this necessarily have constituted a refutation of Euclidean geometry? For there would be nothing to prevent us from continuing to use Euclidean geometry on the assumption that light rays are curved. Nothing could disprove such an assumption. So the highly important conclusion emerges that experience can neither confirm nor refuse a geometry, whichever geometry it be. What geometry one chooses is, for Poincaré, merely a matter of convenience, a convention. We select that system of geometry which enables us to formulate the laws of nature in the simplest way.*" [57]

In conclusion, we can verify whether a space has Euclidean geometry but never if there are deviations from Euclidean geometry, and this is because a real physical process could be responsible for these deviations and not the fact that there is a non-Euclidean geometry.

## 5.7.4. *What Geometry is Realized?*

What geometry one chooses is, for Poincaré, a matter of convention. From this point of view, geometry has obviously no definite reality because one can choose that system of geometry which is convenient and enables us to formulate the laws of nature in the simplest way. What geometry one chooses is therefore a matter of convention, in the opinion of Poincaré.

Within the framework of General Theory of Relativity, the universe has a definite geometry (flat, spherical or hyperbolic) depending on the space contents (usual matter and energy, dark matter, dark energy). Within General Theory of Relativity, we have no choice to select a system of geometry.

What statements are possible within projection theory? Are they close to what General Theory of Relativity says or does projection theory tends to Poincaré's view? Let us discuss this point on the basis of characteristics that are typical for projection theory and on facts in connection with the observer's cognition system. We will come to the conclusion that projection theory tends more or less clear to Poincaré's point of view but also certain elements of General Theory of Relativity are recognizable.

### 5.7.4.1. *The Brain Processes on the Basis of Euclidean Geometry*

There is no space–time in the outside world (see in particular Chapters 2 and 3). Space–time is inside the head of the observer and cannot contain real matter etc. We also have to conclude from the facts that we have discussed in Section 2.3 (Features 1 and 2) that the space–time properties emerge with the objects and processes and these are geometrical positions, trajectories and, in the quantum-theoretical case, probability distributions. All these statements directly follow from the fact that we are principally not able to observe the space–time elements $x$, $y$, $z$ and $t(\tau)$. We can only say something about *distances in connection with masses*, and time intervals in connection with physical processes.

We know nothing about basic reality (instead we construct fictitious realities) but we have only a picture of it that is given in space and time. Due to the principles of evolution, we may definitely say that there is less information in the "picture of reality" than in the outside world (Chapter 2). Let us repeat the main facts. An individual registers situations in the environment in certain patterns which are tailor-made for the particular needs of the species and which are completely free of any compulsion towards precise "objectivity." In particular, we have outlined in Chapter 2 and in [8,9] that we have to assume that an event occurring in the cosmos is portrayed inside a biological system "only" in symbolical form.

In other words, there is a selection process and only that information from the outside world is registered and processed by an individual that is relevant. The brain processes the data from the selected information of the world outside to the "picture of reality" in symbolic form. We know only a little about these processes, but we may assume that the behavior of the things relative to each other is one of the important points. Such relations are obviously produced from the selected outside data and are presented as a picture in space and time on the basis of Euclidean geometry. We can only check it by means of experiments that also take place in reality outside, and the experimental devices are given in the picture of reality as well. Such an experiment was already done by Gauss (Section 5.7.3) on the basis of three points in space and light rays, and, due to the experimental results, Gauss came to the conclusion that the geometry of space is Euclidean. Clearly, from the point of view of projection theory, it is the space inside the head that Gauss investigated and this is Euclidean. However, we have only the three points in the picture and we could in principle also cover this three-point configuration with a non-Euclidean geometry. The brain processes on the basis of Euclidean geometry but this does not mean that we could not connect the final product (picture) with another geometry that is different from Euclidean geometry. In other words, we have the choice to cover the produced picture (three points) by any geometry. It may be Euclidean but also non-Euclidean. This result is in accordance with Poincaré's view that we have discussed above (Section 5.7.3).

### 5.7.4.2. *Geometrical Optics*

The objects in the picture can be manipulated by optical instruments. When we use a certain kind of optical lens we can magnify the objects in the picture. What does that mean within the framework of projection theory? We construct the magnified object by means of the lens and the object which we experience without lens. The one occupies space on the left-hand side of the lens, the other occupies space on the right-hand side of the lens, that is, both objects are in the same space. Therefore, both objects that we see cannot be outside because there is no space outside. Such lens-constructions only allow us to say something about the objects in the picture, without lens and with lens. Nothing can be said about the object and the situation in reality outside. The only thing we can say is the relative changes of the objects in the picture of reality, and nothing else. The investigation of the relationship between the object outside and the object inside is not possible.

### 5.7.4.3. *Experiment with Distorting Glasses*

If one uses glasses which strongly distort the picture in space, the following takes place. After a certain time the observer sees everything in the normal order, that is, the spatial situation is the same as before without distorted glasses and space is again right-angled (Euclidean). In other words, the distorting glasses are effectively ignored after a certain time by the observer's cognition apparatus.

In conclusion, distorting glasses transform the space properties (its metric), which satisfy the axioms of Euclidean geometry, into a space with non-Euclidean geometry. Without changing the physical conditions we observe after a certain time the following strange effect. The cognition apparatus of the observer obviously transforms space with non-Euclidean geometry into usual space which satisfies the axioms of Euclidean geometry. In other words, the cognition apparatus of the observer is able to influence the space and picture, respectively.

Again, we can only check this situation, i.e., the non-Euclidean geometry that appears with distorting glasses, by means of experiments that take place in reality outside which we experience as a space-picture. Such an experiment was already done by Gauss (Section 5.7.3). If we would do it on the basis of three points in space and light rays in analogy to the experiment by Gauss, we would measure that the sum of the angles of the triangle formed by the three points does not amount to two right angles and, due to this experimental results, we have to conclude that the geometry of space is non-Euclidean. Not only are the figures in the picture distorted in the case of distorting glasses but the whole scenario as well, i.e., the rays of light. Therefore, we may conclude that there is a real transformation from Euclidean to non-Euclidean space-metric.

The real physical process outside remains unchanged, only the kind of perception has been changed. We have two geometries for one and the same process outside.

The brain processes temporarily on the basis of non-Euclidean geometry but this does not mean that we cannot connect the final product (the distorted figures in the picture) with another geometry that is different from non-Euclidean geometry. In other words, we have the choice to cover the produced picture (three points, figures) by any geometry. It may be Euclidean and also non-Euclidean. Also this result is in accordance with Poincaré's view (Section 5.7.3).

After a certain time, the non-Euclidean metric transforms back and we get again the picture with Euclidean geometry and with straight rays of light in the three-point experiment. The brain simply ignores the distorting glasses. It is probably better to say that the brain goes back to a Euclidean processing.

Without doubt, the experiment with distorting glasses is basic with respect to the nature of space and the role of the observer. In particular, it demonstrates convincingly that the world in front of us is actually inside the head. A similar experiment has been described in Section 2.5.1 (Chapter 2). The results are in accordance with that what we have concluded in connection with the colour experiment (Section 5.6.1.1).

### 5.7.4.4. *Constancy Phenomena*

How can we explain that the brain transforms the non-Euclidean geometry into a space with Euclidean geometry without changing the physical conditions? It is obviously a process that takes place autonomously in the observers brain without external intervention. Why does the brain do that? The answer is possibly given by evolution in connection with so-called *constancy phenomena*. We already discussed these phenomena in Section 3.16.5 (Chapter 3). Let us briefly repeat this point here.

The recognition of certain situations in the environment requires reliability (reproducible occurrence). How has evolution realized that? It caused organisms to develop physiological mechanisms which register automatically and objectively those signals from the environment that are relevant for the biological system. In this connection, the following is of interest: Of all the information from the environment, only what is relevant for the system is filtered out (selected) and registered. It is most important that an object observed by an individual, is unambiguously recognizable even in the case of large variations in the environment. In order to guarantee this, evolution has developed constancy phenomena whose objectivation performance is managed by a complex physiological apparatus. Konrad Lorenz recognized that very clearly. [18] Because of the fundamental importance let us repeat his statement:

> "*Of special interest to the scientist striving for objectivation is the study of those perceptual functions which convey to us the experience of qualities constantly inherent in certain things in our environment. If, of course, we perceive a certain object (say a sheet of paper) as white, even when different coloured lights, reflecting different wavelengths, are thrown on it; this so-called constancy phenomenon is achieved by the function of a highly complex physiological apparatus which computes, from the colour of the illumination and the colour reflected, the objects constantly inherent property which we call color .... Other neural mechanisms enable us to see that an object which we observe from various sides retains one and the same shape even though the image on our retina assumes a great variety of forms. Other mechanisms make it possible for us to apprehend that an object which we observe from various distances remains the same size, although the size of the retinal image decreases with distance.*" [18]

Such an environmental variation is obviously also given and simulated, respectively, by the use of distorting glasses because they disturb the individual to process the selected information from the outside world within Euclidean geometry, and it is temporarily forced to do that on the basis of non-Euclidean geometry. To process on the basis of non-Euclidean geometry is probably not reliable enough for the individual because this new situation does not possibly allow us to manage certain events in the environment optimal and without risk.

In order to guarantee reliability (reproducible occurrence) also in the case of distorting glasses, the evolution has probably developed a constancy phenomenon. In other words, there should also be a constancy phenomenon that transforms the non-Euclidean space into a Euclidean one.

### 5.7.5. *D'Arcy Wentworth Thompson*

The biologist D'Arcy Wentworth Thompson (1860–1948) investigated changes in biological organisms by means of geometrical considerations. Thompson made important discoveries and was able to describe systematic changes of biological organisms, which occur in the course of evolutionary processes, by means of changes in geometry (its metric). His pioneering discovery can be formulated as follows. The changes during these evolutionary processes can be described by a change in geometry (metric), in most cases from Euclidean to non-Euclidean geometry. In this way, the pictures of one living organism could be transformed into others. Thompson'sfamous book *On Growth and Form*, published in 1917, has had a considerable echo in the scientific community.

Thompson's kind of analysis (see also [58]) can be applied to many organisms and many biological details concerning bones etc. In particular, he throws light on the role of geometry in connection with physically real processes. The discovery by D'Arcy Wentworth Thompson is therefore of particular relevance for what we have analyzed above: The nature of space and time within projection theory. First, we will give some general remarks and after that we will apply that on fishes using the findings of Thompson.

### 5.7.5.1. *Some General Remarks*

Although our cognition apparatus works on the basis of Euclidean geometry, we recognized that there are alternatives in the form of curved (non-Euclidean) spaces. Any pattern drawn on a blank piece of paper can be connected with a curved or a Euclidean coordinate system. It is a matter of convention.

Within projection theory we have a "fictitious realty," defined by $(\mathbf{p}, E)$-space, and we have a "picture of reality," which is represented in $(\mathbf{r}, t)$-space. Both spaces are connected to each other by a Fourier transform and contain exactly the same information. If the process in $(\mathbf{p}, E)$-space is changed we obtain equivalent changes in the picture if the geometry (metric) remains constant. The Fourier transform, defined by Eqs. (5.11) and Eq. (5.12), works within Euclidean geometry.

Let us describe any process, say $A$, by $\psi(\mathbf{p}, E)_A$ and let us change this process. Then, the new process, say $B$, is described by another wavefunction $\psi(\mathbf{p}, E)_B \neq \psi(\mathbf{p}, E)_A$. The corresponding pictures, described in $(\mathbf{r}, t)$-space within Euclidean geometry, are obtained by Fourier transformations, and let us denote the wavefunctions by $\psi(\mathbf{r}, t)_A^{Eucl}$ and $\psi(\mathbf{r}, t)_B^{Eucl}$. Then, we come to the following scheme:

$$\begin{array}{ccc} \psi(\mathbf{p}, E)_A & \rightarrow & \psi(\mathbf{p}, E)_B \\ \downarrow & & \downarrow \\ \psi(\mathbf{r}, t)_A^{Eucl} & & \psi(\mathbf{r}, t)_B^{Eucl}. \end{array} \quad (5.73)$$

Such systematic changes, from $\psi(\mathbf{p}, E)_A$ to $\psi(\mathbf{p}, E)_B$ in the description above, particularly occur in the course of evolutionary developments. Thompson investigated such changes within $(\mathbf{r}, t)$-space. He was of course not familiar with $(\mathbf{p}, E)$-space and the projection principle (projection theory).

### 5.7.5.2. *Fishes*

Figures 5.2 and 5.3 show fishes (*Diodon*, *Orthagoriscus*, *Scarus sp* and *Pomacantus*) in $(\mathbf{r}, t)$-space, that is, these figures show organisms as we experience them in everyday life, and these experiences are made in an Euclidean space (Section 5.7.4.1). Within projection theory, each fish is produced by certain $(\mathbf{p}, E)$-fluctuations in $(\mathbf{p}, E)$-space, and this basic information is projected on $(\mathbf{r}, t)$-space and shows the organism within Euclidean geometry. This is symbolically shown for example,

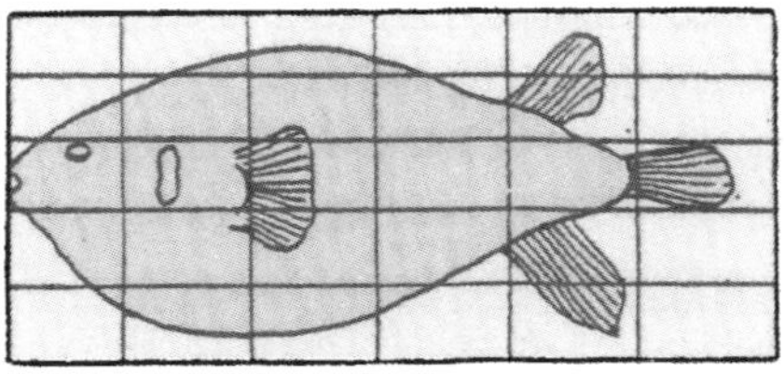

***Diodon***

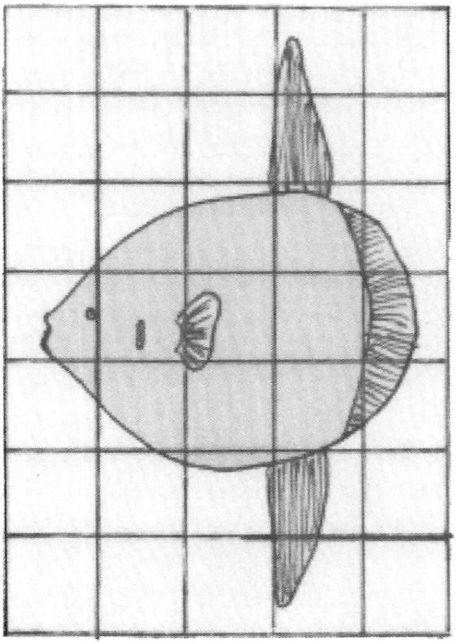

***Orthagoriscus***

Fig. 5.2. Two fish (*Diodon* and *Orthagoriscus*) as we experience them within the observations in everyday life (within Euclidean geometry). *Diodon* gradually changes within an evolutionary process resulting in *Orthagoriscus*. Both fish do really exist in nature.

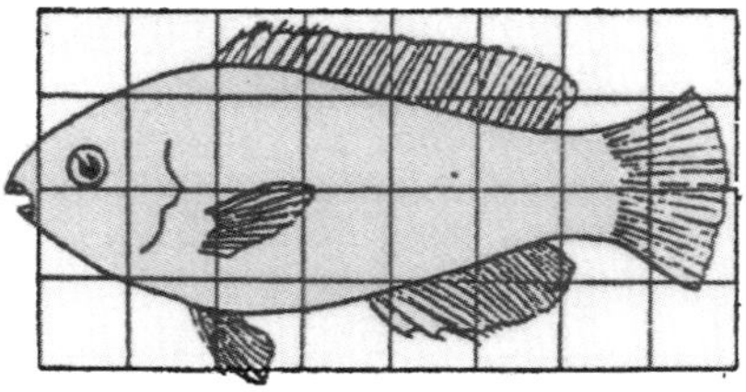

**Scarus sp**

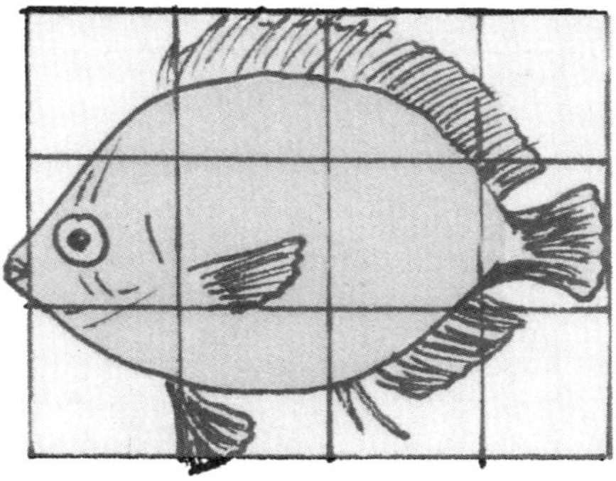

**Pomacanthus**

Fig. 5.3. Two fish (*Scarus sp* and *Pomacantus*) as we experience them within the observations in everyday life (within Euclidean geometry). *Scarus sp* gradually changes within an evolutionary process resulting in *Pomacantus*. Both fish do really exist in nature.

in Fig. 5.4 for *Diodon* and *Orthagoriscus* having, in analogy to Eq. (5.73), in $(\mathbf{p}, E)$-space the wavefunctions $\psi(\mathbf{p}, E)_{Diod}$ and $\psi(\mathbf{p}, E)_{Orth}$ and in $(\mathbf{r}, t)$-space the wavefunctions $\psi(\mathbf{r}, t)_{Diod}^{Eucl}$ and $\psi(\mathbf{r}, t)_{Orth}^{Eucl}$. The wavefunction $\psi(\mathbf{p}, E)_{Diod}$ $[\psi(\mathbf{r}, t)_{Diod}^{Eucl}]$ gradually changes during a continues evolutionary process and we finally obtain $\psi(\mathbf{p}, E)_{Orth}$ $[\psi(\mathbf{r}, t)_{Orth}^{Eucl}]$. These transitions that occur simultaneously can be expressed schematically as follows:

$$
\begin{aligned}
&\psi(\mathbf{p}, E)_{Diod} \xrightarrow{\textit{evolutionary process}} \psi(\mathbf{p}, E)_{Orth}, \\
&\psi(\mathbf{r}, t)_{Diod}^{Eucl} \xrightarrow{\textit{evolutionary process}} \psi(\mathbf{r}, t)_{Orth}^{Eucl}.
\end{aligned}
\tag{5.74}
$$

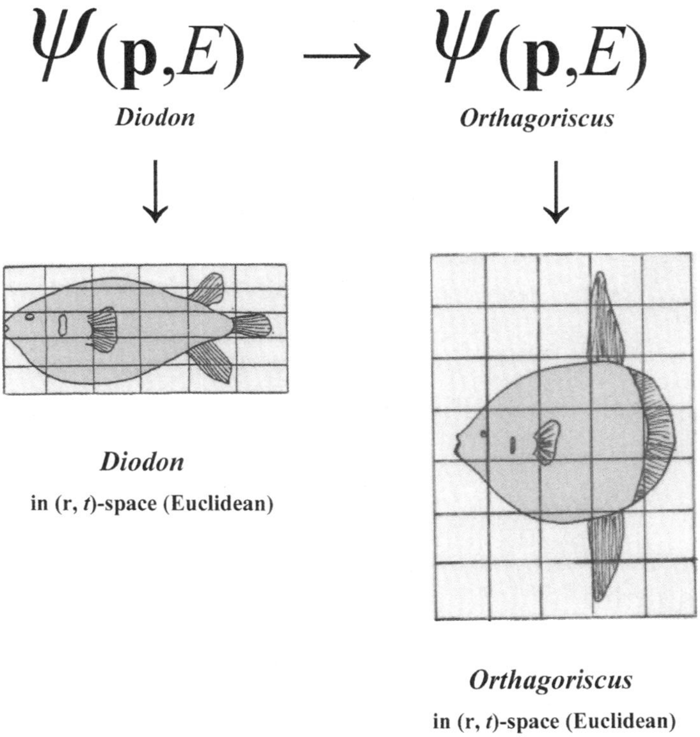


Fig. 5.4. *Projection theory*: *Diodon* (within Euclidean geometry) is produced by $(\mathbf{p}, E)$-fluctuations in $(\mathbf{p}, E)$-space, and we have the projection $\psi(\mathbf{p}, E)_{Diod} \to \psi(\mathbf{r}, t)^{Eucl}_{Diod}$. *Orthagoriscus* (within Euclidean geometry) is produced by $(\mathbf{p}, E)$-fluctuations in $(\mathbf{p}, E)$-space, and we have the projection $\psi(\mathbf{p}, E)_{Orth} \to \psi(\mathbf{r}, t)^{Eucl}_{Orth}$.

What did Thompson say? Thompson explained the evolutionary changes geometrically, that is, he changed the geometry (metric) in order to come from *Diodon* to *Orthagoriscus* and from *Scarus sp* to *Pomacantus*. In other words, he applied to *Diodon* and *Scarus sp* non-Euclidean geometry and got *Orthagoriscus* and *Pomacantus*. These Thompson-transitions [58] can be expressed schematically by

$$\psi(\mathbf{r}, t)^{Eucl}_{Diod} \xrightarrow{\text{evolutionary process}} \psi(\mathbf{r}, t)^{non\text{-}Eucl}_{Orth} \tag{5.75}$$

and

$$\psi(\mathbf{r}, t)^{Eucl}_{Scarus} \xrightarrow{\text{evolutionary process}} \psi(\mathbf{r}, t)^{non\text{-}Eucl}_{Poma}. \tag{5.76}$$

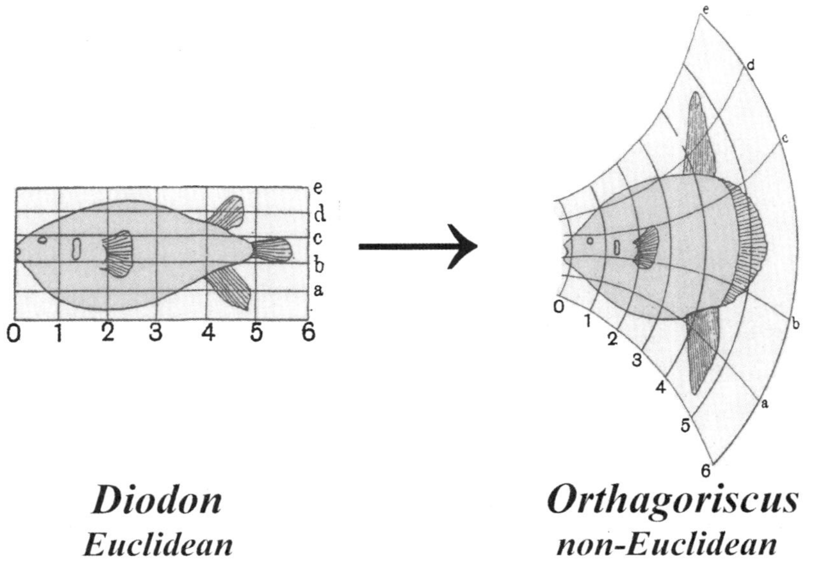


Fig. 5.5. Thompson-transition: *Diodon* → *Orthagoriscus* [Eq. (5.75)]. Reprinted with permission from [58].

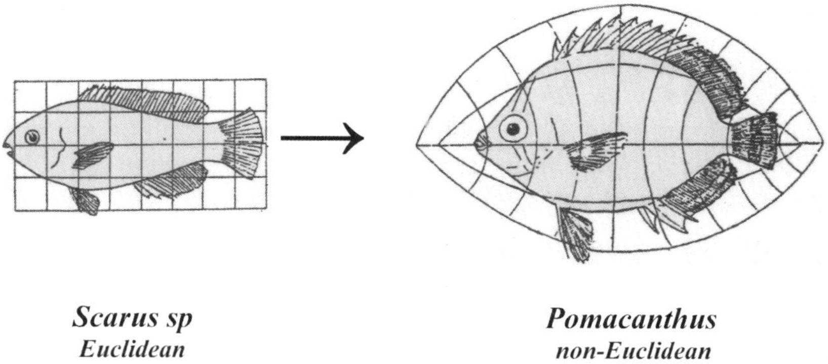


Fig. 5.6. Thompson-transition: *Scarus sp* → *Pomacantus* [Eq. (5.76)]. Reprinted with permission from [58].

The same facts are also represented in Figs. 5.5 and 5.6.

This discovery made by Thompson, summarized in (5.75), (5.76) and Figs. 5.5 and 5.6, is exactly what we expect from projection theory. Let us briefly discuss why.

In Figs. 5.4, the processes in $(\mathbf{p}, E)$-space are changed in order to get the right pictures of the fishes (within Euclidean geometry). In order to get *Orthagoriscus* from *Diodon*, the processes in $(\mathbf{p}, E)$-space are changed and we have the transition $\psi(\mathbf{p}, E)_{Diod} \rightarrow \psi(\mathbf{p}, E)_{Orth}$. This path is represented in Fig. 5.7.

However, because $(\mathbf{p}, E)$-space and $(\mathbf{r}, t)$-space are completely equivalent, we can also use another path, namely that over geometry

$$\psi(\mathbf{p},E) \rightarrow \psi(\mathbf{p},E)$$

***Diodon*** ***Orthagoriscus***

$$\downarrow$$

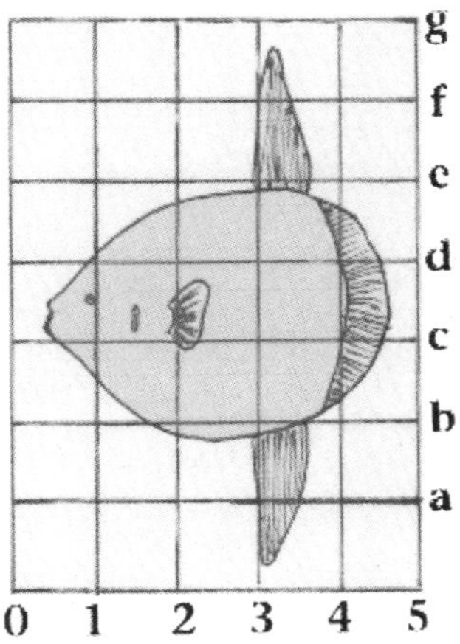


***Orthagoriscus***

(non-Euclidean)

Fig. 5.7. In the description of the production of *Orthagoriscus* from *Diodon* by evolutionary processes, we can work within $(\mathbf{p}, E)$-space, and we have the transition $\psi(\mathbf{p}, E)_{Diod} \rightarrow \psi(\mathbf{p}, E)_{Orth}$. Here there is a change of $(\mathbf{p}, E)$-fluctuations. This is one path of two.

changes. That is, we produce the picture of *Diodon* [$\psi(\mathbf{r},t)^{Eucl}_{Diod}$] by $\psi(\mathbf{p},E)_{Diod}$ within Euclidean geometry and after that we represent the same *Diodon* within non-Euclidean geometry and we get *Orthagoriscus* [$\psi(\mathbf{r},t)^{non\text{-}Eucl}_{Poma}$]. This second path (Fig. 5.8) is identical with the Thompson-transition (5.75).

Both paths are completely equivalent because $(\mathbf{p},E)$-space and $(\mathbf{r},t)$-space have exactly the same information contents. Note that the choice of geometry (Euclidean, non-Euclidean) is only a matter of convention. We have outlined this point in Section 5.7.3.

It is important to note that changes in the $(\mathbf{p},E)$-distribution are equivalent to changes in the geometry (metric). This behavior

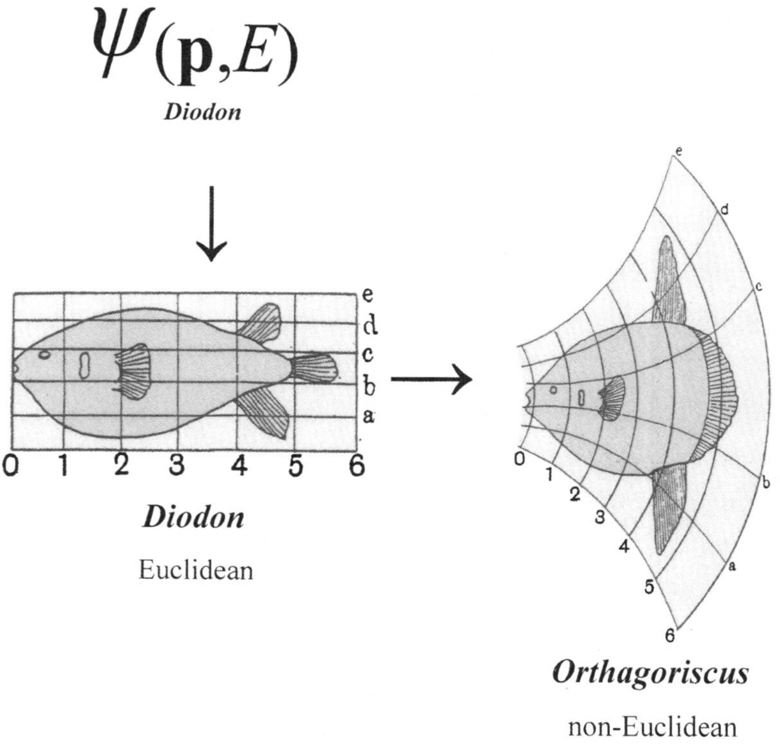


Fig. 5.8. In the description of the production of *Orthagoriscus* from *Diodon* by evolutionary processes we can also work, as alternative, within $(\mathbf{r},t)$-space, and we have the transition $\psi(\mathbf{r},t)^{Eucl}_{Scarus} \rightarrow \psi(\mathbf{r},t)^{non\text{-}Eucl}_{Poma}$. Here there is a change of geometry (metric). This path within $(\mathbf{r},t)$-space is identical with the Thompson-transition. This second path is completely equivalent to the first path discussed in Fig. 5.7.

is also observed within General Theory of Relativity (Section 5.7.1). Changes in the momentum — energy tensor $T_{\mu\nu}$ lead to changes in the geometry ($g_{\mu\nu}$).

It is obvious that these evolutionary processes together with Thompson's discovery of geometrical transitions support projection theory and all the details of it, in particular, what we have stated about space–time.

## 5.8. Flatness and Horizon Problem

### 5.8.1. *Flatness Problem*

Gauss found that the geometry of the space of our experience is Euclidean (Section 5.7.3). He surveyed a triangle formed by three mountains with sides measuring 69, 85, and 107 km. But our conception (projection theory) of space–time strictly requires that all space-regions, independent of their size, must be Euclidean. There is no space–time outside, and the space inside the observer is Euclidean for all processes that take place outside (Section 5.7.4), and this is because our cognition apparatus works on the basis of Euclidean geometry.

In fact, large-scale observations of the cosmic microwave background demonstrated that the geometry of the cosmos is Euclidean, i.e., it is flat to an accuracy of a few percent. This result supports convincingly projection theory, in particular, its space–time conception.

In connection with the flat geometry of the cosmos, General Theory of Relativity has a certain problem and needs *inflation theory* together with dark matter and dark energy but nobody knows what it is and where it is. Projection theory does not need such contested theories like inflation.

### 5.8.2. *Horizon Problem*

Inflation theory is an *ad hoc* solution not only for the explanation of the flatness of space but also for the *horizon problem*. On a large scale the cosmos appears statistically isotropic and homogeneous, and the whole universe is obviously in equilibrium. However, in the big bang model without inflation, the gravitational expansion does not give the early universe enough time to equilibrate. Different regions in space, that are widely separated, have not contacted each other because of the large distances between them. The value of the velocity of light is not large enough for an exchange of energy, heat, etc. Therefore these regions should not have the same temperature and other physical properties, but they have. This problem can be solved with the help of inflation theory, which suggests that there was a period of exponential expansion in the very early cosmos.

The situation within projection theory is quite different from the cosmological standard theory of usual physics. Let us demonstrate that by means of a simple one-dimensional model that describes the interaction between a specific system with its environment. Again, interaction exclusively means that there is an exchange of momentum **p** and energy $E$ between the system and the environment (momentum and energy are conserved during the interaction process). Let us assume that these fluctuations are given by

$$\begin{aligned} \psi(\mathbf{p}, E) &= A \exp\{-\alpha(E - E_0)^2\} \exp\{-bp_x^2\} \cos\left\{\frac{\alpha_0}{\hbar} p_x\right\} \\ &= A \exp\left\{-\alpha(E - E_0)^2\right\} \Psi(p_x). \end{aligned} \tag{5.77}$$

The Fourier transform with respect to $p_x$ leads to the equivalent information in $(\mathbf{r}, E)$-space (instead of $(\mathbf{r}, t)$-space we may also use $(\mathbf{r}, E)$-space, see [12]):

$$\begin{aligned} \psi(x, E) &= A \exp\{-\alpha(E - E_0)^2\} \left[ C \exp\left\{-\frac{(x - \alpha_0)^2}{4b\hbar^2}\right\} \right. \\ &\quad \left. + C \exp\left\{-\frac{(x + \alpha_0)^2}{4b\hbar^2}\right\} \right] \\ &= A \exp\{-\alpha(E - E_0)^2\} \Psi(x). \end{aligned} \tag{5.78}$$

The probability densities $\psi^*(p_x)\psi(p_x)$ and $\psi^*(x)\psi(x)$, based on Eqs. (5.77) and (5.78), are represented in Fig. 5.9.

The picture $\psi^*(x)\psi(x)$ consists of two parts (particle 1 and particle 2), and both parts have exactly the same properties: shape, width, height. There is no exchange of any physical signals through

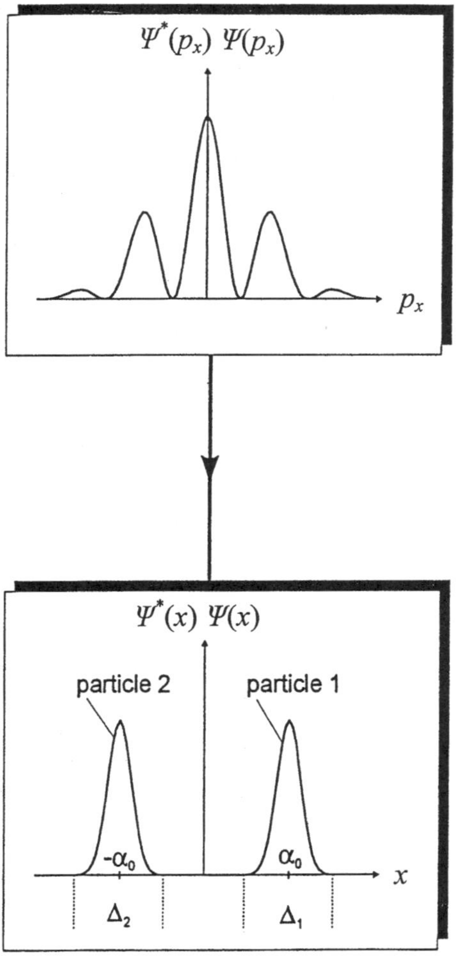


Fig. 5.9. On the cosmological horizon problem. Its solution within projection theory. More detail is given in the text.

space. Nevertheless they are identical, and this is due to the fact that particle 1 and particle 2 are produced synchronously by only *one process* in $(\mathbf{p}, E)$-space on the basis of $\psi(p_x)$ described by Eq. (5.77), but this single process produces two separate regions (geometrical pictures) in $(\mathbf{r}, t)$-space that are strongly correlated. In other words, particle 1 and particle 2 are not independent of each other but have to be considered as *one* system because they have been produced by only *one* process in

$(\mathbf{p}, E)$-space. In other words, we observe *non-local effects* in $(\mathbf{r}, t)$-space. More details concerning non-local effects within projection theory are given in Chapter 3 and [12].

At time $\tau$ we measure with a detector either in the space-region $\Delta_1$ (see Fig. 5.9) a certain momentum $p_x$ or we measure with a detector in the space-region $\Delta_2 = \Delta_1$ a certain momentum $p_x$, but we have never two events at time $\tau$, one in connection with particle 1 and another in connection particle 2. The time interval $\Delta\tau$ between such measurements, at different positions $\alpha_0$ and $-\alpha_0$, can be infinitesimal small even when the distance $d = 2\alpha_0$ between particle 1 and particle 2 are as large as the universe. In the case of independent particles we would have at time $\tau$ two events and of course two independent processes in $(\mathbf{p}, E)$-space.

All the properties just discussed are completely independent on the distance $d = 2\alpha_0$ between particle 1 and particle 2 and $d$ might be as large as the universe. (Note, that $\alpha_0$ plays the role of a frequency in $(\mathbf{p}, E)$-space where the processes take place.) Furthermore, the distance $d = 2\alpha_0$ can be dependent on $\tau$, i.e., we can have $d = d(\tau)$. Thus, we may have an expansion, a contraction with certain velocities or accelerations. In particular, an expansion process might start for $d \gg \Delta = \Delta_1 = \Delta_2$, and both particles will never be in contact.

In general, we have the following situation in projection theory: These *non-local effects* are entirely due to the projection process from reality $[(\mathbf{p}, E)$-space$]$ onto $(\mathbf{r}, t)$-space, like a flashlight on a screen. There is no preferred point in $(\mathbf{r}, t)$-space, say $\mathbf{r}_1$ and $t_1$, which would influence instantaneously all the other points $\mathbf{r}$, $t$ for which $\psi^*(\mathbf{r}, t)\psi(\mathbf{r}, t) \neq 0$ by actions through space–time. Since there are only projections, no information or material objects move through space–time, and no exchange of physically real signals through $(\mathbf{r}, t)$-space takes place. Exactly the same is the case for the simplified case $\psi^*(x)\psi(x)$.

## 5.9. Summary

Physical statements about the universe are presently formulated on the basis of the General Theory of Relativity and on quantum theory.

Quantum theory and the General Theory of Relativity lead to sets of laws that work fantastically. Quantum field theory being accurate to about one part in $10^{11}$, and general relativity has been tested to be correct to one part in $10^{14}$. However, if we put them together we inevitably obtain irreconcilable differences. Both theories are mutually exclusive. In this chapter, the situation is analyzed in connection with the cosmological constant because this constant connects quantum phenomena with space–time features of the General Theory of Relativity. It turned out that both theories have a serious vacuum (space–time) problem. Within quantum field theory the energy density of the vacuum is obviously much too large and, on the other hand, General Theory of Relativity does not fulfil Mach's principle.

It is usually assumed that the quantum theoretical value for the vacuum energy is wrong because the cosmological constant $\Lambda$ becomes unacceptable large in this case. Within the framework of these considerations we work within quantum field theory. Here the laws of usual quantum theory are combined with the laws of Special Theory of Relativity. In this connection it is important to note that within quantum field theory, no mechanism could be found and no reasonable procedure could be introduced to reduce the value for the vacuum energy. The search for a solution of this big problem already lasts approximately two decades but, up to now, without any success. Therefore, we should conclude that the vacuum energy problem is tightly linked with the very foundations of physics. Due to the unsuccessful investigations on lower physical levels makes the step to the top level probably necessary. Then, the following question arises. Are our conceptions of basic physics really realistic enough for the solutions of problems of this calibre? Just in connection with usual quantum theory a lot of questions have not been answered up to the present day. We discussed this point, in particular, in connection with time and we came to the following conclusion.

The required symmetry between space and time (Special Theory of Relativity) is definitely not fulfilled within the framework of usual quantum theory (in its relativistic and its non-relativistic version). In particular, time remains a simple parameter in conventional quantum mechanics. Thus, we may state that time is still a *classical* quantity within conventional quantum theory. From this point of view, usual quantum theory is not complete, and the relation $\Delta E \Delta \tau \geq \hbar/2$ is quite questionable. In Bunge's opinion, it is a total stranger in usual quantum theory. However, just this equation is responsible for the appearance of the vacuum energy leading to the problem in connection with the cosmological constant $\Lambda$.

In Section 5.3 we quoted some further critical points which appear in connection with usual quantum theory. The main facts can be summarized as follows: (1) Schrödinger's equation could not be deduced in usual quantum theory. (2) The *collapse of the wavefunction* cannot be explained in usual quantum theory. (3) The use of a particle defined as a local existent (which is assumed within the frame of usual quantum theory) seem also to be problematic.

To overcome this situation we came to the following conclusion: The real world cannot be embedded in space–time (*container principle*) but should be projected on it (*projection principle*), and we experience these projections as "pictures of reality."

The fundamentally new nature of time within projection theory is particularly reflected in the fact that the time $t$ becomes a system-specific property. Within conventional quantum theory, time $t = \tau$ is always a simple *external* parameter and has nothing to do with the process under investigation. Within projection theory we obtain a new uncertainty relation for the energy $E$ and time $t$ ($\delta E \delta t \geq \hbar/2$). In contrast to $\Delta E \Delta \tau \geq \hbar/2$, relation $\delta E \delta t \geq \hbar/2$ agrees completely in its physical content with the uncertainty relation for the momentum and the coordinate. The required symmetry between space and time that is required by Special Theory of Relativity is definitely fulfilled within projection theory but not in usual quantum theory (in its relativistic as well as in its non-relativistic version). In contrast to $\Delta E \Delta \tau \geq \hbar/2$, equation $\delta E \delta t \geq \hbar/2$ does not lead to temporary violations of the conservation law for the energy.

In connection with the role of the cosmological constant $\Lambda$, we have the following situation within projection theory. From specific estimations for the cosmological constant we may conclude that the theoretical prediction deviates from the observed value by a factor of $10^{122}$. The basis for the quantum field theoretical calculation of

$\Lambda$ is equation $\Delta E \Delta \tau \geq \hbar/2$, and this equation allows that the conservation law for the energy can be temporarily violated leading to the unacceptable (wrong) value for the cosmological constant $\Lambda$. Within projection theory we have, instead of $\Delta E \Delta \tau \geq \hbar/2$, the law $\delta E \delta t \geq \hbar/2$ and this changes the situation completely. Strictly speaking, a cosmological constant $\Lambda$ in the sense of General Theory of Relativity gives no sense within projection theory. The following points are relevant:

1. Within projection theory, space–time is not a container for the integration of physically real objects etc. Nothing can be inside space–time, no matter, no energy. In particular, there can be no vacuum energy. Space–time exclusively play the role of an auxiliary element for the representation of the picture of reality. Space–time can only contain geometrical positions, trajectories etc. as in the case of a sheet of paper.
2. The interactions exclusively take place in $(\mathbf{p}, E)$-space in form of $(\mathbf{p}, E)$-fluctuations where $\mathbf{p}$ and $E$ are strictly conserved. Again, a temporary violation of these conservation laws gives no sense within projection theory. In contrast to $\Delta E \Delta \tau \geq \hbar/2$, equation $\delta E \delta t \geq \hbar/2$ does not lead to temporary violations of the conservation law for the energy.

With respect to usual quantum theory the following question arise: Are the zero point energies of quantum fields real? The Casimir effect has often been used to demonstrate that the zero point energies of quantum fields are real. However, it is shown in [30] that the Casimir forces can be determined without reference to such zero point energies. "*In discussion with the cosmological constant, the Casimir effect is often invoked as decisive evidence that the zero point energies of quantum fields are 'real.' On the contrary, Casimir effects can be formulated and Casimir forces can be computed without reference to zero point energies.*" [30]

In Section 5.5, we discussed Mach's principle in connection with General Theory of Relativity. After that we made some principal remarks on the features of space and time. Here the role of the observer is relevant, and we have to distinguish between the *outside world* and the *inside world*. We came to the conclusion that space and time should be not elements of the outside world but are obviously located inside the observer's brain.

Reality is projected on the observer's cognition system leading to species-specific world views ("pictures of reality"). The observer constructs (consciously or unconsciously) such a picture on the basis of a "frame" which is in the case of a human observer given by our well-experienced space and time, but, in principle, this frame can be different for different biological systems. This view is confirmed by Poincaré. What geometry one chooses is, for Poincaré, a matter of convention. From this point of view, geometry has obviously no definite reality because one can choose that system of geometry which is convenient and enables to formulate the laws of nature in the simplest way. This is in perfect agreement with projection theory and is supported by the Thompson-transitions that we have discussed in Section 5.7.5.

The conception of space–time within projection theory strictly requires that all spaces-regions, independent of their size, must be flat (Euclidean): There is no space–time outside, and the space inside the observer is *flat* (Euclidean) for all processes that take place outside, and this is because our cognition apparatus works on the basis of Euclidean geometry. In fact, large-scale observations of the cosmic microwave background demonstrated that the geometry of the cosmos is Euclidean, i.e., it is flat to an accuracy of a few percent. Inflation theory or similar approaches are not needed within projection theory.

The *horizon problem* is also solved within projection theory by *non-local effects*. These non-local effects are entirely due to the projection process from reality [$(\mathbf{p}, E)$-space] onto $(\mathbf{r}, t)$-space, like a flashlight on a screen. There is no preferred point in $(\mathbf{r}, t)$-space, say $\mathbf{r}_1$ and $t_1$, which would influence instantaneously all the other points $\mathbf{r}, t$ for which $\psi^*(\mathbf{r}, t)\psi(\mathbf{r}, t) \neq 0$ by actions through space–time. Since there are only projections, no information or material objects move through space–time. In other words, there is no exchange of any physically real signal *through* space.

6

# Final Remarks

1. First, we have asked the following important question. Can the progress in science, in particular in physics, lead to a final truth of the physical world? It cannot! The absolute truth, which is so to speak embedded in basic reality, can never be observed and, therefore, theoretical considerations about it make no sense.

   The perception of complete reality in the sense of a precise reproduction implies that with growing fine structures, increasing information of the outside world is needed. Then, the evolution would have furnished the sense organs with the property to transmit as much information from the outside world as possible. But the opposite is correct: The strategy of nature is to take up as little information from reality as possible. Reality outside is not assessed by "complete" and "incomplete" but by "favorable towards life" and "hostile towards life".
2. The common or naive point of view assumes the following: the inside world which we feel to be outside us, actually exists in the outside world in exactly the same form as we perceive. According to this view there is only one difference between the inside world and the outside world. Inside there are only geometrical positions, whereas outside there are the real material bodies instead of the geometrical positions. In other words, it is normally assumed that the geometrical positions are merely replaced by material objects.

   But why should events in nature occur, so to speak, twice, one outside of us, and again in the form of a picture? This would be against the "principles of evolution" and the "principle of usefulness", respectively. It would make not much sense if events in a world, which is tailored to fulfil the principle of usefulness, would take place twice. There should be no one-to-one correspondence between the inside world (picture) and the outside world. Therefore, a visitor of a cinema does not get at the pay desk a small but true model of the cinema (a precise reproduction of the cinema) in order to find a certain place in the cinema. A simple cinema ticket with the essential information is more appropriate. In this respect, the cinema ticket is the picture of the cinema.

3. The facts from the theory of science have taught us that we will never be able to make final and complete statements about the physical universe, i.e., from the point of view of theory of science, the "absolute truth" can never be obtained. Why? Because a metric for the measurement of the "distance" between bodies of knowledge is not definable, and there is no possibility to express certain peculiarities of the absolute truth and basic (absolute) reality, respectively. An absolute reality may exist within theory of science but it is not accessible to man.

   The results that follow from the theory of science strongly confirm the statements, which we have worked out in connection with the principles of evolution and the principle of usefulness, respectively.

4. The consequent analysis showed that the material objects, which we observe in everyday life, are not embedded in space (space–time). Space and time have to be considered as auxiliary elements for the representation of the selected information about the outside world (basic reality). The chick experiment demonstrated that these pictures must be species-dependent, and this is confirmed by the principles of evolution. In other words, the physical reality is projected onto space and time, and the space–time elements $x, y, z, \tau$ cannot be seized with our five senses or with specific measuring instruments. These features of space and time must be considered in the formulation of physical theories. The projection principle has been applied to quantum phenomena, and the results of Chapter 3 can be summarized as follows.

5. The outside world (basic reality) is not accessible to the human observer. Therefore, the pictures in front of us are the most direct and essential event. Thus, we may state the following. Within the memory of man, all essential things are represented within

the frame of pictures that appear spontaneously in front of us. This is the most basic statement within projection theory. These pictures of reality are represented in $(\mathbf{r}, t)$-space and are described on the basis of processes which take place in fictitious reality. Each process is characterized by $(\mathbf{p}, E)$-fluctuations in $(\mathbf{p}, E)$-space (fictitious reality). The processes in $(\mathbf{p}, E)$-space are projected onto $(\mathbf{r}, t)$-space.

6. Each process and system, respectively, defines its own time-structure where the letter $t$ is the variable for describing these system-specific time structures. The quantity $t$ does not replace the time $\tau$, but the external parameter $\tau$ also exists in projection theory as reference time. In other words, projection theory extends the frame of description by $t$. Instead of $\mathbf{r}, \tau$ (conventional physics), within projection theory we have $\mathbf{r}, t, \tau$.
7. Not only does $\mathbf{r}$ act as a statistical variable but also $t$. Thus, the wavefunction of a system is defined in terms of the variables $\mathbf{r}, t$, i.e., we have $\Psi = \Psi(\mathbf{r}, t)$, and $\Psi(\mathbf{r}, t)$ can be dependent on $\tau$ or not. The probability distributions for $\mathbf{r}$ and $t$ can only be determined in connection with $\mathbf{p}$ and $E$ [the variables of $(\mathbf{p}, E)$-space)], and we come to the following statement: "*The measurement of one of the possible values for* $\mathbf{p}$ *and for* $E$ *at time* $\tau$ *is done in the space–time intervals* $\mathbf{r}$, $\mathbf{r} + d\mathbf{r}$ *and* $t$, $t + dt$ *with the probability density of* $\Psi^*(\mathbf{r}, t)\Psi(\mathbf{r}, t)$." Clearly, this instruction is more general than Born's probability description.
8. The system-specific time $t$ behaves statistically. In general, $t$ jumps arbitrarily from one $t$-position to another, and we cannot know when we observe an event in the past, present or in the future. In contrast to that, the reference time $\tau$, measured with our clocks, goes by definition always strictly from the past to the future. This feature of time $t$ is completely new. Even the notion "system-specific time" $t$ is not known in conventional quantum theory. Here only the reference time $\tau$ appears which however only plays the role of an external parameter.
9. Point-like particles with definite energy and definite momentum are not definable within the frame of projection theory. This "point-model", used in usual quantum theory, is obviously a

hold-over from classical mechanics and is mainly based on our observations in everyday life. Furthermore, there are non-local effects, that is, we observe in $(\mathbf{r}, t)$-space global instantaneous changes in the structure, and it is a feature of the projection that all space–time points $\mathbf{r}, t$ $(-\infty \leq \mathbf{r} \leq \infty, \ -\infty \leq t \leq \infty)$ are instantaneously involved. This effect is comparable with a flashlight on a screen.

10. The systematic $(\mathbf{p}, E)$-fluctuations in $(\mathbf{p}, E)$-space reflect processes and we have to consider $(\mathbf{p}, E)$-space as reality. However, because the variables $\mathbf{p}$ and $E$ are auxiliary elements, this reality is a "fictitious reality". Therefore, the $\mathbf{p}, E$-fluctuations (the interactions of the systems with other systems) are fictitious in character and merely simulate the real, but unknown processes in basic reality.

    Due to the conservation laws for the momentum $\mathbf{p}$ and the energy $E$, we must have systematic interaction processes [$(\mathbf{p}, E)$-fluctuations] of the system under investigation with other systems of the environment and, therefore, we have in fact to consider $(\mathbf{p}, E)$-space as (fictitious) reality. However, there do not exist such conservation laws for the variables $\mathbf{r}$ and $t$, i.e., the realized $(\mathbf{r}, t)$-point at time $\tau$ of the system under investigation is independent of the realized $(\mathbf{r}, t)$-points at the same time $\tau$ of the other systems of the environment. There is no exchange of certain "space–time pieces" between the pictures. In other words, there are no correlations between the $(\mathbf{r}, t)$-points at time $\tau$. This indicates that $\mathbf{r}$ and $t$ are in fact elements of a picture. In other words, the information of fictitious reality [$(\mathbf{p}, E)$-space] is projected onto $(\mathbf{r}, t)$-space by a Fourier transform and we get the "picture of reality".

11. Projection theory requires that non-interacting systems cannot exist in nature. Why does nature not admit such systems, i.e., free elementary systems without internal structure? Such free systems with constant momentum $\mathbf{p}_0$ and constant energy $E_0$ are in a certain sense "dead systems" because they are not involved in the processes in nature. There are no $(\mathbf{p}, E)$-fluctuations between a free system and its surroundings, and such systems would be useless.

Only interacting systems with non-vanishing $(\mathbf{p}, E)$-fluctuations $(\Delta\mathbf{p} \neq 0,\ \Delta E \neq 0)$ are relevant when we base our considerations on the "principle of usefulness". This feature is also a requirement of projection theory because within this theory, free (non-interacting) systems without substructure cannot exist. Thus, projection theory obviously contains the "principle of usefulness". Only processes are relevant because static systems are dead systems and cannot be involved in the scenario in nature. Like Whitehead and Bergson, we argue for the primacy of process.

12. The features of projection theory directly justify the introduction of "distance-independent communications". That is, there should also be interactions between systems possible that are not dependent on their mutual space-time distance. Such kind of interaction describes the form of objects (systems).

    Objects that do not communicate by distance-dependent interactions but only by distance-independent communications behave statistically. We showed that the space–time positions of the probability densities, which define the systems in $(\mathbf{r}, t)$-space, must behave statistically, i.e., these peaks jump arbitrarily through space and time. These jumps are independent from each other because the interaction between the systems is not dependent on their distance but we assumed that this interaction is only distance-independent and creates the form of the systems.

13. The physical picture within Newton's theory is quite different from that of projection theory. In classical mechanics, the $(\mathbf{p}, E)$-states are definitely changed when a body moves through space (relative to space) with varying velocity, and for the explanation of this scenario space–effects (inertia) are essential in Newton's mechanics. In particular, here a non-interacting body may move through space with a certain constant velocity, and this property directly comes into play by unrealistic space–effects (inertia) in Newton's theory. However, such kind of space–effect has to be considered as unphysical and has been strongly criticized by Ernst Mach and, in particular, also by Albert Einstein. Nevertheless, also within Special Theory of Relativity, and General Theory of Relativity the situation concerning space-effects could not really be

improved. More detail concerning Theory of Relativity are given in Chapter 5.

The situation within projection theory is quite different. Here Mach's principle is fulfilled and this is a quite general statement. The reason is obvious because "only" geometrical structures (objects) are embedded in $(\mathbf{r}, t)$-space, and $(\mathbf{r}, t)$-space plays the role of an auxiliary element for the representation of the world, that is, to form pictures of it. In other words, since there can be no material objects embedded in $(\mathbf{r}, t)$-space, material objects cannot be influenced by space and time $[(\mathbf{r}, t)$-space$]$. Therefore, Mach's principle is from the beginning fulfilled and must be strictly valid for all physical situations. In particular, we have recognized that in this case (Mach's principle is fulfilled) the geometrical objects may perform arbitrary jumps through space and time $[(\mathbf{r}, t)$-space$]$.

14. Within projection theory the interaction potential can be interpreted at best as an auxiliary element without any imaginable background. (There is no information or energy transfer through $(\mathbf{r}, t)$-space.) But within conventional physics, the notion "interaction" as an imaginable quality becomes questionable when we analyze this term in more detail.

    In contrast to projection theory, within conventional physics there exist some descriptive interpretations for the interaction (action-at-a-distance, proximity effect, principle of least action), which is here expressed by $U(x, y, z)$ and not by $V(x, y, z, t)$. We have briefly discussed these interpretations for $U(x, y, z)$ and their relevance in connection with the force laws as, for example, for the gravitational law $m_1 m_2/r^2$. However, this force law cannot be derived from one of the interpretations. This is a critical point.

    We came to the conclusion that these interpretations for the notion "interaction" cannot help us further to understand the mutual influence in connection with two or more interacting bodies. In particular, all these interpretations (action-at-a-distance, proximity effect, principle of least action) are not able to deduce the force law $m_1 m_2/r^2$. As in projection theory, we have to conclude that also in conventional physics the interaction law described by $m_1 m_2 / r^2$ is an abstract law.

In particular, there is no possibility to explain gravity by mechanical-corpuscular effects and, therefore, gravity has to be considered as a metaphysical element when the scientific standards are based on the mechanical-corpuscular world view. Gravity can be qualified as an occult element from this point of view. In this connection Kuhn remarked [10]: "*Gravity, interpreted as an innate attraction between every pair of particles of matter, was an occult quality in the same sense as the scholastics' "tendencies to fall" had been.*"

The situation in projection theory is quite different. Here we have no material objects in space–time [$(\mathbf{r}, t)$-space] but only geometrical structures. These geometrical structures are not static elements. There is an exchange of momentum and energy in $(\mathbf{p}, E)$-space with other systems, and an object comes into existence by these $(\mathbf{p}, E)$-fluctuations, which are projected onto $(\mathbf{r}, t)$-space, leading to the shape of the systems and to relative structural configurations between them.

There are $(\mathbf{r}, t)$-correlations in $(\mathbf{r}, t)$-space, but there is no exchange of information between the various geometrical structures, that is, nothing acts across $(\mathbf{r}, t)$-space as in the case of conventional physics ("proximity effect" and the "forces acting-at-the-distance"); the $(\mathbf{r}, t)$-correlations come exclusively into existence by the $(\mathbf{p}, E)$-fluctuations in $(\mathbf{p}, E)$-space. The $(\mathbf{r}, t)$-structures in $(\mathbf{r}, t)$-space are exclusively the result of projection processes and no geometrical part of $(\mathbf{r}, t)$-space is influenced by the others by actions through $(\mathbf{r}, t)$-space. From the point of view of the projection principle, such actions through space and time make no sense since $(\mathbf{r}, t)$-space merely contains pictures.

15. In contrast to usual physics, within projection theory a system-specific time $t$ is defined, that is, each system produces its own time scale. The time $\tau$ (measured with our clocks in everyday life) serves as reference time and has nothing to do with the system under investigation. Despite the statistical fluctuations, at time $\tau$ the whole of time $t$ — past, present and future — is laid out

frozen before us: $\Psi^*(\mathbf{r}, t)\Psi(\mathbf{r}, t)$, $-\infty < \mathbf{r}, t < \infty$. However, we do not observe the whole scenario $\Psi^*(\mathbf{r}, t)\Psi(\mathbf{r}, t)$ at time $\tau$, but only selected configurations of reality: $\Psi^*(\mathbf{r}, t_0)\Psi(\mathbf{r}, t_0)$, $-\infty < \mathbf{r} < \infty$, for $t_0 = \tau$. Each photograph represents such a configuration in space, at a certain time $\tau = t_0$. The transition from $\Psi^*(\mathbf{r}, t)\Psi(\mathbf{r}, t)$ to $\Psi^*(\mathbf{r}, t_0)\Psi(\mathbf{r}, t_0)$ must be due to an interplay between the system under investigation, described by $\Psi(\mathbf{r}, t)$, and the observers' function that is characterized by the reference time structure $\Psi_{ref}(t)$.

16. Within General theory of Relativity, paradoxes (contradictions) can appear in connection with time travel to the past. It turned out that the proposed physical solutions (many-world theory, argument of self-consistency, chronology protection conjecture) for eliminating the contradictions are not convincing.

    On the other hand, within projection theory time travel is principally possible without using relativistic elements. Due to the relationship between reference time $\tau$ and system-specific time $t$, the principal potentiality for time travel is given, i.e., for time travel to the future as well as to the past. The grandmother paradox and similar situations do not appear in projection theory. Here time travel becomes possible if we are able to decouple an individual reference time structure from the others. Time shifts with respect to the individual reference time is connected to specific changes of the individual brain functions.

    The reason why General Theory of Relativity leads to paradoxes in connection with time travel to the past is possibly due to the fact that this theory is not able to connect the biological structure of the time traveller with the basic elements of the relativistic framework.

17. In Chapter 4, we introduced levels of reality, and we could order the elements of reality as follows. Basic reality exists independently of the observer, and is in principle not accessible in a direct way. Basic reality is exclusively observable or describable by means of pictures on various levels, i.e., levels of reality. A picture-independent view is obviously not possible for a human observer. Space and time do not belong to basic reality but have to be considered as auxiliary elements for the representation

of the pictures and are, within projection theory, elements of the brain.

18. The hard objects (trees, stones, etc.) in front of us are states of the observers brain; we have only the impression that all these things are in front of us, i.e., outside of us. The hard objects, that we feel to be positioned in front of us, and the physical laws are equally states of the observer's brain. Nevertheless, they are different from each other and are therefore positioned on different levels.
19. All hard objects (material objects like trees, cars, atoms, molecules, etc.) are described by specific $(\mathbf{p}, E)$-processes in $(\mathbf{p}, E)$-space and always produce specific pictures (pattern) in $(\mathbf{r}, t)$-space. Such systems have to be considered as "states of matter". On the other hand, a general physical law [as, for example, $i\hbar\partial/\partial t\Psi(\mathbf{r}, t) = -\hbar^2/2m_0\Delta\Psi(\mathbf{r}, t) + V(x, y, z, t)\Psi(\mathbf{r}, t)$] is not depictable in $(\mathbf{r}, t)$-space. Therefore, such a law cannot be produced by $(\mathbf{p}, E)$-processes in $(\mathbf{p}, E)$-space. We have defined these general physical laws as "states of mind".
20. Like the hard (material) objects, the general physical laws have to considered as *real objects*. Physical laws are here *objects* in generalized form, more general than the hard objects. As we have already remarked above, both, the hard objects as well as the physical laws are within projection theory "states of the observer's brain".
21. The general physical law itself is not depicable in $(\mathbf{r}, t)$-space [$(\mathbf{p}, E)$-space)]. Only specific solutions (in connection with modelling) for $\Psi(\mathbf{r}, t)^{\text{specific}}$ [$\Psi(\mathbf{p}, E)^{\text{specific}}$] can be pictured. Another kind of representation should be considered for the abstract general physical law $i\hbar\partial/\partial t\Psi(\mathbf{r}, t) = -\hbar^2/(2m_0)\Delta\Psi(\mathbf{r}, t) + V(x, y, z, t)\Psi(\mathbf{r}, t)$ which contains potentially the characteristics of all possible physical systems, i.e., all possible functions $V(\mathbf{r}, t)$, and this feature can be expressed symbolically by $\Psi = \Psi(V)$.
22. Any specific model is captured in $(\mathbf{r}, t)$-space [$(\mathbf{p}, E)$-space], that is, within the framework of a specific model one cannot leave $(\mathbf{r}, t)$-space [$(\mathbf{p}, E)$-space] in order to describe a general physical law which is not despictable in $(\mathbf{r}, t)$-space. Therefore, both levels can principally not be joined. However, we may conclude a specific

model from the general physical law, but we cannot conclude the general physical law from the specific model. The "concept of induction" is not valid. Projection theory leads to the "principle of deduction".

23. Space–time $[(\mathbf{r}, t)$-space$]$ is tightly-linked to $(\mathbf{p}, E)$-space (and vice versa), and the statements, formulated above, particularly mean that the "object" physical law cannot be the outcome of $(\mathbf{p}, E)$-fluctuations as in the case of the hard objects in front of us. While the specific $(\mathbf{p}, E)$-processes in $(\mathbf{p}, E)$-space produce specific pictures in $(\mathbf{r}, t)$-space with exactly the same information, the physical law is not depictable in $(\mathbf{r}, t)$-space and, therefore, a general physical law cannot be produced by $(\mathbf{p}, E)$-processes, i.e., in connection with physical (material) brain functions.
24. Within projection theory, there is the following strict law. If there are no $(\mathbf{p}, E)$-fluctuations in $(\mathbf{p}, E)$-space, there cannot be a physical picture in $(\mathbf{r}, t)$-space. A picture for a physically real process is not defined without $(\mathbf{p}, E)$-fluctuations and vice versa! Then, by definition, systems that are not depictable in $(\mathbf{r}, t)$-space are not physical systems (are not material objects).
25. Clearly, we can have pictures in $(\mathbf{r}, t)$-space without $(\mathbf{p}, E)$-fluctuations, but these kind of pictures are not physical representations but merely drawings and belong to the "products of imagination". Such drawings (for example, an art work) may be arbitrary and must not satisfy physical laws.
26. Because a physical law is more general than the specific $(\mathbf{p}, E)$-processes (the characteristics of all possible $(\mathbf{p}, E)$-processes are equally dictated by the physical law), we may conclude that this physical law is positioned at a higher level than that on which the models for these specific $(\mathbf{p}, E)$-processes are positioned.
27. Therefore, a physical law (formula) cannot be described by specific models (physical processes). The attempt to describe the products of mind by certain physical brain functions (neuronal activities) has no basis within projection theory. The reason for that is simple and has been worked out in Chapter 4. Any physical process, also those in connection with brain functions, is described by $(\mathbf{p}, E)$-fluctuations in $(\mathbf{p}, E)$-space) and produces a concrete

picture in $(\mathbf{r}, t)$-space. In the *reverse case*, we may argue as follows. Since the physical law $i\hbar\partial/\partial t\Psi(\mathbf{r}, t) = -\hbar^2/(2m_0)\Delta\Psi(\mathbf{r}, t) + V(x, y, z, t)\Psi(\mathbf{r}, t)$ (or other states of the mind) does not produce a concrete picture in $(\mathbf{r}, t)$-space, the general physical law cannot be the result of physical brain functions [described by $(\mathbf{p}, E)$-fluctuations].

28. Since the origin of feelings and thinking or the products of mind cannot be described as pictures in $(\mathbf{r}, t)$-space, their origin cannot be based on physical laws. The physical laws exclusively describe the hard objects (the objects in front of us but also atoms, molecules, etc.) in $(\mathbf{r}, t)$-space and exclusively come into play by the interactions [$(\mathbf{p}, E)$-fluctuations] in $(\mathbf{p}, E)$-space. However, the physical law itself is not depictable in $(\mathbf{r}, t)$-space, as we have outlined in detail.
29. In Chapter 5, we discussed the cosmological constant within conventional physics as well as within projection theory. Here is a summary of the main results: It is usually assumed that the quantum theoretical value for the vacuum energy is wrong because the cosmological constant $\Lambda$ becomes unacceptably large in this case. Within the framework of these considerations, we work within quantum field theory. Here the laws of usual quantum theory are combined with the laws of the Special Theory of Relativity. In this connection it is important to note that within quantum field theory no mechanism could be found and no reasonable procedure could be introduced to reduce the value for the vacuum energy. The search for a solution of this big problem already lasts approximately two decades but, up to now, without any success. Therefore, we should conclude that the vacuum energy problem is tightly linked with the very foundations of physics. The unsuccessful investigations on lower physical levels makes the step to the top level probably necessary.

    Then, the following question arises: are our conceptions of basic physics really realistic enough for the solution of problems of this calibre? It turned out that quantum theory and General Theory of Relativity have a serious vacuum (space–time) problem. Within quantum field theory the energy density of the vacuum is obviously much too large and, on the other hand, General Theory of Relativity does not fulfil Mach's principle. Just in connection with usual quantum theory, a lot of questions have not been answered up to the present day. We discussed this point, in particular, in connection with time and we came to the following conclusion.

The required symmetry between space and time (Special Theory of Relativity) is definitely not fulfilled within the framework of usual quantum theory (in its relativistic and its non-relativistic version). In particular, time remains a simple parameter in conventional quantum mechanics. Thus, we may state that time is still a *classical* quantity within conventional quantum theory. From this point of view, usual quantum theory is not complete, and the relation $\Delta E \Delta \tau \geq \hbar/2$ is quite questionable. In Bunge's opinion, it is a total stranger in usual quantum theory. However, just this equation is responsible for the appearance of the vacuum energy leading to the problem in connection with the cosmological constant $\Lambda$.

30. The fundamentally new nature of time within projection theory is particularly reflected in the fact that the time $t$ becomes a system-specific property. Within conventional quantum theory time $t = \tau$ is always a simple *external* parameter and has nothing to do with the process under investigation.

Within projection theory, we obtain a new uncertainty relation for the energy $E$ and time $t$ ($\delta E \delta t \geq \hbar/2$). In contrast to $\Delta E \Delta \tau \geq \hbar/2$, relation $\delta E \delta t \geq \hbar/2$ agrees completely in its physical content with the uncertainty relation for the momentum and the coordinate. The required symmetry between space and time that is required by Special Theory of Relativity is definitely fulfilled within projection theory but not in usual quantum theory (in its relativistic as well as in its non-relativistic version). In contrast to $\Delta E \Delta \tau \geq \hbar/2$, equation $\delta E \delta t \geq \hbar/2$ does not lead to temporary violations of the conservation law for the energy. Within projection theory, we have the following situation.

31. In connection with the role of the cosmological constant $\Lambda$ within projection theory we have the following situation. From specific estimations for the cosmological constant, we may conclude that

the theoretical prediction deviates from the observed value by a factor of $10^{122}$. The basis for the quantum field theoretical calculation of $\Lambda$ is equation $\Delta E \Delta \tau \geq \hbar/2$, and this equation allows that the conservation law for the energy can be temporarily violated leading to the unacceptable (wrong) value for the cosmological constant $\Lambda$. Within projection theory we have, instead of $\Delta E \Delta \tau \geq \hbar/2$, the law $\delta E \delta t \geq \hbar/2$ and this changes the situation completely. Strictly speaking, a cosmological constant $\Lambda$ in the sense of General Theory of Relativity gives no sense within projection theory. The following points are relevant.

(i) Within projection theory, space–time is not a container for the integration of physically real objects etc. Nothing can be inside space–time, no matter, no energy. In particular, there can be no vacuum energy. Space–time exclusively plays the role of an auxiliary element for the representation of the picture of reality. Space–time can only contain geometrical positions, trajectories etc. as in the case of a sheet of paper.

(ii) The interactions exclusively take place in $(\mathbf{p}, E)$-space in form of $(\mathbf{p}, E)$-fluctuations where $\mathbf{p}$ and $E$ are strictly conserved. Again, a temporary violation of these conservation laws gives no sense within projection theory. In contrast to $\Delta E \Delta \tau \geq \hbar/2$, equation $\delta E \delta t \geq \hbar/2$ does not lead to temporary violations of the conservation law for the energy.

32. Reality is projected on the observer's cognition system leading to species-specific world views ("pictures of reality"). The observer constructs (consciously or unconsciously) such a picture on the basis of a "frame" which is in the case of a human observer given by our well-experienced space and time, but, in principle, this frame can be different for different biological systems. This view is confirmed by Poincaré. What geometry one chooses is, for Poincaré, a matter of convention. From this point of view, geometry has obviously no definite reality because one can choose that system of geometry which is convenient and enables us to formulate the laws of nature in the simplest way. This is in perfect agreement with projection theory and is supported by the Thompson-transitions that we have discussed in Sec. 5.7.5.

33. The conception of space–time within projection theory strictly requires that all space-regions, independent of their size, must be flat (Euclidean): There is no space–time outside, and the space inside the observer is *flat* (Euclidean) for all processes that take place outside, and this is because our cognition apparatus works on the basis of Euclidean geometry. In fact, large-scale observations of the cosmic microwave background demonstrated that the geometry of the cosmos is Euclidean, i.e., it is flat to an accuracy of a few percent. Inflation theory or similar approaches are not needed within projection theory.

    The *horizon problem* is also solved within projection theory by *non-local effects.* These non-local effects are entirely due to the projection process from reality [$(\mathbf{p}, E)$-space] onto $(\mathbf{r}, t)$-space, like a flashlight on a screen. There is no preferred point in $(\mathbf{r}, t)$-space, say $\mathbf{r}_1$ and $t_1$, which would influence instantaneously all the other points $(\mathbf{r}, t)$ for which $\psi^*(\mathbf{r}, t)\psi(\mathbf{r}, t) \neq 0$ by actions through space–time. Since there are only projections, no information or material objects move through space–time. In other words, there is no exchange of any physically real signal *through* space.

# Appendix A
# On the Registration of Events within Projection Theory

Let us suppose that there is an interaction between the system under investigation and the measuring instrument (detector) defined. How can we understand this measuring process within projection theory? Here we would like to study the principal features.

## A.1. Equivalencies

For the understanding of the measurement of certain events in connection with the system under investigation, two features are relevant:

Feature 1:

Both spaces, $(\mathbf{r}, t)$-space and $(\mathbf{p}, E)$-space, are equivalent concerning their information content. A certain situation in $(\mathbf{r}, t)$-space has its correspondence in $(\mathbf{p}, E)$-space.

Feature 2:

In connection with the Fourier transform, which transforms the information from $(\mathbf{r}, t)$-space to $(\mathbf{p}, E)$-space and vice versa, the following feature is relevant. In order to get the information of the system under investigation at a certain space–time point, say $\mathbf{r}, t$, we need all the information of $(\mathbf{p}, E)$-space. This has been symbolically expressed in Section 3.3 by

$$\mathbf{r}, t \leftarrow (\mathbf{p}, E). \tag{A.1}$$

There is an equivalence between the point $\mathbf{r}, t$ and the distribution $(\mathbf{p}, E)$, which can be expressed by

$$\mathbf{r}, t \Leftrightarrow (\mathbf{p}, E). \tag{A.2}$$

Features 1 and 2 are needed in connection with the registration of detector-events. Let us assume that the system is described by $\Psi(\mathbf{r}, t)$ in $(\mathbf{r}, t)$-space and, on the other hand, by $\Psi(\mathbf{p}, E)$ in $(\mathbf{p}, E)$-space. Then we may argue as follows.

The position of the detector at the point $\mathbf{r}, t$ in $(\mathbf{r}, t)$-space reflects a certain relationship between the detector and the system under investigation which is characterized by $\Psi(\mathbf{r}, t)$ $[\Psi^*(\mathbf{r}, t)\Psi(\mathbf{r}, t)]$ at the space–time position $\mathbf{r}, t$; the detector must be adjusted to the $(\mathbf{r}, t)$-configuration of $\Psi(\mathbf{r}, t)[\Psi^*(\mathbf{r}, r)\Psi(\mathbf{r}, t)]$. Due to Feature 1, this picture corresponds to an equivalent situation in $(\mathbf{p}, E)$-space. There is a relationship between the detector [represented as a $(\mathbf{p}, E)$-structure] and a specific distribution $(\mathbf{p}, E)$, which corresponds to the position $\mathbf{r}, t$ of the detector in $(\mathbf{r}, t)$-space. See in particular Feature 2 and Fig. A.1. The range of the distribution $(\mathbf{p}, E)$ is defined by the range of the function $\Psi^*(\mathbf{p}, E)\Psi(\mathbf{p}, E)$ with $-\infty \leq \mathbf{p}, E \leq \infty$ that characterizes the system.

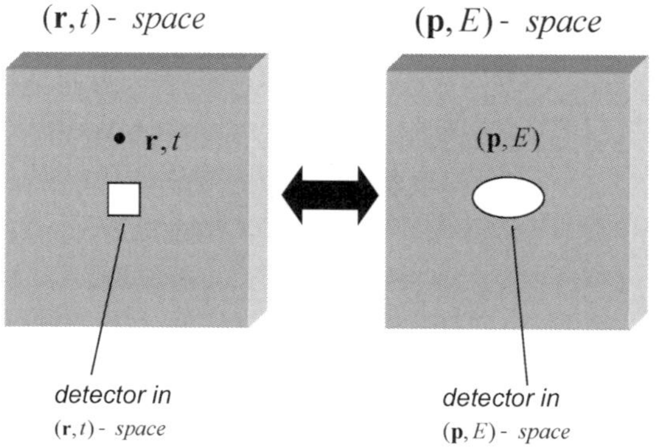


Fig. A.1. Both spaces, $(\mathbf{r}, t)$-space and $(\mathbf{p}, E)$-space, are equivalent concerning their physical information. The situation in $(\mathbf{r}, t)$-space is equivalent to that in $(\mathbf{p}, E)$-space. Therefore, the relationship between the system and the detector, positioned at the point $\mathbf{r}, t$ in $(\mathbf{r}, t)$-space, is equivalent to an interaction of the detector [represented as a $(\mathbf{p}, E)$-structure] with a distribution $(\mathbf{p}, E)$ in $(\mathbf{p}, E)$-space [Eqs. (A.1) and (A.2)].

This relationship in $(\mathbf{p}, E)$-space means that there is an "interaction" between the detector and the system under investigation described in $(\mathbf{p}, E)$-space by $\Psi(\mathbf{p}, E)$. In projection theory, the interaction is expressed by $(\mathbf{p}, E)$-fluctuations, that is, there is an exchange of momentum and energy between the detector and the system, which can be symbolized by $\Delta\mathbf{p}_{\text{det}}, \Delta E_{\text{det}}$. Due to the interaction between the detector and the system, the state of the detector is changed.

In conclusion, at reference time $\tau$ the detector is positioned in $(\mathbf{r}, t)$-space at the point $\mathbf{r}, t$, and there is a relation between the detector and the system which is characterized by $\Psi(\mathbf{r}, t)$ $[\Psi^*(\mathbf{r}, t)\Psi(\mathbf{r}, t)]$ at this space–time position $\mathbf{r}, t$; the detector must be adjusted to the $(\mathbf{r}, t)$-configuration of $\Psi(\mathbf{r}, t)[\Psi^*(\mathbf{r}, r)\Psi(\mathbf{r}, t)]$. An equivalent situation to this $(\mathbf{r}, t)$-space configuration is given in $(\mathbf{p}, E)$-space having the following characteristics (see also Fig. A.1). The same detector, here represented as structure, has a certain relation to the distribution $(\mathbf{p}, E)$ that reflects an interaction $[(\mathbf{p}, E)$-fluctuations$]$ between the detector and the system with $\Psi(\mathbf{p}, E)$, which takes at reference time $\tau$ one of the possible values for the momentum $\mathbf{p}$ and the energy $E$ of the distribution $(\mathbf{p}, E)$ that is given with a certain probability density $\Psi^*(\mathbf{p}, E)\Psi(\mathbf{p}, E)$.

This situation, that is, the detector and its changes due to the interaction of the detector with the system under investigation, is projected onto $(\mathbf{r}, t)$-space. In this way the observer registers in $(\mathbf{r}, t)$-space the changes of the detector due to its interaction with the system under investigation in $(\mathbf{p}, E)$-space.

## A.2. No Material Objects in Space (Space–Time)

It must be emphasized that we do not need in projection theory for the interpretation of the wavefunction $\Psi(\mathbf{r}, t)$ a point-like material particle that is embedded in space and time $[(\mathbf{r}, t)$-space$]$. Such kind of particle may not exist in projection theory because all processes take place in $(\mathbf{p}, E)$-space. Born's interpretation of conventional quantum theory is based on this assumption, that is, a material object is needed which is embedded in space. However, Born's construction is not free of problems. In [34], a brief critical discussion is given.

Let us repeat the situation in projection theory. Within projection theory, measurement means that the properties of $(\mathbf{r}, t)$-space and $(\mathbf{p}, E)$-space are connected and, as we have already outlined in Section 3.2.2, this leads to the following picture. The measurement of one of the possible values for $\mathbf{p}$ and for $E$ is done at time $\tau$ in the space–time intervals $\mathbf{r}, \mathbf{r} + d\mathbf{r}$ and $t, t + dt$ with the probability density of $\Psi^*(\mathbf{r}, t)\Psi(\mathbf{r}, t)$ (Section 3.10.1). Again, this statement means that we are working simultaneously on the basis of two spaces [$(\mathbf{r}, t)$-space and $(\mathbf{p}, E)$-space].

# Appendix B
# The Classical Case

Let us define the classical case as follows. The variables $\mathbf{p}, E, \mathbf{r}, t$ are given at each time $\tau$ without any uncertainty. That is, instead of the quantum-theoretical probability distributions $(\mathbf{p}), (E), (\mathbf{r}), (t)$, we have simultaneously the definite values $\mathbf{p}, E, \mathbf{r}, t$ at time $\tau$, and this feature defines the classical case. In other words, the transition from the quantum-theoretical description to classical considerations is given by

$$\tau : (\mathbf{p}), (E), (\mathbf{r}), (t) \rightarrow \mathbf{p}, E, \mathbf{r}, t. \tag{B.1}$$

For example, at $\tau_0$ we have

$$\begin{aligned} \mathbf{p} = \mathbf{p}_0, E = E_0, \\ \mathbf{r} = \mathbf{r}_0, t = t_0. \end{aligned} \tag{B.2}$$

## B.1. Probability Densities

At each time $\tau$ the normalization integral for $(\mathbf{p}, E)$-space is defined through [12]

$$\int_{-\infty}^{\infty} \Psi^*(\mathbf{p}, E)\Psi(\mathbf{p}, E) dp_x dp_y dp_z dE = 1. \tag{B.3}$$

Using (B.2) we get

$$\int_{-\infty}^{\infty} \Psi^*(\mathbf{p}_0, E_0)\Psi(\mathbf{p}_0, E_0) dp_x dp_y dp_z dE = 1 \tag{B.4}$$

with

$$\Psi^*(\mathbf{p}_0, E_0)\Psi(\mathbf{p}_0, E_0)\, C = 1 \tag{B.5}$$

and

$$C = \int_{\mathbf{p}_0, E_0} dp_x dp_y dp_z dE$$

$$= \lim_{\Delta_\mathbf{p}, \Delta_E \to 0} \int_{\mathbf{p}_0, E_0}^{\mathbf{p}_0 + \Delta_\mathbf{p}, E + \Delta_E} dp_x dp_y dp_z dE \to 0, \tag{B.6}$$

$$\Delta_\mathbf{p} = (\Delta_{p_x}, \Delta_{p_y}, \Delta_{p_z}). \tag{B.7}$$

Then, we have with (B.5)

$$\Psi^*(\mathbf{p}_0, E_0)\Psi(\mathbf{p}_0, E_0) \to \infty \tag{B.8}$$

and

$$\Psi(\mathbf{p}_0, E_0) \to \infty. \tag{B.9}$$

Exactly the same procedure holds for $(\mathbf{r}, t)$-space. Here the normalization integral is defined at each time $\tau$ by [12]

$$\int_{-\infty}^{\infty} \Psi^*(\mathbf{r}, t)\Psi(\mathbf{p}, E) dxdydzdt = 1. \tag{B.10}$$

Using (B.2) we obtain

$$\int_{-\infty}^{\infty} \Psi^*(\mathbf{r}_0, t_0)\Psi(\mathbf{p}_0, E_0) dxdydzdt = 1 \tag{B.11}$$

with

$$\Psi^*(\mathbf{r}_0, t_0)\Psi(\mathbf{p}_0, E_0) D = 1 \tag{B.12}$$

and

$$D = \int_{\mathbf{r}_0, t_0} dxdydzdt = \lim_{\Delta_\mathbf{r}, \Delta_t \to 0} \int_{\mathbf{r}_0, t_0}^{\mathbf{r}_0 + \Delta_\mathbf{r}, t + \Delta_t} dxdydzdt \to 0, \tag{B.13}$$

$$\Delta_\mathbf{r} = (\Delta_x, \Delta_y, \Delta_z). \tag{B.14}$$

Then, we have with (B.11)

$$\Psi^*(\mathbf{r}_0, t_0)\Psi(\mathbf{p}_0, E_0) \to \infty \tag{B.15}$$

and

$$\Psi(\mathbf{r}_0, t_0) \to \infty. \tag{B.16}$$

In other words, both functions, $\Psi(\mathbf{p}_0, E_0)$ and $\Psi(\mathbf{r}_0, t_0)$, have to be simultaneously infinite.

## B.2. Probabilities

The probability $\Delta W$ of finding the system at time $\tau_0$ in the state $\mathbf{p} = \mathbf{p}_0, E = E_0$ is expressed by

$$\Delta W = \Psi^*(\mathbf{p}_0, E_0)\Psi(\mathbf{p}_0, E_0) \lim_{\Delta_\mathbf{p}, \Delta_E \to 0} \Delta_{p_x}\Delta_{p_y}\Delta_{p_z}\Delta_E = 1, \tag{B.17}$$

where (B.5) and (B.6) have been used. In other words, at time $\tau_0$ the system is with certainty in the state $\mathbf{p} = \mathbf{p}_0, E = E_0$. This is the case for any time $\tau$.

On the other hand, the probability $\Delta W$ of finding the system at time $\tau_0$ in the state $\mathbf{r} = \mathbf{r}_0, t = t_0$ is expressed by

$$\Delta W = \Psi^*(\mathbf{r}_0, t_0)\Psi(\mathbf{r}_0, t_0) \lim_{\Delta_\mathbf{r}, \Delta_t \to 0} \Delta_x\Delta_y\Delta_z\Delta_t = 1, \tag{B.18}$$

where (B.12) and (B.13) have been used. From Eq. (B18) directly follows that the system is at time $\tau_0$ with certainty in the state $\mathbf{r} = \mathbf{r}_0, t = t_0$. This is the case for any time $\tau$.

## B.3. Transformations in the Classical Limit

The relationship between $\Psi(\mathbf{r}, t)$ and $\Psi(\mathbf{p}, E)$ are given by the Fourier-transform (Chapter 3)

$$\Psi(\mathbf{r}, t) = \frac{1}{(2\pi\hbar)^2} \int_{-\infty}^{\infty} \Psi(\mathbf{p}, E) \exp\left\{ i\left[\frac{\mathbf{p}}{\hbar}\cdot\mathbf{r} - \frac{E}{\hbar}t\right]\right\} dp_x dp_y dp_z dE. \tag{B.18}$$

Using Eq. (B.2) we get

$$\Psi(\mathbf{r}_0, t_0) = \frac{1}{(2\pi\hbar)^2}\Psi(\mathbf{p}_0, E_0) \exp\left\{ i\left[\frac{\mathbf{p}_0}{\hbar}\cdot\mathbf{r}_0 - \frac{E_0}{\hbar}t_0\right]\right\} C, \tag{B.19}$$

where the parameter $C$ is expressed by (B.6).

On the other hand, the inverse transformation

$$\Psi(\mathbf{p}, E) = \frac{1}{(2\pi\hbar)^2} \int_{-\infty}^{\infty} \Psi(\mathbf{r}, t) \exp\left\{-i\left[\frac{\mathbf{p}}{\hbar}\cdot\mathbf{r} - \frac{E}{\hbar}t\right]\right\} dtdxdydz \tag{B.20}$$

leads to

$$\Psi(\mathbf{p}_0, E_0) = \frac{1}{(2\pi\hbar)^2} \Psi(\mathbf{r}_0, t_0) \exp\left\{-i\left[\frac{\mathbf{p}_0}{\hbar}\cdot\mathbf{r}_0 - \frac{E_0}{\hbar}t_0\right]\right\} D, \tag{B.21}$$

where the parameter $D$ is expressed by (B.13).

Discussion

Let us study (B.19) and (B.21) for $\hbar \rightarrow 0$. We will recognize that only in this case the conditions (B.9) and (B.16) can be fulfilled.

1. The terms

$$\exp\left\{i\left[\frac{\mathbf{p}_0}{\hbar}\cdot\mathbf{r}_0 - \frac{E_0}{\hbar}t_0\right]\right\} \tag{B.22}$$

and

$$\exp\left\{-i\left[\frac{\mathbf{p}_0}{\hbar}\cdot\mathbf{r}_0 - \frac{E_0}{\hbar}t_0\right]\right\} \tag{B.23}$$

[see (B.19) and B.21)] remain uncertain in the case of $\hbar \rightarrow 0$ and can take values between 1 and $-1$ for both, the real and the imaginary part.

2. Only in the case $\hbar \rightarrow 0$ the function $\Psi(\mathbf{r}_0, t_0)$ becomes infinite when $\Psi(\mathbf{p}_0, E_0)$ is infinite [see (B.19)], as is required by Eq. (B.16). Otherwise (if $\hbar$ is not zero) the function $\Psi(\mathbf{r}_0, t_0)$ remains uncertain: $\Psi(\mathbf{r}_0, t_0) = 0 \cdot \infty$.

   Furthermore, only in the case $\hbar \rightarrow 0$, the function $\Psi(\mathbf{p}_0, E_0)$ becomes infinite when $\Psi(\mathbf{r}_0, t_0)$ is infinite [see (B.21)], as is required by Eq. (B.9). If $\hbar$ is not zero the function $\Psi(\mathbf{p}_0, E_0)$ remains uncertain $\Psi(\mathbf{p}_0, E_0) = 0 \cdot \infty$.

In conclusion, the Eqs. (B.16) and (B.9) are fulfilled if

$$\frac{1}{(2\pi\hbar)^2}\exp\left\{i\left[\frac{\mathbf{p}_0}{\hbar}\cdot\mathbf{r}_0-\frac{E_0}{\hbar}t_0\right]\right\}C=1 \tag{B.24}$$

and

$$\frac{1}{(2\pi\hbar)^2}\exp\left\{-i\left[\frac{\mathbf{p}_0}{\hbar}\cdot\mathbf{r}_0-\frac{E_0}{\hbar}t_0\right]\right\}D=1 \tag{B.25}$$

are fulfilled [see in particular (B.19) and (B.21)]. Only in the case of (B.24) and (B.25) we get

$$\hbar=\left[\frac{1}{(2\pi)^2}\exp\left\{i\left[\frac{\mathbf{p}_0}{\hbar}\cdot\mathbf{r}_0-\frac{E_0}{\hbar}t_0\right]\right\}C\right]^{1/2}\rightarrow 0 \tag{B.26}$$

and

$$\hbar=\left[\frac{1}{(2\pi)^2}\exp\left\{-i\left[\frac{\mathbf{p}_0}{\hbar}\cdot\mathbf{r}_0-\frac{E_0}{\hbar}t_0\right]\right\}D\right]^{1/2}\rightarrow 0. \tag{B.27}$$

The constants $C$ and $D$ are zero [see (B.6) and (B.13)].

3. Thus, instead of (B.18) and (B.20), at time $\tau_0$ the following transformation formula is valid in the classical limit

$$\Psi(\mathbf{r}_0,t_0)=\Psi(\mathbf{p}_0,E_0). \tag{B.28}$$

Equation (B.28) is even valid when the values of both functions $\Psi(\mathbf{r}_0,t_0)$ and $\Psi(\mathbf{p}_0,E_0)$ are finite: $\Psi(\mathbf{r}_0,t_0)\neq\infty$ and $\Psi(\mathbf{p}_0,E_0)\neq\infty$. In the next section, we will discuss this equation further.

## B.4. Classical Projections

What does the system-specific time $t_0$ mean? We know from our analysis in Chapter 3 that the system-specific time $t$ is a quantum-theoretical quantity and is not defined in the classical case. The probability density $\Psi^*\Psi$ is a sharp function for all times $\tau$ if the variables $\mathbf{p}, E, \mathbf{r}, t$ take definite values. In the case of $\tau_0$, we have the definite values $\mathbf{p}_0, E_0, \mathbf{r}_0, t_0$, and $\Psi^*\Psi$ is a sharp function at $t_0$ (Fig. B.1). As we have outlined in Chapter 3, $\Psi^*\Psi$ can only be observed by the reference time system $\Psi^*_{\text{ref}}(\tau-t)\Psi_{\text{ref}}(\tau-t)$, which takes the form

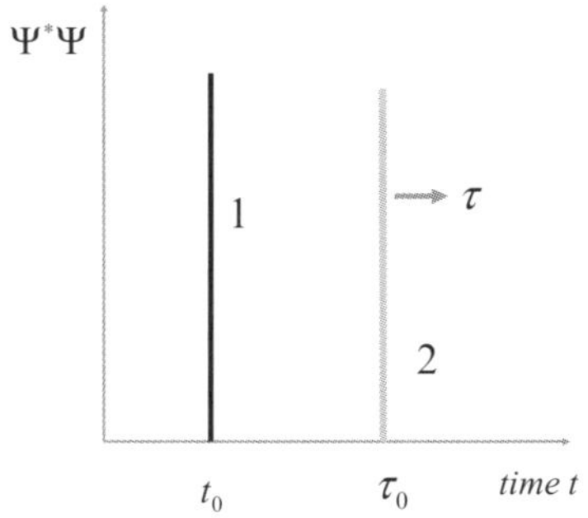


Fig. B.1. $\Psi^*\Psi$ as a function of time **t**. The classical system under investigation (line 1) and the reference time system (line 2). In this case ($t_0 \neq \tau_0$) a system-specific time (here $t_0$) is still defined, which reflects quantum-behavior (see Chapter 3 and [12]). Thus, the system under investigation behaves not completely classically for $t_0 \neq \tau_0$.

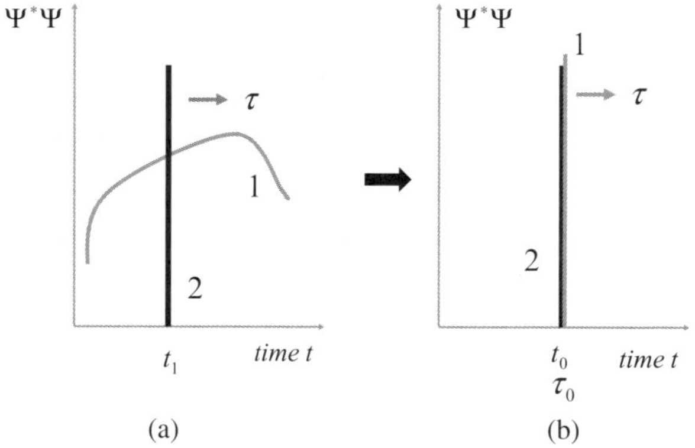


Fig. B.2. $\Psi^*\Psi$ as a function of time $t$. For a quantum system we have a probability distribution ((a), curve 1: system under investigation, curve 2: reference system). More details concerning quantum systems are given in Chapter 3 and Ref. 12). In the case of a classical system $\Psi^*\Psi$ must be a sharp function at $t_0$ [see (b)], and we have $t_0 = \tau_0$. Otherwise the system-specific time (here $t_0$) is not eliminated, and the system under investigation would still show quantum-behavior (see also Fig. B.1).

$\Psi^*_{\text{ref}}(\tau_0 - t)\Psi_{\text{ref}}(\tau_0 - t)$ at $\tau_0$. In the case of $t_0 \neq \tau_0$, we get Fig. B.1. This means that the system could not be observed. The existence of $t_0$ still reflects quantum-behavior, but $t_0$ is automatically eliminated if we require that only those systems are relevant that are observable. This is the case for $t_0 = \tau_0$ and we come to Fig. B.2, that is, in the classical case the system-specific time must be identified with the reference time.

In the quantum-theoretical case we have probability distributions for the variables in $(\mathbf{r}, t)$-space as well as for those in $(\mathbf{p}, E)$-space. In particular, we have the following feature (Chapter 3). The measurement of one of the possible values for $\mathbf{p}$ and for $E$ at time $\tau$ is done in the space–time intervals $\mathbf{r}, \mathbf{r} + d\mathbf{r}$ and $t, t + dt$ with the probability density of $\Psi^*(\mathbf{r}, t)\Psi(\mathbf{r}, t)$, where $\mathbf{p}$ and $E$ are statistical variables too and are given with the probability density $\Psi^*(\mathbf{p}, E)\Psi(\mathbf{p}, E)$.

In the classical case, there are no longer probability distributions, but we have definite values for $\mathbf{r}$ and $t$ in $(\mathbf{r}, t)$-space and also definite values for $\mathbf{p}$ and $E$ in $(\mathbf{p}, E)$-space: $\mathbf{p}_0, E_0, \mathbf{r}_0, t_0$ with $t_0 = \tau_0$. We measure with certainty the definite values $\mathbf{p}_0, E_0$ at time $\tau_0$ at the definite space–time point $\mathbf{r}_0, t_0$. Thus, there is a strict correlation between $\mathbf{r}_0, t_0 = \tau_0$ and $\mathbf{p}_0, E_0$:

$$\tau_0 : \mathbf{p}_0 = \mathbf{p}_0(\mathbf{r}_0, \tau_0), \;\; E_0 = E_0(\mathbf{r}_0, \tau_0). \tag{B.29}$$

The values $\mathbf{p}_0, E_0$ belong to $(\mathbf{p}, E)$-space, and the values $\mathbf{r}_0, t_0 = \tau_0$ belong to $(\mathbf{r}, t)$-space (see Fig. B.3).

Also in the classical case the projection principle must be valid, and our analysis showed that this is in fact realizable. We have not to recourse to a particle, which is embedded in space, having at time $\tau_0$ the position $\mathbf{r}_0$, the energy $E_0$ and the momentum $\mathbf{p}_0$ there.

## B.5. Information Content of the Different Spaces

Clearly, transformation (B.28) holds for each time $\tau$, and this is because condition (B.2) for classical behaviour is fulfilled at each time $\tau$, that is, we have $\mathbf{p}, E, \mathbf{r}, t$ at time $\tau$ with

$$\Psi(\mathbf{r}, t) = \Psi(\mathbf{p}, E) \tag{B.30}$$

if $\mathbf{p}, E, \mathbf{r}, t = \tau$ take definite values without any uncertainty.

In analogy to (B.29), there is a strict correlation between $\mathbf{r}, t = \tau$ and $\mathbf{p}, E$:

$$\tau : \mathbf{p} = \mathbf{p}(\mathbf{r}, \tau), \;\; E = E(\mathbf{r}, \tau). \tag{B.31}$$

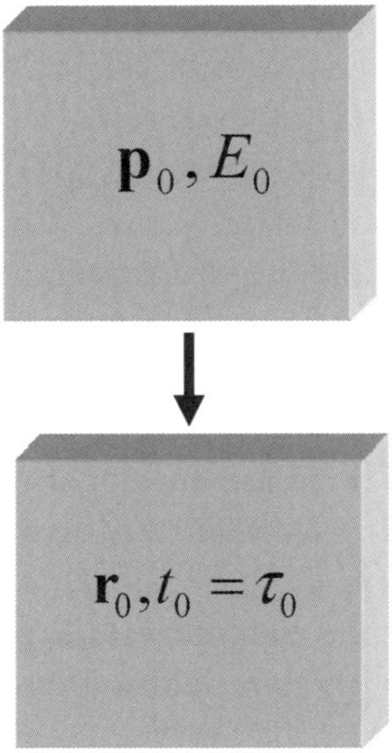


Fig. B.3. Also for a classical system the projection principle is valid. It has the values $\mathbf{p}_0, E_0$ in $(\mathbf{p}, E)$-space, and the values $\mathbf{r}_0, t_0 = \tau_0$ in $(\mathbf{r}, t)$-space. The (classical) processes take place within $(\mathbf{p}, E)$-space, and these processes are projected onto $(\mathbf{r}, t)$-space.

The values $\mathbf{p}, E$ belong to $(\mathbf{p}, E)$-space, and the values $\mathbf{r}, t = \tau$ belong to $(\mathbf{r}, t = \tau)$-space (see Fig. B.3).

Also here we can state that Eq. (B.30) is even valid when the values of both functions $\Psi(\mathbf{r}, t)$ and $\Psi(\mathbf{p}, E)$ are finite: $\Psi(\mathbf{r}, t) \neq \infty$ and $\Psi(\mathbf{p}, E) \neq \infty$.

In the quantum-theoretical case, the wavefunctions $\Psi(\mathbf{r}, t)$ and $\Psi(\mathbf{p}, E)$ are connected by a Fourier-transform and we have the equivalence

$$\Psi(\mathbf{r}, t) \leftrightarrow \Psi(\mathbf{p}, E) \tag{B.32}$$

indicating that the information in both spaces, $(\mathbf{r}, t)$-space and $(\mathbf{p}, E)$-space, is exactly the same. In the classical limit, instead of (B.32) relation (B.30) holds. Also here the information is exactly the same in both spaces. In other words, we always have

$$\mathbf{p}, E\text{-information} = \mathbf{r}, t(\tau)\text{-information}. \tag{B.33}$$

## B.6. The Relationship between the Classical Variables $\mathbf{r}_0, t_0, \mathbf{p}_0, E_0$

Let us use the following equations [12], which can directly be verified on the basis of (B.18):

$$i\hbar\frac{\partial}{\partial t}\Psi(\mathbf{r},t) = \frac{1}{(2\pi\hbar)^2}\int_{-\infty}^{\infty} E\,\Psi(\mathbf{p},E) \times \exp\left\{i\left[\frac{\mathbf{p}}{\hbar}\cdot\mathbf{r} - \frac{E}{\hbar}t\right]\right\} dp_x dp_y dp_z dE, \quad \text{(B.34)}$$

$$-\frac{\hbar^2}{2m_0}\Delta\Psi(\mathbf{r},t) = \frac{1}{(2\pi\hbar)^2}\int_{-\infty}^{\infty} -\frac{\mathbf{p}^2}{2m_0}\Psi(\mathbf{p},E) \times \exp\left\{i\left[\frac{\mathbf{p}}{\hbar}\cdot\mathbf{r} - \frac{E}{\hbar}t\right]\right\} dp_x dp_y dp_z dE. \quad \text{(B.35)}$$

In the case of condition (B.2), we immediately obtain with (B.6)

$$\left[i\hbar\frac{\partial}{\partial t}\Psi(\mathbf{r},t)\right]_{\substack{\mathbf{r}=\mathbf{r}_0\\ t=t_0}} = \frac{1}{(2\pi\hbar)^2}E_0\Psi(\mathbf{p}_0,E_0) \times \exp\left\{i\left[\frac{\mathbf{p}_0}{\hbar}\cdot\mathbf{r}_0 - \frac{E_0}{\hbar}t_0\right]\right\} C \quad \text{(B.36)}$$

and

$$\left[\frac{\hbar^2}{2m_0}\Delta\Psi(\mathbf{r},t)\right]_{\substack{\mathbf{r}=\mathbf{r}_0\\ t=t_0}} = -\frac{1}{(2\pi\hbar)^2}\frac{\mathbf{p}_0^2}{2m_0}\Psi(\mathbf{p}_0,E_0) \times \exp\left\{i\left[\frac{\mathbf{p}_0}{\hbar}\cdot\mathbf{r}_0 - \frac{E_0}{\hbar}t_0\right]\right\} C. \quad \text{(B.37)}$$

Application of (B.24) finally leads to

$$\left[i\hbar\frac{\partial}{\partial t}\Psi(\mathbf{r},t)\right]_{\substack{\mathbf{r}=\mathbf{r}_0\\ t=t_0}} = E_0\Psi(\mathbf{p}_0,E_0) \quad \text{(B.38)}$$

and

$$\left[\frac{\hbar^2}{2m_0}\Delta\Psi(\mathbf{r},t)\right]_{\substack{\mathbf{r}=\mathbf{r}_0\\ t=t_0}} = -\frac{\mathbf{p}_0^2}{2m_0}\Psi(\mathbf{p}_0,E_0). \quad \text{(B.39)}$$

Furthermore, in [12] we showed that the following relationship exists:

$$f(\mathbf{p}, E) = \frac{1}{(2\pi\hbar)^2} \int_{-\infty}^{\infty} \times \left\{ i\hbar \frac{\partial}{\partial t} \Psi(\mathbf{r}, t) + \frac{\hbar^2}{2m_0} \Delta\Psi(\mathbf{r}, t) - U(x, y, z)\Psi(\mathbf{r}, t) \right\} \Psi(\mathbf{r}, t) \times \exp\left\{ -i \left[ \frac{\mathbf{p}}{\hbar} \cdot \mathbf{r} - \frac{E}{\hbar} t \right] \right\} dt dx dy dz, \tag{B.40}$$

where $f(\mathbf{p}, E)$ is unknown but can be considered as part of the quantum-theoretical interaction potential.

We can formulate an equation in $(\mathbf{r}, t)$-space, which is equivalent to (B.40). [12] That leads to a function $f(\mathbf{r}, t)$, where $f(\mathbf{r}, t)$ is connected to $f(\mathbf{p}, E)$ by a Fourier-transform [12]:

$$f(\mathbf{p}, E) = \frac{1}{(2\pi\hbar)^2} \int_{-\infty}^{\infty} f(\mathbf{r}, t) \exp\left\{ -i \left[ \frac{\mathbf{p}}{\hbar} \cdot \mathbf{r} - \frac{E}{\hbar} t \right] \right\} dt dx dy dz. \tag{B.41}$$

Applying again condition (B.2) and Eq. (B.40) we obtain

$$f(\mathbf{p}_0, E_0) = \frac{1}{(2\pi\hbar)^2} \left\{ \left[ i\hbar \frac{\partial}{\partial t} \Psi(\mathbf{r}, t) \right]_{\substack{\mathbf{r} = \mathbf{r}_0 \\ t = t_0}} + \left[ \frac{\hbar^2}{2m_0} \Delta\Psi(\mathbf{r}, t) \right]_{\substack{\mathbf{r} = \mathbf{r}_0 \\ t = t_0}} - U(x_0, y_0, z_0)\Psi(\mathbf{r}_0, t_0) \right\} \times \exp\left\{ -i \left[ \frac{\mathbf{p}_0}{\hbar} \cdot \mathbf{r}_0 - \frac{E_0}{\hbar} t_0 \right] \right\} \int_{\mathbf{r}_0, t_0} dt dx dy dz, \tag{B.42}$$

$$f(\mathbf{p}_0, E_0) = \frac{1}{(2\pi\hbar)^2} f(\mathbf{r}_0, t_0) \times \exp\left\{ -i \left[ \frac{\mathbf{p}_0}{\hbar} \cdot \mathbf{r}_0 - \frac{E_0}{\hbar} t_0 \right] \right\} \int_{\mathbf{r}_0, t_0} dt dx dy dz. \tag{B.43}$$

With (B.25) we have

$$f(\mathbf{p}_0, E_0) = \left[ i\hbar \frac{\partial}{\partial t} \Psi(\mathbf{r}, t) \right]_{\substack{\mathbf{r} = \mathbf{r}_0 \\ t = t_0}} + \left[ \frac{\hbar^2}{2m_0} \Delta \Psi(\mathbf{r}, t) \right]_{\substack{\mathbf{r} = \mathbf{r}_0 \\ t = t_0}}$$
$$- U(x_0, y_0, z_0)\Psi(\mathbf{r}_0, t_0) \tag{B.44}$$

and

$$f(\mathbf{p}_0, E_0) = f(\mathbf{r}_0, t_0). \tag{B.45}$$

Using (B.38), (B.39), (B.28) and (B.45), (B.44) takes the form

$$f(\mathbf{r}_0, t_0) = \lim_{\Psi(\mathbf{r}_0, t_0) \to \infty} \left[ E_0 \Psi(\mathbf{r}_0, t_0) - \frac{\mathbf{p}_0^2}{2m_0} \Psi(\mathbf{r}_0, t_0) \right.$$
$$\left. - U(x_0, y_0, z_0)\Psi(\mathbf{r}_0, t_0) \right] \tag{B.46}$$

and we get the

$$E_0 = \frac{\mathbf{p}_0^2}{2m_0} + V(x_0, y_0, z_0) \tag{B.47}$$

with

$$V(x_0, y_0, z_0) = U(x_0, y_0, z_0) + \lim_{\Psi(\mathbf{r}_0, t_0) \to \infty} \frac{f(\mathbf{r}_0, t_0)}{\Psi(\mathbf{r}_0, t_0)}, \tag{B.48}$$

where $V(x_0, y_0, z_0)$ is the quantum-theoretical potential and $U(x_0, y_0, z_0)$ its classical limit. Since we are investigating here the classical limit, we must have

$$V(x_0, y_0, z_0) = U(x_0, y_0, z_0). \tag{B.49}$$

Because of $\Psi(\mathbf{r}_0, t_0) \to \infty$ [classical limit, see (B.16)] the term $f(\mathbf{r}_0, t_0)/\Psi(\mathbf{r}_0, t_0)$ in (B.48) is always zero if $f(\mathbf{r}_0, t_0)$ remains finite: $f(\mathbf{r}_0, t_0) \neq \infty$. Therefore, w must have

$$\lim_{\Psi(\mathbf{r}_0, t_0) \to \infty} \frac{f(\mathbf{r}_0, t_0)}{\Psi(\mathbf{r}_0, t_0)} = 0 \tag{B.50}$$

for $f(\mathbf{p}_0, E_0) \neq \infty$.

Then, we obtain for the classical case the well-known expression

$$E_0 = \frac{\mathbf{p}_0^2}{2m_0} + U(x_0, y_0, z_0) \tag{B.51}$$

for a configuration at time $\tau_0$, or we may also write

$$E = \frac{\mathbf{p}^2}{2m_0} + U(x, y, z) \tag{B.52}$$

for each time $\tau$. The system-specific time does not appear in (B.51) and (B.52). As in the analysis given in Section B.4, the system-specific time is eliminated.

Equations (B.51) and (B.52), respectively, is a result of projection theory, that is, we have two spaces, $(\mathbf{r}, \tau)$-space and $(\mathbf{p}, E)$-space, and the information of $(\mathbf{p}, E)$-space is projected onto $(\mathbf{r}, \tau)$-space. In contrast to that, within conventional physics (B.51) is interpreted in connection with a particle of mass $m_0$, which is embedded in space, having the momentum $\mathbf{p}_0$ and the energy $E_0$ in space.

The classical function $U(x, y, z)$ has nothing to do with an interaction process in $(\mathbf{r}, \tau)$-space. $U(x, y, z)$ is an energy and, therefore, it belongs to $(\mathbf{p}, E)$-space, which is not defined in conventional physics. $U(x, y, z)$ has two tasks:

1. It describes the energy fluctuations in $(\mathbf{p}, E)$-space with other systems (environment).
2. Due to (B.31) there must be a correlation between the values $\mathbf{r}, t = \tau$ and $\mathbf{p}, E$. Thus, also in projection theory we need a function like $U(x, y, z)$. It correlates the geometrical position $\mathbf{r}$ at time $\tau$ of the system in $(\mathbf{r}, t)$-space with the energy value in $(\mathbf{p}, E)$-space at the same time $\tau$. However the meaning of $U(x, y, z)$ is quite different from that of conventional physics (see Section 3.15).

Thus, the appearance of the (abstract) function $V(x, y, z, t)$ is a logical consequence within the theoretical structures of projection theory (see in particular [12]).

Let us briefly repeat the main features of $U(x, y, z)$ that is has within Newton' mechanics, where everything is embedded in $(\mathbf{r}, \tau)$. How does the mutual influence between two bodies (as, for example, the earth and the sun) come into existence? What mechanism can explain this mutual influence within Newton' mechanics? In this connection, two "pictures" turned out to be of particular relevance (see in particular Section 3.15): the "proximity effect" and the "action-at-the-distance".

Is it possible to deduce the forces expressed by the well-known formula $m_1 m_2/r^2$ on the basis of these pictures. As already mentioned in Section 3.15, the ideas "proximity effect" and "action-at-the-distance" cannot be used for that, since they interpret the relation $m_1 m_2/r^2$, but are not able to explain the mathematical structure of this force law. As we have stated in Section 3.15 there is no possibility within Newton's mechanics to explain by a mechanism how the mutual influence between two bodies (as, for example, between the earth and the sun) comes into existence.

In conclusion, the interpretations for the notion "interaction" cannot help further to understand the mutual influence in connection with two or more interacting bodies. In particular, all these interpretations are not able at all to deduce the force law $m_1 m_2/r^2$. As in projection theory, we also come in conventional physics to the conclusion that the interaction law, described by $m_1 m_2/r^2$, is an abstract law.

## B.7. Applications

Within conventional classical mechanics reality is embedded in space and time. Let us consider a particle of mass $m_0$ in an external field, and let us treat this problem only in one dimension using the coordinate $x$. Then, the total energy $E_{\text{total}}$ is given by the kinetic energy $E_k$ and the potential $U(x)$, and instead of (B.52) we obtain

$$E_{\text{total}} = \frac{p_x^2}{2m_0} + U(x), \qquad \text{(B.53)}$$

where $E_{\text{total}}$ remains constant, that is, when $U(x)$ increases the kinetic energy $E_k$ must decrease and vice versa. However, within projection theory there can be no potential energy because any energy can only be an element of $(\mathbf{p}, E)$-space. Thus, we need another argumentation.

We demonstrated in Section 5.7.1 that a re-interpretation is possible. Instead to assume that the world is embedded in space (the view of conventional physics), we may interpret Eq. (B.52) [Eq. (B.53)]] in terms of the projection principle (see in particular Section B.4): The processes take place within $(\mathbf{p}, E)$-space and are projected onto $(\mathbf{r}, t)$-space (Fig. B.3). However, the term potential energy, that is, the function $U(x)$, may no longer appear in connection with $(\mathbf{r}, t)$-space, and we have to shift the interaction problem on $(\mathbf{p}, E)$-space as is necessary within projection theory.

In $(\mathbf{p}, E)$-space the system with mass $m_0$ exchanges momentum and energy with another system. In other words, the kinetic energy $E_k = E$ is the energy of the system $[E = E_{\text{total}} - U(x)]$ and we have $p_x$-fluctuations $\Delta p_x$ and $E$-fluctuations $\Delta E$ [instead of the potential $U(x)$] changing the momentum $p_x$ and the energy $E$ of the system in the course of time $\tau$. Therefore, the energy $E$ and the momentum $p_x$ vary with time $\tau$ and the following relation is valid:

$$\tau : E = \frac{p_x^2}{2m_0}. \tag{B.54}$$

This is the situation in $(\mathbf{p}, E)$-space. But what about the situation in $(\mathbf{r}, \tau)$-space? Let us briefly investigate this point.

There is no momentum, no energy, no mass $m_0$ in $(\mathbf{r}, \tau)$-space, but we may "only" have geometrical positions in $(\mathbf{r}, \tau)$-space. From (B.54), we immediately find

$$\frac{dE}{dp_x} = v = \frac{dx}{d\tau} \tag{B.55}$$

with $p_x = m_0 v$, where $v$ is the velocity of the system. The term $dE/dp_x$ in Eq. (B.55) belongs to $(\mathbf{p}, E)$-space, and the term $dx/d\tau$ in Eq. (B.55) describes the system in $(\mathbf{r}, \tau)$-space. Then, the term $dE/dp_x$ corresponds to "reality" and $dx/d\tau$ is an element of the "picture of reality" (Fig. B.4):

$$\frac{dE}{dp_x}(\text{reality}) \rightarrow \frac{dx}{d\tau}(\text{picture of reality, geomertical positions}). \tag{B.56}$$

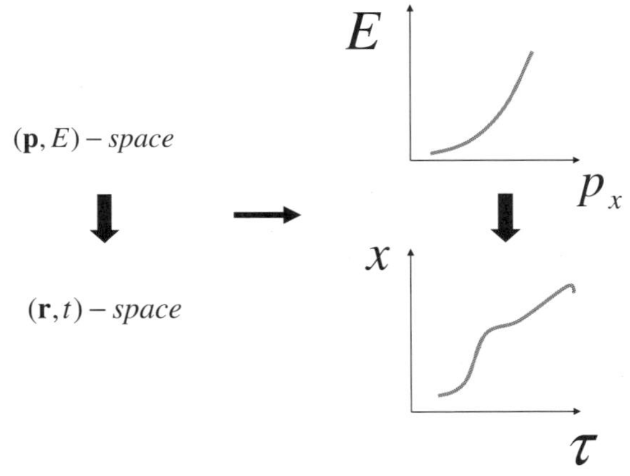


Fig. B.4. The energy $E$ as a function of momentum $p_x$ [according to (B.54)]. Its projection onto $(\mathbf{r}, \tau)$-space leads to $x = x(\tau)$, whereby condition (B.55) must be fulfilled.

From Eq. (B.55) directly follows that both spaces, $(\mathbf{p}, E)$-space and $(\mathbf{r}, \tau)$-space, contain exactly the same information, as required above [Section B.5, Eq. (B.33)]. Both spaces are connected by the function $U(x)$: $U$ has the dimensions of energy and belongs to $(\mathbf{p}, E)$-space and $x$ is a geometrical position and belongs to $(\mathbf{r}, \tau)$-space. Also this is compatible with what is outlined in Section B.5.

Equation (B.55) must be fulfilled at each time $\tau_i$, that is, we have a strict relation between the values $p_{xi}$, $E_i$ and $x_i$ at time $\tau_i$. For each time $\tau_i$, the condition

$$\frac{dE_i}{dp_{xi}} = \upsilon = \frac{dx_i}{d\tau_i} \tag{B.57}$$

must be valid. In Fig. B.5, two examples are given.

## B.8. On the Classical Equations of Motion: Some General Remarks

Again, Fig. B.5 only expresses the relation of the events in $(\mathbf{p}, E)$-space and those in $(\mathbf{r}, \tau)$-space. However, we cannot say something about the trajectory $x = x(\tau)$, that is, a description of $x = x(\tau)$ is not possible on the basis of condition (B.57). In order to obtain

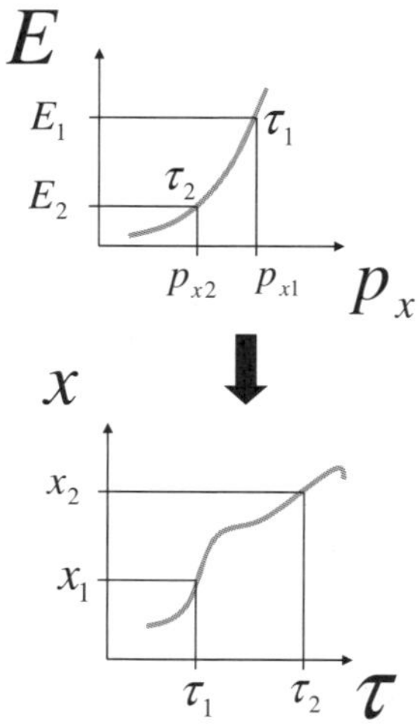


Fig. B.5. There is a strict relation between the values $p_{xi}$, $E_i$ and $x_i$ at time $\tau_i$. For each time $\tau_i$ the condition (B.57) must be fulfilled. This is demonstrated in the figure (identical with Fig. B.4) for $i = 1, 2$ with $\tau_1 < \tau_2$.

a trajectory $x = x(\tau)$, we need "equations of motion" in analogy to Newton's equations of motion. The corresponding relations of projection theory must however be different from Newton's equations. There are some features in projection theory in the case of free, non-interacting quantum systems, which should also be valid in the classical limit. What features are meant here?

### B.8.1. *Free, non-interacting Systems*

It can be shown within projection theory that free (non-interacting) systems cannot exist if the system is elementary in character. [12] Elementary in character means that the system cannot be divided into subsystems. A system can be considered as free if the potential function becomes zero: $V(x, y, z, t) = 0$. Then, also the operator $V(i\hbar\partial/\partial p_x, \ldots, -i\hbar\partial/\partial E)$ (Section 3.7) is not definable. In this case we have $\Psi^*(\mathbf{r}, t)\Psi(\mathbf{r}, t) = 0$, $\Psi^*(\mathbf{p}_0, E_0)\Psi(\mathbf{p}_0, E_0) = 0$ with $\mathbf{p}_0 = 0, E_0 = 0$ for all times $\tau$ (see Section 3.9.1 and [12]). This is the quantum-theoretical result. However, it is easy to recognize that a quantum free system is identical with a classical free system. Thus, also in the classical limit the $(\mathbf{p}, E)$-space is empty and, consequently, this "non-existing" system in $(\mathbf{p}, E)$-space cannot be observable in $(\mathbf{r}, \tau)$-space because of $\Psi^*(\mathbf{r}, t = \tau)\Psi(\mathbf{r}, t = \tau) = 0$, that is, it can also not exist as geometrical structure in $(\mathbf{r}, \tau)$-space, which is identical to the fact that there can be no event in $(\mathbf{r}, \tau)$-space.

In contrast to projection theory, from Newton's equations of motion it follows that in the case of $\mathbf{p}_0 = 0$, $E_0 = 0$, a free, non-interacting system can exist. It is a resting system ($\upsilon = 0$) with $m_0 \neq 0$. The property $\upsilon = 0$ reflects "inertia" leading to the effect that the free system interacts with space [$(\mathbf{r}, \tau)$-space] within Newton's mechanics. In other words, Mach's principle is not fulfilled here (see also Chapter 3). Such space-effects (violation of Mach's principle) do not take place in projection theory.

It is shown in [12] that relativistic rest mass effects does not influence the result of projection theory for free, non-interacting system. It is demonstrated that in the case of $\mathbf{p}_0 = 0$, $E_0 = 0$ also the mass must also be zero: $m_0 = 0$.

## B.8.2. *The Parameter $m_0$*

1. The "mass" $m_0$ in the re-interpreted formula (B.52) has to be treated as a simple parameter with any physical background. In projection theory it is forbidden to connect with $m_0$ a certain point-like real system in $(\mathbf{r}, \tau)$-space. We have one geometrical structure described by $\Psi^*(\mathbf{r}, t = \tau)\Psi(\mathbf{r}, t = \tau)$ when we base our considerations on (B.52) alone. Then, there can be no additional geometrical structure due to $m_0$.
2. The term "free system" in projection theory means that there are no distance-dependent interactions between the system under investigation and other systems. Such a free system can in principle have a certain structure in $(\mathbf{r}, t)$-space defining its form (shape), and this form can only be the result of distance-independent interactions leading to the mass $m_0 \neq 0$. However, this does change the situation. Also such a system cannot move with $\upsilon = 0$ in $(\mathbf{r}, \tau)$-space, but must also here behave strictly statistically (Chapter 3). Clearly, this is in contrast to Newton's mechanics, where $\upsilon = 0$ is possible for $\mathbf{p}_0 = 0$, $E_0 = 0$.

### B.8.3. *Complex Systems*

In the case of a complex free system consisting of various interacting parts, which does not interact with external other systems, an observer that is positioned within the system can of course investigate it experimentally. Let us denote the frame of reference in which the whole system together with the observer is resting by $S'$. How does this system $S'$ move relative to a frame of reference $S$ in which another observer is resting?

Since we assumed that system $S'$ does not interact with another external system, the interaction between $S$ and $S'$ must be zero. Furthermore, within projection theory there can be no interaction between the space $[(\mathbf{r}, \tau)$-space$]$ and system $S'$ and, therefore, the motion relative to the resting frame of reference $S$ must be strictly statistical.

### B.8.4. *Concluding Remarks*

In Sections B.8.1 and B.8.2, we have shown that Mach's principle is fulfilled also in the classical limit of projection theory. This is definitely not the case within Newton's mechanics. This important result has to be taken into account in the formulation of the classical equations of motion. Thus, the equations of motion within projection theory must necessarily be different from those of classical conventional mechanics.

# References


1. Stephen W. Hawking, *A Brief History of Time* (Bantam Press, London, 1988).
2. John Horgan, *The End of Science* (Addison and Wesley, USA, 1996).
3. Richard P. Feynman, *The Character of Physical Law* (M.I.T. Press, Cambridge/ Massachusetts and London, 1977).
4. Nicholas Rescher, *The Limits of Science* (University of California Press, Berkeley, Los Angeles, London, 1984).
5. E. Haeckel, *Die Welträtsel* (Alfred Kröner, Stuttgart, 1984).
6. Hoimar von Ditfurth, *Der Geist fiel nicht vom Himmel* (Deutscher Taschenbuch Verlag, München, 1980).
7. C. G. Jung, *Synchronizität, Akausalität and Okkultismus* (Deutscher Taschenbuch Verlag, München, 1990).
8. W. Schommers, *Space and Time, Matter and Mind* (World Scientific, New Jersey, London, Singapore, 1994).
9. W. Schommers, *The Visible and the Invisible* (World Scientific, New Jersey, London, Singapore, 1998).
10. Thomas Kuhn, *The Structure of Scientific Revolutions* (University of Chicago Press, London, 1962).
11. Wolfram Schommers, *Advanced Science Letters* **1** (2008) 59.
12. Wolfram Schommers, *Quantum Processes* (World Scientific, New Jersey, London, Singapore).
13. K. Gödel, *Rev. Mod. Phys.* **21** (1947) 447.
14. R. Brückner, *Das schielende Kind* (Schwabe Verlag, Basel, Stuttgart, 1977).
15. M. Jammer, *The Philosophy of Quantum Mechanics* (Wiley, New York, 1974).
16. B. de Witt, *Physics Today*, September 1970.
17. Herwig Schopper, *Materie und Antimaterie* (Piper, München, Zürich, 1989).
18. Konrad Lorenz, *Die Rückseite des Spiegels* (Piper, München, 1973).
19. Jim Al-Khalili, *Black Holes, Worm Holes and Time Machines* (Institute of Physics, Bristol, 1999).
20. Kenneth W. Ford, *Elementary Particles* (Blaisdell Publishing Company, New York, Toronto, London, 1963).
21. J. Richard Gott, *Time Travel in Einstein's Universe* (A Martiner Book Houghton Mifflin Company, Boston and New York, 2001).
22. J. Gribbin, In search of Schrödinger's cat, in *Quantum Physics and Reality* (Bantam Books, Toronto, New York, London, Sydney, Auckland, 1984).
23. Julius Fraser, *Die Zeit* (Deutscher Taschenbuch Verlag, München, 1991).
24. Michael Talbot, *Mystik und neue Physik* (Heyne Verlag, München, 1989).

25. Karl Popper/Franz Kreuzer, *Offene Gesellschaft* (Offenes Universum, Piper, München, 1986).
26. Brandon Carter, in *Physical Cosmology and Philosophy*, ed. John Leslies (Macmillan, New York, 1990), p. 129.
27. Wolfram Schommers, *Advanced Science Letters* **1** (2008) 59.
28. Brian Greene, *The Elegant Universe* (Vintage Books, A Division of Random House Inc., New York, 1999).
29. K. C. Cole, *The Hole in the Universe* (A Harvest Book, Hartcourt Inc, San Diego, New York, London, 2001).
30. R. L. Jaffe, *Phys. Rev. D* **72** (2005) 021301.
31. Charles Seife, *Alpha and Omega* (Penguin Books, New York, 2003).
32. Edmund J. Copeland, M. Sami and Shinji Tsujikawa, *International Journal of Modern Physics D* **15** (2006) 1753.
33. Henning Genz, in *Lexikon der Physik* (Spektrum Akademischer Verlag, Heidelberg, Berlin, 1998).
34. W. Schommers, in *Quantum Theory and Pictures of Reality* (Springer-Verlag, Berlin, Heidelberg, 1989), Chaps. 2, 5.
35. P. Ghose, D. Home, G. S. Agarwal, *Phys. Lett. A* **153** (1991) 403.
36. P. Ghose, *Testing Quantum Mechanics on a New Ground* (Cambridge University Press, 1999).
37. Y. Mizobuchi, Y. Othake, *Phys. Lett. A* **168** (1992) 1.
38. C. S. Unnikrishnan, S. A. Murphy, *Phys. Lett. A* **221** (1996) 1.
39. G. Brida, M. Genovese, M. Gramegna and E. Predazzi, *Phys. Lett. A* **328** (1996) 313.
40. Wolfram Schommers, *J. Comp. Theor. Nanoscience* **4** (2007) 992.
41. W. Finkelnburg, *Einführung in die Atomphysik* (Springer-Verlag, Berlin, New York, Heidelberg, 1962).
42. A. Rubinowics, *Quantum Mechanics* (Elsevier, Amsterdam, 1968).
43. Wolfram Schommers, in *Handbook of Theoretical and Computational Nanoscience*, Vol. 1: Basic Concepts, Nanomachines, and Medical Nanodevices (American Scientific Publishers, Los Angeles, 2006), 137–219.
44. L. D. Landau and E. M. Lifschitz, *Quantum Mechanics* (Pergamon, Oxford, 1965).
45. L. de Broglie, *Die Elementarteilchen* (Goverts, Hamburg, 1943).
46. M. Bunge, *Canadian Journal of Physics* **48** (1970) 1410.
47. Banesh Hoffman, *Albert Einstein–Creator and Rebel* (The Viking Press, New York, 1972).
48. Peter Graneau and Neal Graneau, *In the Grip of the Distance Universe* (World Scientific, New Jersey, London, Singapore, 2006).
49. A. Einstein, *Grundzüge der Relativitätstheorie* (Verlag Viehweg und Sohn, Braunscheig, 1973).
50. O. Heckmann, *Sterne, Kosmos, Weltmodelle* (Deutscher Taschenbuch Verlag GmbH & Co. KG, München, 1980).
51. M. Born, *Die Relativitätstheorie Einsteins* (Springer-Verlag, Berlin, Heidelberg, 1964).

52. H. Dehnen, in *Philosophie und Physik der Raum-Zeit*, eds. J. Audretsch und K. Mainzer (Wissenschaftsverlag, Mannheim, Wien, Zürich, 1988).
53. B. Kanitscheider, *Kosmologie* (Philipp Reclam jun., Stuttgart, 1984).
54. G. Falk and W. Ruppel, *Mechanik, Relativität, Gravitation* (Springer-Verlag, Berlin, Heidelberg, 1983).
55. E. Dammann, *Erkenntnisse jenseits von Raum und Zeit* (Droemersche Verlagsanstalt Th. Knaur Nachf., München, 1990).
56. John D. Barrow, *The Artful Universe* (Little, Brown and Company, Boston, New York, Toronto, London, 1995).
57. M. Jammer, *Concepts of Space* (Dover Publications Inc., New York, 1993).
58. D'Arcy Wentworth Thompson, *On Growth and Form* (Dover Publications, Inc., New York, 1992).

# Index